HANDBOOK OF THIN-FILM DEPOSITION PROCESSES AND TECHNIQUES

MATERIALS SCIENCE AND PROCESS TECHNOLOGY SERIES

Editors

Rointan F. Bunshah, University of California, Los Angeles *(Materials Science and Technology)*

Gary E. McGuire, Microelectronics Center of North Carolina *(Electronic Materials and Processing)*

DEPOSITION TECHNOLOGIES FOR FILMS AND COATINGS; Developments and Applications: by *Rointan F. Bunshah et al*

CHEMICAL VAPOR DEPOSITION FOR MICROELECTRONICS; Principles, Technology, and Applications: by *Arthur Sherman*

SEMICONDUCTOR MATERIALS AND PROCESS TECHNOLOGY HANDBOOK; For Very Large Scale Integration (VLSI) and Ultra Large Scale Integration (ULSI): edited by *Gary E. McGuire*

SOL-GEL TECHNOLOGY FOR THIN FILMS, FIBERS, PREFORMS, ELECTRONICS, AND SPECIALTY SHAPES: edited by *Lisa C. Klein*

HYBRID MICROCIRCUIT TECHNOLOGY HANDBOOK; Materials, Processes, Design, Testing and Production: by *James J. Licari* and *Leonard R. Enlow*

HANDBOOK OF THIN FILM DEPOSITION PROCESSES AND TECHNIQUES; Principles, Methods, Equipment and Applications: edited by *Klaus K. Schuegraf*

DIFFUSION PHENOMENA IN THIN FILMS AND MICROELECTRONIC MATERIALS: edited by *Devendra Gupta* and *Paul S. Ho*

IONIZED-CLUSTER BEAM DEPOSITION AND EPITAXY: by *Toshinori Takagi*

Related Titles

ADHESIVES TECHNOLOGY HANDBOOK: by *Arthur H. Landrock*

HANDBOOK OF THERMOSET PLASTICS: edited by *Sidney H. Goodman*

HANDBOOK OF CONTAMINATION CONTROL IN MICROELECTRONICS; Principles, Applications and Technology: edited by *Donald L. Tolliver*

HANDBOOK OF THIN-FILM DEPOSITION PROCESSES AND TECHNIQUES

Principles, Methods, Equipment and Applications

Edited by

Klaus K. Schuegraf

Tylan Corporation
Carson, California

NOYES PUBLICATIONS
Park Ridge, New Jersey, U.S.A.

Copyright © 1988 by Noyes Publications
No part of this book may be reproduced or utilized in any form or by any means, electronic or mechanical, including photocopying, recording or by any information storage and retrieval system, without permission in writing from the Publisher.
Library of Congress Catalog Card Number: 87-34702
ISBN: 0-8155-1153-1
Printed in the United States

Published in the United States of America by
Noyes Publications
Mill Road, Park Ridge, New Jersey 07656

10 9 8 7 6 5 4 3 2 1

Library of Congress Cataloging-in-Publication Data

Handbook of thin-film deposition processes and techniques :
  principles, methods, equipment, and applications / edited by Klaus K. Schuegraf.
        p. cm.
    Bibliography: p.
    Includes index.
    ISBN 0-8155-1153-1
    1. Thin film devices--Design and construction--Handbooks, manuals, etc. I. Schuegraf, Klaus K. II. Title: Handbook of thin-film deposition processes and techniques.
  TK7872.T55H36   1988
  621.381'72--dc19   87-34702
                                                  CIP

Preface

The technology of thin film deposition has advanced dramatically during the past 30 years. This advancement was driven primarily by the need for new products and devices in the electronics and optical industries. The rapid progress in solid-state electronic devices would not have been possible without the development of new thin film deposition processes, improved film characteristics and superior film qualities. Thin film deposition technology is still undergoing rapid changes which will lead to even more complex and advanced electronic devices in the future. The economic impact of this technology can best be characterized by the world-wide sales of semiconductor devices, which exceeded $40 billion in 1987.

This book is intended to serve as a handbook and guide for the practitioner in the field, as a review and overview of this rapidly evolving technology for the engineer and scientist, and as an introduction for the student in several branches of science and engineering.

This handbook is a review of 13 different deposition technologies, each authored by experts in their particular field. It gives a concise reference and description of the processes, methods, and equipment for the deposition of technologically important materials. Emphasis is placed on recently developed film deposition processes for application in advanced microelectronic device fabrications that require the most demanding approaches. The discussions of the principles of operation for the deposition equipment and its suitability, performance, controls, capabilities and limitations for production applications are intended to provide the reader with basic understanding and appreciation of these systems. Key properties and areas of application of industrially important materials created by thin film deposition processes are described. Extensive use of references, reviews and bibliographies provides source material for specific use and more detailed study.

The topics covered in each chapter of this book have been carefully selected to include advanced and emerging deposition technologies with potential for manufacturing applications. An attempt was made to compare competing technologies and to project a scenario for the most likely future developments. Several other deposition technologies have been excluded since adequate recent reviews are already available. In addition, the technology for deposition or coating of films exceeding 10 microns in thickness was excluded, since these films have different applications and are in general based on quite different deposition techniques.

Many people contributed and assisted in the preparation of this handbook. My thanks go to the individual authors and their employers, who provided detailed work and support. I am especially indebted to Werner Kern, who provided many valuable suggestions and assisted in co-authoring several sections of this book. Last but not least, my special thanks go to George Narita, Executive Editor of Noyes Publications, for providing continued encouragement and patience for the completion of all the tasks involved.

Torrance, California Klaus K. Schuegraf
July, 1988

Contributors

Russell L. Abber
Hughes Aircraft Company
El Segundo, California

Brian Chapman
Lucas Labs
Santa Clara, California

Robert Chow
Varian Associates
Thin Film Division
Santa Clara, California

George J. Collins
Colorado State University
Department of Electrical Engineering
Fort Collins, Colorado

Martin L. Hammond
Tetron/Gemini Systems
Freemont, California

Werner Kern
RCA Laboratories
David Sarnoff Research Center
Princeton, New Jersey

Walter S. Knodle
High Yield Technology, Inc.
Mountain View, California

Stefano Mangano
Lucas Labs
Santa Clara, California

Seitaro Matsuo
NTT Atsugi ECL
Kanagawa, Japan

James J. McNally
Air Force Academy
Colorado Springs, Colorado

John R. McNeil
University of New Mexico
Department of Electrical Engineering
Albuquerque, New Mexico

Cameron A. Moore
Colorado State University
Department of Electrical Engineering
Fort Collins, Colorado

Son Van Nguyen
IBM Corporation
General Technology Division
Essex Junction, Vermont

Paul D. Reader
Ion Tech, Inc.
Fort Collins, Colorado

Arnold Reisman
Microelectronics Center of
 North Carolina
Research Triangle Park,
 North Carolina

Ronald C. Rossi
Tylan Corporation
Carson, California

Klaus K. Schuegraf
Tylan Corporation
Carson, California

Toshinori Takagi
Kyoto University
Ion Beam Engineering Experimental
 Laboratory
Kyoto, Japan

Lance R. Thompson
Sandia National Labs
Albuquerque, New Mexico

Isao Yamada
Kyoto University
Ion Beam Engineering Experimental
 Laboratory
Kyoto, Japan

Peter Younger
Eaton Corporation
Beverly, Massachusetts

Zeng-Qi Yu
Colorado State University
Fort Collins, Colorado

John L. Zilko
AT&T Bell Laboratories
Murray Hill, New Jersey

NOTICE

To the best of the Publisher's knowledge the information contained in this book is accurate; however, the Publisher assumes no responsibility nor liability for errors or any consequences arising from the use of the information contained herein. Mention of trade names or commercial products does not constitute endorsement or recommendation for use by the Publisher.

Final determination of the suitability of any information, procedure, or equipment for use contemplated by any user, and the manner of that use, is the sole responsibility of the user. The book is intended for informational purposes only. Expert advice should be obtained at all times when implementation is being considered. Due caution should be exercised in the handling of equipment or materials which might be considered hazardous.

Contents

1. **DEPOSITION TECHNOLOGIES AND APPLICATIONS:
   INTRODUCTION AND OVERVIEW** 1
   Werner Kern and Klaus K. Schuegraf
   Objective and Scope of This Book 1
   **Importance of Deposition Technology in Modern Fabrication
   Processes**..2
   Classification of Deposition Technologies. 3
   Overview of Various Thin-Film Deposition Technologies. 3
   Evaporative Technologies 3
   Molecular Beam Epitaxy............................5
   Glow-Discharge Technologies............................5
   Sputtering.......................................5
   Plasma Processes.................................6
   Cluster Beam Deposition...........................7
   Gas-Phase Chemical Processes 8
   Reactors 9
   Vapor-Phase Epitaxy 10
   Photo-Enhanced Chemical Vapor Deposition (PHCVD) 10
   Laser-Induced Chemical Vapor Deposition (LCVD)........10
   Ion Implantation..................................11
   Thermal Oxidation 11
   Oxidation of Silicon................................11
   Other Gas-Phase Oxidations.........................11
   Liquid-Phase Chemical Formation 11
   Electrolytic Anodization.............................12
   Electroplating....................................12
   Chemical Reduction Plating..........................12
   Electroless Plating.................................12
   Electrophoretic Deposition 12

        Immersion Plating.................................13
        Mechanical Methods..............................13
        Liquid-Phase Epitaxy.............................13
   Criteria for the Selection of a Deposition Technology for
   Specific Applications...............................14
      Thin-Film Applications..............................14
        Electronic Components..........................14
        Electronic Displays.............................14
        Optical Coatings...............................14
        Magnetic Films for Data Storage....................14
        Optical Data Storage Devices.....................15
        Antistatic Coatings.............................15
        Hard Surface Coatings..........................15
      Material Characteristics.............................15
      Process Technology................................17
      Thin-Film Manufacturing Equipment.....................19
   Summary and Perspective for the Future....................20
   References.......................................22

2. SILICON EPITAXY BY CHEMICAL VAPOR DEPOSITION..........26
Martin L. Hammond

   Introduction......................................26
        Applications of Silicon Epitaxy.........................27
   Theory of Silicon Epitaxy by CVD........................29
   Silicon Epitaxy Process Chemistry.......................31
   Commercial Reactor Geometries.......................33
      Horizontal Reactor...............................34
      Cylinder Reactor................................34
      Vertical Reactor.................................35
      New Reactor Geometry...........................35
   Theory of Chemical Vapor Deposition....................36
   Process Adjustments...............................38
      Horizontal Reactor...............................39
      Cylinder Reactor................................39
      Vertical Reactor.................................41
      Control of Variables..............................43
   Equipment Considerations for Silicon Epitaxy...............44
      Gas Control System..............................44
      Leak Testing...................................45
      Gas Flow Control...............................46
      Dopant Flow Control.............................47
   Other Equipment Considerations.......................51
      Heating Power Supplies...........................51
      Effect of Pressure...............................52
      Temperature Measurement........................53
      Backside Transfer...............................55
      Intrinsic Resistivity...............................56
      Phantom p-Type Layer...........................56

Contents

Defects in Epitaxy Layers.................................56
    Haze...57
    Pits..57
    Orange Peel..57
    Faceted Growth......................................57
    Edge Crown..57
    Etch Pits..58
    Slips...58
    Stacking Faults......................................58
    Spikes and Hillocks..................................58
    Shallow Pits...58
Safety..59
Key Technical Issues..................................59
    Productivity/Cost....................................59
    Uniformity/Quality...................................62
    Buried Layer Pattern Transfer........................62
    Autodoping..66
        Dopant Transitions.............................69
New Materials Technology for Silicon Epitaxy.............74
Low Temperature Epitaxy...............................75
Conclusions...76
References..76

3. LOW PRESSURE CHEMICAL VAPOR DEPOSITION........80
Ronald C. Rossi

Introduction...80
Equipment..81
    Horizontal Reactor...................................83
        Gas Control Systems...........................84
        Vacuum Systems...............................84
        Process Control Systems.......................85
    Vertical Reactors....................................85
    Bell Jar Reactors....................................85
    Single Wafer Reactors................................86
Principles of Low-Pressure CVD........................87
LPCVD Processing....................................88
    Polysilicon...90
    Silicon Nitride.......................................95
    Low-Temperature Oxide (LTO)........................100
    Other LPCVD Processes.............................106
        Tetraethylorthosilicate (TEOS).................106
        Diacetoxyditertiarybutoxysilane (DADBS).......106
        Phosphorus-Doped Silicon.....................106
        Doped LPLTO................................107
        Tungsten....................................107
        Tungsten Silicide.............................108
        Semi-Insulating Polysilicon (SIPOS)............108
        Aluminum and Aluminum Silicon Alloys.........108

xii Contents

        Boron Nitride................................108
    Summary.....................................108
    References..................................109

4. **PLASMA-ASSISTED CHEMICAL VAPOR DEPOSITION**...........112
   V.S. Nguyen
    Introduction.................................112
    General Principles..........................113
        Nature of Plasma...........................113
        Reaction Kinetics in Plasma..................117
        Deposition Mechanism.......................120
            Radical Mechanism......................122
            Ionic Mechanism........................123
    The Deposited Films.........................124
        Silicon Nitride.............................124
        Silicon Oxynitride..........................126
        Silicon Oxide..............................128
        Silicon Films..............................130
        Other Conductor and Semiconductor Films......130
    Equipment for Plasma Deposition..............132
    Effects of Operating Parameters..............138
    Future Research and Development.............140
    References..................................141

5. **MICROWAVE ELECTRON CYCLOTRON RESONANCE PLASMA CHEMICAL VAPOR DEPOSITION**..........................147
   Seitaro Matsuo
    Introduction.................................147
        ECR Plasma Deposition Apparatus.............148
    Divergent Magnetic Field Plasma Extraction...150
    Deposition Characteristics....................154
        Silicon Nitride Deposition....................155
        Silicon Dioxide.............................159
        Ion Incidence Effects.......................159
        Material Supply By Sputtering................163
    ECR Plasma CVD System......................166
    Conclusions.................................168
    References..................................168

6. **MOLECULAR BEAM EPITAXY: EQUIPMENT AND PRACTICE**....170
   Walter S. Knodle and Robert Chow
    The Basic MBE Process......................170
    Competing Deposition Technologies............173
        Liquid Phase Epitaxy........................173
        Vapor Phase Epitaxy and MOCVD.............174
    MBE-Grown Devices.........................176
        Transistors.................................177
        Microwave and Millimeter Wave Devices.......178

Optoelectronic Devices........................ 180
Integrated Circuits........................... 181
MBE Deposition Equipment 183
   Vacuum System Construction 183
      Construction Practices 183
      Multi-Chamber Systems 184
      Pumping Considerations....................... 186
      Sample Transfer Techniques 186
   Sources................................. 186
      Thermal Evaporation Sources 187
      Electron Beam Heated Sources................... 191
      Implantation Sources......................... 191
      Gas Sources 191
      Source Shutters and the Source Flange 192
   Sample Manipulation 192
      Sample Mounting 192
      Sample Temperature Control..................... 192
      Sample Rotation Control 193
   System Automation........................... 194
   Performance Parameters........................ 194
Principles of Operation........................... 194
   Substrate Preparation.......................... 198
      III-V Substrate Cleaning 198
      Silicon Substrate Cleaning...................... 198
      II-VI Substrate Cleaning 199
   Growth Procedure............................ 199
      Thermal Transient........................... 200
      Doping Control............................. 200
      Compositional Control........................ 201
      Interrupted Growth.......................... 203
      In Situ Metallization 203
   In Situ Analysis.............................. 204
      Reflection High Energy Electron Diffraction 204
      X-ray Photoelectron Spectroscopy 204
      Auger Electron and Secondary Ion Mass Spectroscopy..... 204
      Residual Gas Analysis......................... 205
   Materials Evaluation........................... 205
      Optical Microscopy 205
      Hall Effect................................ 206
      Capacitance-Voltage.......................... 206
      Photoluminescence Spectroscopy................... 207
      Deep Level Transient Spectroscopy.................. 207
   Safety.................................... 208
Recent Advances.............................. 208
   RHEED Oscillation Control...................... 209
   GaAs on Silicon 210
   Oval Defect Reduction......................... 210
   Chemical Beam Epitaxy/Gas Source MBE 212

xiv Contents

        Hydride MBE..212
        Metalorganic MBE..................................212
        Superlattice Structures..............................213
            Strained-Layer Superlattices....................213
            Superlattice Buffer Layers......................214
            Superlattice Device Structures.................214
    Future Developments................................214
        Production Equipment.............................214
        In Situ Processing.................................216
        Process Developments.............................217
            Ionized Cluster Beam Epitaxy..................217
            Vacuum Chemical Epitaxy......................217
            Irradiation Assisted MBE.......................218
        Toxic Gases and Environmental Concerns..........218
    References...218

7. METAL-ORGANIC CHEMICAL VAPOR DEPOSITION: TECHNOLOGY AND EQUIPMENT........................234
J. L. Zilko

    Introduction...234
    Physical and Chemical Properties of Sources Used in MOCVD....237
        Physical and Chemical Properties of Organometallic
            Compounds....................................238
        Organometallic Source Packaging..................241
        Hydride Sources and Packaging....................243
    Growth Conditions, Mechanisms and Chemistry.............244
        Growth Conditions and Materials Purity.............245
        Growth Mechanisms...............................249
        Gas Phase Chemical Reactions.....................251
    System Design and Construction......................252
        Leak Integrity and Cleanliness.....................252
        Oxygen Gettering Techniques.....................253
        Gas Manifold Design.............................254
        Reaction Chamber...............................256
        Exhaust and Low Pressure MO-CVD................260
    Future Developments................................261
    References...265

8. PHOTOCHEMICAL VAPOR DEPOSITION....................270
Russell L. Abber

    Introduction...270
    Theory..271
    Review of Photo-CVD Applications....................273
        Silicon...273
        Dielectrics and Insulators.........................276
        Metals...277
        Compound Semiconductors.......................279
        Miscellaneous...................................280

Photo-CVD Equipment.	280
Commercial Equipment	280
Reactor Design	282
Summary.	286
References	286

9. INTRODUCTION TO SPUTTERING ... 291
Brian Chapman and Stefano Mangano

Principle and Implementation of Sputtering	291
Introduction.	291
What Is Sputtering?	294
Applications of Sputtering	295
Sources of Sputtering 'Bullets'	295
Ion Beam Sputtering	295
Ions From Plasmas.	296
Glow Discharge DC Sputtering.	297
Practical DC Sputtering Systems.	298
Challenges in Sputter Deposition	299
High Rate Sputtering	299
DC Magnetrons.	299
Sputtering of Insulators	300
RF Sputtering.	300
RF Magnetrons.	301
Reactive Processes.	301
Reactive Sputter Deposition	302
Control of Stoichiometry	302
Bias Sputtering	303
Properties of Bias Sputtered Films	303
Topography Control With Bias.	304
DC or RF Bias?	304
Sputter Deposition Equipment.	305
Variety of Equipment	305
Semiconductor Deposition Equipment	306
Static Systems	307
Planar Rotation Systems.	308
In-Line System.	308
Modules of a Sputter Deposition System	308
Etching	311
Sputter Etching.	311
Patterning By Sputter Etching	311
Glow Discharge Etching	312
Ion Beam Etching	312
Limitations of Sputter Etching.	312
Plasma Etching.	313
Patterning By Plasma Etching	313
Patterning By Lift-Off	314
Future of Sputtering	315
Sputtering For Step Coverage	315

xvi Contents

   Sputtering or CVD?................................ 316
   Sputter-Assisted Processes........................... 316
   Conclusions...................................... 317
  References...................................... 317

10. LASER AND ELECTRON BEAM ASSISTED PROCESSING 318
 Cameron A. Moore, Zeng-qu Yu, Lance R. Thompson, and
 George J. Collins
  Introduction..................................... 318
  Beam Assisted CVD of Thin Films 319
   Conventional CVD Methods 319
   Electron Beam Assisted CVD......................... 320
   Laser Assisted CVD............................... 320
   Experimental Apparati of Beam Assisted CVD 320
   Comparison of Beam Deposited Film Properties 322
    Laser-Deposited Dielectric Films 322
    Laser-Deposited Metallic Films..................... 325
    Electron-Beam Deposited Dielectric Films............. 327
  Submicron Pattern Delineation With Large Area Glow
  Discharge Pulsed Electron-Beams 330
  Beam Induced Thermal Processes....................... 333
   Overview....................................... 333
   Electron Beam Annealing of Ion-Implanted Silicon 334
   Electron Beam Alloying of Silicides 336
   Laser and Electron Beam Recrystallization of Silicon on SiO_2 .. 338
  Summary and Conclusions 340
  References...................................... 341

11. IONIZED CLUSTER BEAM DEPOSITION 344
 Isao Yamada, Toshinori Takagi, and Peter Younger
  Introduction..................................... 344
  Formation of Clusters and Properties of the Cluster........... 345
  Ionized Cluster Beam Deposition Equipment 349
  Film Formation Kinetics............................. 351
  Film Properties................................... 354
   Metals... 357
   Metal-Insulator-Semiconductor Structures 358
   Semiconductors 360
   Oxides, Nitrides and Others......................... 361
  Conclusions 362
  References...................................... 362

12. ION BEAM DEPOSITION............................... 364
 John R. McNeil, James J. McNally and Paul D. Reader
  Introduction..................................... 364
  Overview of Ion Beam Applications 364
   Categories of Kaufman Ion Sources..................... 365
   Operational Considerations.......................... 367

    Ion Beam Probing..367
    Substrate Cleaning With Ion Beams........................370
    Applications...373
        Ion Beam Sputtering....................................373
            Aspects of Sputtering..............................373
            Advantages/Disadvantages of Ion Beams for Sputtering.....376
            Aspects of Ion Beam Sputter Apparatus...............376
            Properties of Ion Beam Sputtered Films..............379
        Ion Assisted Deposition...............................380
            Equipment..380
            Procedures.......................................382
            Examples of Applications of IAD to Optical Coatings......383
            IAD Results......................................383
            Application Summary.............................390
    Concluding Comments.....................................391
    References...391

13. PLASMA AND ELEVATED PRESSURE OXIDATION IN VERY LARGE SCALE INTEGRATION AND ULTRA LARGE SCALE INTEGRATION.....................................393
    Arnold Reisman
    Introduction..393
    Plasma Assisted Oxidation Processes.......................396
    Elevated Pressure Oxidation................................402
    Conclusions..405
    References...406

INDEX..409

1

Deposition Technologies and Applications: Introduction and Overview

Werner Kern and Klaus K. Schuegraf

OBJECTIVE AND SCOPE OF THIS BOOK

The aim of this book is to provide a concise reference and description of the processes, methods, and equipment for depositing technologically important materials. Emphasis is placed on the most recently developed processes and techniques of film deposition for applications in high technology, in particular advanced microelectronic device fabrication that requires the most sophisticated and demanding approaches. The volume is intended to serve as a handbook and guide for the practitioner in the field, as a review and overview of this rapidly evolving technology for the engineer and scientist, and as an introduction to the student in several branches of science and engineering.

The discussions of the principles of operation of deposition equipment and its suitability, performance, control, capabilities, and limitations for production applications are intended to provide the reader with a basic understanding and appreciation of these systems. Key properties and areas of application of industrially important materials created by thin-film deposition processes will be described. Extensive use of references, reviews, and bibliographies is made to provide source material for specific use and more detailed study.

The chapter topics have been carefully selected to include primarily advanced and emerging deposition technologies in this rapidly evolving field. Many other important deposition technologies have not been included if adequate recent reviews are already available, or if the technologies are primarily intended for forming thick films or coatings that generally exceed a thickness of about ten micrometers, but the importance of these technologies is nevertheless recognized. Finally, an attempt has been made to compare competing technologies and to project a scenario for the most likely developments in the future.

IMPORTANCE OF DEPOSITION TECHNOLOGY IN MODERN FABRICATION PROCESSES

Deposition technology can well be regarded as the major key to the creation of devices such as computers, since microelectronic solid-state devices are all based on material structures created by thin-film deposition. Electronic engineers have continuously demanded films of improved quality and sophistication for solid-state devices, requiring a rapid evolution of deposition technology. Equipment manufacturers have made successful efforts to meet the requirements for improved and more economical deposition systems and for *in situ* process monitors and controls for measuring film parameters. Another important reason for the rapid growth of deposition technology is the improved understanding of the physics and chemistry of films, surfaces, interfaces, and microstructures made possible by the remarkable advances in analytical instrumentation during the past 20 years. A better fundamental understanding of materials leads to expanded applications and new designs of devices that incorporate these materials.

A good example of the crucial importance of deposition technology is the fabrication of semiconductor devices, an industry that is totally dependent on the formation of thin solid films of a variety of materials by deposition from the gas, vapor, liquid, or solid phase. The starting material, epitaxial films of semiconductors, are usually grown from the gas phase. Chemical vapor deposition of a single-crystal silicon film on a single-crystal silicon substrate of the same crystallographic orientation, a process known as homoepitaxy, is accomplished by hydrogen reduction of dichlorosilane vapor. If a single-crystal film of silicon is deposited on a nonsilicon crystal substrate, the process is termed heteroepitaxy. Layers of single-crystal compound semiconductors are created to a thickness of a few atom layers by molecular beam epitaxy.

Subsequent steps in the fabrication process create electrical structures that require the deposition of an insulating or dielectric layer such as an oxide, glass, or nitride, by one of several types of chemical vapor deposition (CVD) processes, by plasma-enhanced chemical vapor deposition (PECVD), or by any one of a number of sputtering deposition methods. The deposition of conductor films for contact formation and interconnections can be accomplished by vacuum evaporation or sputtering. CVD processes are especially suitable if polysilicon, polycides, or refractory metals are to be deposited.

The deposition of subsequent levels of insulator is repeated to build multilevel structures. Deposition may be complemented by spin-on techniques of organic polymeric materials, such as a polyimide, or of organometallic-based glass forming solutions. Spin-on deposition is especially useful if planarization of the device topography is required, as in the case of most high-density, multilevel conductor, VLSI circuits. The sequence of alternate film deposition of metals and insulators may be repeated several more times, with repetitive spin-on deposition of photopolymer masking solution for the delineation of contact openings, grid lines, and other pattern features by etching operations and lift-off techniques. Film formation by methods other than deposition are used in a few steps of the fabrication sequence; these include thermal oxidation of the substrate; ion implantation, nitridation, silicide formation; electrolytic and electroless metal deposition, and spray deposition (e.g., of organometallic solutions for forming antireflection coatings). These examples attest to the fact that the vast majority of material formation processes in semiconductor device technology (and in other areas of electronic device fabrication) are crucially dependent on film deposition technology.

CLASSIFICATION OF DEPOSITION TECHNOLOGIES

There are many dozens of deposition technologies for material formation.[1-4] Since the concern here is with *thin-film* deposition methods for forming layers in the thickness range of a few nanometers to about ten micrometers, the task of classifying the technologies is made simpler by limiting the number of technologies to be considered.

Basically, thin-film deposition technologies are either purely physical, such as evaporative methods, or purely chemical, such as gas- and liquid-phase chemical processes. A considerable number of processes that are based on glow discharges and reactive sputtering combine both physical and chemical reactions; these overlapping processes can be categorized as physical-chemical methods.

A classification scheme is presented in Table 1, where we have grouped thin-film deposition technologies according to evaporative glow-discharge, gas-phase chemical, and liquid-phase chemical processes. The respective pertinent chapter numbers in this book have also been indicated. Certain film formation processes such as oxidation which, strictly speaking, are not deposition processes, have been included because of their great importance in solid-state technology.

OVERVIEW OF VARIOUS THIN-FILM DEPOSITION TECHNOLOGIES

The following is a brief description of the principles, salient features, applications, and selected literature references of the more important technologies for thin-film deposition and formation categorized in Table 1. Technologies that are not covered in the chapters of this book have also been included in this discussion to present a more comprehensive overview of the entire field.

Evaporative Technologies

Although one of the oldest techniques used for depositing thin films, *thermal evaporation or vacuum evaporation*[5-7] is still widely used in the laboratory and in industry for depositing metal and metal alloys. The following sequential basic steps take place: (1) A vapor is generated by boiling or subliming a source material; (2) the vapor is transported from the source to the substrate; and (3) the vapor is condensed to a solid film on the substrate surface. Although deceptively simple in principle, the skilled practitioner must be well versed in vacuum physics, material science, mechanical and electrical engineering as well as in elements of thermodynamics, kinetic theory of gases, surface mobility, and condensation phenomena.

Evaporants cover an extraordinary range of varying chemical reactivity and vapor pressures. This variety leads to a large diversity of source components including resistance-heated filaments, electron beams; crucibles heated

4 Thin-Film Deposition Processes and Techniques

Table 1: Survey and Classification of Thin-Film Deposition Technologies

EVAPORATIVE METHODS

- Vacuum Evaporation
    - Conventional vacuum evaporation
    - Electron-beam evaporation
    - Molecular-beam epitaxy (MBE) (Chapt. 6)
    - Reactive evaporation

GLOW-DISCHARGE PROCESSES

- Sputtering (Chapt. 9)
    - Diode sputtering
    - Reactive sputtering
    - Bias sputtering (ion plating)
    - Magnetron sputtering
    - Ion beam deposition (Chapt. 12)
    - Ion beam sputter deposition
    - Reactive ion plating
    - Cluster beam deposition (CBD) (Chapt. 11)

- Plasma Processes
    - Plasma-enhanced CVD (Chapt. 4)
    - Plasma oxidation (Chapt. 13)
    - Plasma anodization
    - Plasma polymerization
    - Plasma nitridation
    - Plasma reduction
    - Microwave ECR plasma CVD (Chapt. 5)
    - Cathodic arc deposition

GAS-PHASE CHEMICAL PROCESSES

- Chemical vapor deposition (CVD)
    - CVD epitaxy (Chapt. 2)
    - Atmospheric-pressure CVD (APCVD)
    - Low-pressure CVD (LPCVD) (Chapt. 3)
    - Metalorganic CVD (MOCVD) (Chapt. 7)
    - Photo-enhanced CVD (PHCVD) (Chapt. 8)
    - Laser-induced CVD (PCVD) (Chapt. 10)
    - Electron-enhanced CVD (Chapt. 10)

- Thermal Forming Processes
    - Thermal oxidation (Chapt. 13)
    - Thermal nitridation
    - Thermal polymerization

    - Ion implantation

LIQUID-PHASE CHEMICAL TECHNIQUES

- Electro processes
    - Electroplating
    - Electroless plating
    - Electrolytic anodization
    - Chemical reduction plating
    - Chemical displacement plating
    - Electrophoretic deposition

- Mechanical techniques
    - Spray pyrolysis
    - Spray-on techniques
    - Spin-on techniques

    - Liquid phase epitaxy

by conduction, radiation, or rf-induction; arcs, exploding wires, and lasers. Additional complications include source-container interactions, requirements for high vacuum, precise substrate motion (to ensure uniformity) and the need for process monitoring and control.

Molecular Beam Epitaxy. (MBE)[8-16] is a sophisticated, finely controlled method for growing single-crystal epitaxial films in a high vacuum (10^{-11} torr). The films are formed on single-crystal substrates by slowly evaporating the elemental or molecular constituents of the film from separate Knudsen effusion source cells (deep crucibles in furnaces with cooled shrouds) onto substrates held at a temperature appropriate for chemical reaction, epitaxy, and re-evaporation of excess reactants. The furnaces produce atomic or molecular beams of relatively small diameter, which are directed at the heated substrate, usually silicon or gallium arsenide. Fast shutters are interposed between the sources and the substrates. By controlling these shutters, one can grow superlattices with precisely controlled uniformity, lattice match, composition, dopant concentrations, thicknesses, and interfaces down to the level of atomic layers.

The most widely studied materials are epitaxial layers of III-V semiconductor compounds, but silicon, metals, silicides, and insulators can also be deposited as single-crystal films by this versatile and uniquely precise method. Complex layer structures and superlattices for fabricating gallium arsenide heterojunction solid-state lasers, discrete microwave devices, optoelectronic devices, waveguides, monolithic integrated optic circuits, and totally new devices, have been created. An additional important advantage of MBE is the low temperature requirement for epitaxy, which for silicon is in the range of 400°C to 800°C,[15] and for gallium arsenide, 500°C to 600°C.[9] Several production systems with associated analytic equipment are now available.[16] The extremely limited product throughput, the complex operation, and the expensive equipment are, at present, the major limitations of this promising deposition technology for production applications.

Glow-Discharge Technologies

The electrode and gas-phase phenomena in various kinds of glow discharges (especially rf discharges) represent a rich source of processes used to deposit and etch thin films. Creative exploitation of these phenomena has resulted in the development of many useful processes for film deposition (as well as etching), as listed in Table 1.

Sputtering. The most basic and well known of these processes is sputtering,[17-25] the ejection of surface atoms from an electrode surface by momentum transfer from bombarding ions to surface atoms. From this definition, sputtering is clearly an etching process, and is, in fact, used as such for surface cleaning and for pattern delineation. Since sputtering produces a vapor of electrode material, it is also (and more frequently) used as a method of film deposition similar to evaporative deposition. Sputter-deposition has become a generic name for a variety of processes.

Diode Sputtering. Diode sputtering uses a plate of the material to be deposited as the cathode (or rf-powered) electrode (target) in a glow discharge. Material can thus be transported from the target to a substrate to form a film. Films of pure metals or alloys can be deposited when using noble gas discharges (typically Ar) with metal targets.

Reactive Sputtering. Compounds can be synthesized by reactive sputtering, that is, sputtering elemental or alloy targets in reactive gases; alternatively, they can be deposited directly from compound targets.

Bias Sputtering. Bias sputtering or ion-plating[25] is a variant of diode sputtering in which the substrates are ion bombarded during deposition and prior to film deposition to clean them. Ion bombardment during film deposition can produce one or more desirable effects, such as re-sputtering of loosely-bonded film material, low-energy ion implantation, desorption of gases, conformal coverage of contoured surface, or modification of a large number of film properties. The source material need not originate from a sputtering target, but can be an evaporation source, a reactive gas with condensable constituents, or a mixture of reactive gases with condensable constituents, and other gases that react with the condensed constituents to form compounds.

It should be noted that *all* glow discharge processes involve sputtering in one form or another, since it is impossible to sustain a glow discharge without an electrode at which these processes occur. In "electrodeless" discharges, rf power is capacitively coupled through the insulating wall of a tubular reactor. In this case, the inside wall of the tube is the main electrode of the discharge. However, sputtering can also lead to undesirable artifacts in this and other glow discharge processes.

Magnetron Sputtering. Another variant in sputtering sources uses magnetic fields transverse to the electric fields at sputtering-target surfaces. This class of processes is known as magnetron sputtering.[20-22] Sputtering with a transverse magnetic field produces several important modifications of the basic processes. Target-generated secondary electrons do not bombard substrates because they are trapped in cycloidal trajectories near the target, and thus do not contribute to increased substrate temperature and radiation damage. This allows the use of substrates that are temperature-sensitive (for example, plastic materials) and surface-sensitive (for example, metal-oxides-semiconductor devices) with minimal adverse effects. In addition, this class of sputtering sources produces higher deposition rates than conventional sources and lends itself to economic, large-area industrial application. There are cylindrical, conical, and planar magnetron sources, all with particular advantages and disadvantages for specific applications. As with other forms of sputtering, magnetron sources can be used in a reactive sputtering mode. Alternatively, one can forego the low-temperature and low radiation-damage features and utilize magnetron sources as high-rate sources by operating them in a bias-sputtering mode.

Ion-Beam Sputtering. Ion beams, produced in, and extracted from glow discharges in a differentially pumped system, are important to scientific investigations of sputtering, and are proving to be useful as practical film-deposition systems for special materials on relatively small substrate areas. There are several advantages of ion-beam sputtering deposition.[23] The target and substrate are situated in a high-vacuum environment rather than in a high-pressure glow discharge. Glow discharge artifacts are thereby avoided, and higher-purity films usually result. Reactive sputtering and bias sputtering with a separate ion gun can be used.

Plasma Processes. The fact that some chemical reactions are accelerated at a given temperature in the presence of energetic reactive-ion bombardment is the basis of processes for surface treatments such as plasma oxidation, plasma nitriding, and plasma carburizing.[26-28] A metal to be oxidized, nitrided or carburized is made the cathode of a glow discharge and is simultaneously heated by radiant or rf-induction means. The discharge gas is either O_2, N_2 plus H_2, or CH_4. Very thick (0.1-2 mm) protective coatings on a variety of metals can be produced in this way to render surfaces hard and/or corrosion resistant.

Anodization. Plasma anodization[26,28,29] is a technique for producing thin oxide films (less than 100 nm) on metals such as aluminum, tantalum, titanium, and zirconium, collectively referred as "valve metals." In this case, a dc discharge is set up in an oxygen atmosphere and the substrates (shielded from the cathode to avoid sputter deposition) are biased positively with respect to the anode. This bias extracts negative oxygen ions from the discharge to the surface, which is also bombarded with electrons that assist the reaction. The process produces very dense, defect-free, amorphous oxide films that are of interest as gate material in III-V compound semiconductor devices such as in microwave field-effect transistors.

Deposition of Inorganic/Organic Films. Plasma deposition of inorganic films[27,29-37] and plasma polymerization of organic reactants to produce films of organic polymers[38] involve the introduction of a volatile reactant into a glow discharge which is usually generated by an rf force. The reactant gases or vapors are decomposed by the glow discharge mainly at surfaces (substrate, electrodes, walls), leaving the desired reaction product as a thin solid film. Plasma deposition is a combination of a glow-discharge process and low-pressure chemical vapor deposition, and can be classified in either category. Since the plasma *assists* or *enhances* the chemical vapor deposition reaction, the process is usually denoted as PACVD or PECVD. The possibilities for producing films of various materials and for tailoring their properties by judicious manipulation of reactant gases or vapors and glow-discharge parameters are very extensive. Plasma deposition processes are used widely to produce films at lower substrate temperatures and in more energy-efficient fashion than can be produced by other techniques. For example, they are widely used to form secondary-passivation films of plasma silicon nitride on semiconductor devices, and to deposit hydrogenated, amorphous silicon layers for thin-film solar cells.

Microwave Electron Cyclotron Resonance Deposition. ECR plasma deposition[39,40] employs an electron cyclotron resonance (ECR) ion source to create a high-density plasma. The plasma is generated by resonance of microwaves and electrons through a microwave discharge across a magnetic field. The main feature of this recently introduced process is the high rate of deposition obtained at a low temperature of deposition.

Cluster Beam Deposition. Ionized cluster beam deposition (ICB) or cluster beam deposition[41-45] is one of the most recent emerging technologies for the deposition of thin films with growth-control capabilities not attainable by other processes. ICB depositions is one of several techniques classified as ion-assisted thin-film formation. The material to be deposited emerges and expands into a vacuum environment from a small nozzle of a heated confinement crucible, usually constructed of high-purity graphite. The vapor pressure within the crucible is several orders of magnitude higher than the pressure of the vacuum chamber so that the expanding vapor supercools. Homogeneous nucleation results in the generation of atomic aggregates or clusters of up to a few thousand atoms held together by weak interatomic forces. The clusters passing through the vacuum towards the substrate can in part be positively charged by impact ionization with electron beam irradiation. Closely controlled accelerating voltages add energy to the ionized clusters which then impinge on the substrate, diffuse or migrate along the plane of the surface, and finally form a thin film of exceptional purity. The complete and detailed process is extremely complex but offers unprecedented possibilities of film formation once the fundamentals and engineering technology are fully understood and exploited. Plasma deposition (and plasma etching) processes represent cases in which technology is leading science. The detailed interactions of plasma chemistry, plasma physics, and possible synergistic effects are still largely unexplained. In view of the technological importance of these processes, much more research and process modeling is required to obtain an adequate understanding of these deposition mechanisms.

Gas-Phase Chemical Processes

Methods of film formation by purely chemical processes in the gas or vapor phases include chemical vapor deposition and thermal oxidation. Chemical vapor deposition (CVD)[26,33,46-55] is a materials synthesis process whereby constituents of the vapor phase react chemically near or on a substrate surface to form a solid product. The deposition technology has become one of the most important means for creating thin films and coatings of a very large variety of materials essential to advanced technology, particularly solid-state electronics where some of the most sophisticated purity and composition requirements must be met. The main features of CVD are its versatility for synthesizing both simple and complex compounds with relative ease and at generally low temperatures. Both chemical composition and physical structure can be tailored by control of the reaction chemistry and deposition conditions. Fundamental principles of CVD encompass an interdisciplinary range of gas-phase reaction chemistry, theremodynamics, kinetics, transport mechanisms, film growth phenomena, and reactor engineering.

Chemical reaction types basic to CVD include pyrolysis (thermal decomposition), oxidation, reduction, hydrolysis, nitride and carbide formation, synthesis reactions, disproportionation, and chemical transport. A sequence of several reaction types may be involved in more complex situations to create a particular end product. Deposition variables such as temperature, pressure, input concentrations, gas flow rates and reactor geometry and operating principle determine the deposition rate and the properties of the film deposit. Most CVD processes are chosen to be heterogeneous reactions. That is, they take place at the substrate surface rather than in the gas phase. Undesirable homogeneous reactions in the gas phase nucleate particles that may form powdery deposits and lead to particle contamination instead of clean and uniform coatings. The reaction feasibility (other than reaction rate) of a CVD process under specified conditions can be predicted by thermodynamic calculations, provided reliable thermodynamic data (especially the free energy of formation) are available. Kinetics control the rate of reactions and depends on temperature and factors such as substrate orientation. Considerations relating to heat, mass, and momentum transport phenomena are especially important in designing CVD reactors of maximum efficiency. Since important physical properties of a given film material are critically influenced by the structure (such as crystallinity), control of the factors governing the nucleation and structure of a growing film is necessary.

Thin-film materials that can be prepared by CVD cover a tremendous range of elements and compounds. Both inorganic, organometallic, and organic reactants are used as starting materials. Gases are preferred because they can be readily metered and distributed to the reactor. Liquid and solid reactants must be vaporized without decomposition at suitable temperatures and transported with a carrier gas through heated tubes to the reaction chamber, which complicates processing, especially in the case of reduced-pressure systems. Materials deposited at low temperatures (e.g., below 600°C for silicon) are generally amorphous. Higher temperatures tend to lead to polycrystalline phases. Very high temperatures (typically 900°C to 1100°C in the case of silicon) are necessary for growing single-crystal films. These films are oriented according to the structure of the substrate crystal; this phenomenon, known as epitaxy, is of crucial practical importance in solid-state device technology.

CVD has become an important process technology in several industrial fields. As noted, applications in solid-state microelectronics are of prime importance. Thin CVD films of insulators, dielectrics (oxides, silicates, nitrides), elemental and compound semiconductors (silicon, gallium arsenide, etc.), and conductors (tungsten, molybdenum, aluminum, refractory metal silicides) are extensively utilized in the fabrication of solid-state devices. Hard and wear-resistant coatings of materials such as boron, diamond-like carbon, borides, carbides and nitrides have found important applications in tool technology. Corrosion resistant coatings, especially oxides and nitrides, are used for metal protection in metallurgical applications. Numerous other types of materials, including vitreous graphite and refractory metals, have been deposited mainly in bulk form or as thick coatings. Many of these CVD reactions have long been used for coating of substrates at reduced pressure, often at high temperatures.

Reactors. The reactor system (comprising the reaction chamber and all associated equipment) for carrying out CVD processes must provide several basic functions common to all types of systems. It must allow transport of the reactant and diluent gases to the reaction site, provide activation energy to the reactants (heat, radiation, plasma), maintain a specific system pressure and temperature, allow the chemical processes for film deposition to proceed optimally, and remove the by-product gases and vapors. These functions must be implemented with adequate control, maximal effectiveness, and complete safety.

The most sophisticated CVD reactors are those used for the deposition of electronic materials. Low-temperature (below 600°C) production reactors for normal- or atmospheric-pressure CVD (APCVD) include rotary vertical-flow reactors and continuous, in-line conveyorized reactors with various gas distribution features. They are used primarily for depositing oxides and binary and ternary silicate glass coatings for solid-state devices. Reactors for mid-temperature (600°C to 900°C) and high-temperature (900°C to 1300°C) operation are either hot-wall or cold-wall types constructed of fused quartz. Hot-wall reactors, usually tubular in shape, are used for exothermic processes where the high wall temperature avoids deposition on the reactor walls. They have been used for synthesizing complex layer structures of compound semiconductors for microelectronic devices. Cold-wall reactors, usually bell-jar shaped, are used for endothermic processes, such as the deposition of silicon from the halides or the hydrides.

Heating is accomplished by rf induction or by high-intensity radiation lamps. Substrate susceptors of silicon carbide-coated graphite slabs are used for rf-heated systems.

Reactors operating at low pressure (typically 0.1–10 torr) for low-pressure CVD (LPCVD) in the low-, mid-, or high-temperature ranges are resistance-heated hot-wall reactors of tubular, bell-jar, or close-spaced design. In the horizontal tubular design the substrate slices (silicon device wafers) stand up in a carrier sled and gas flow is horizontal. The reduced operating pressure increases the mean-free path of the reactant molecules, which allows a closely spaced wafer stacking. The very high packing density achieved (typically 100 to 200 wafers per tube) allows a greatly increased throughput, hence substantially lower product cost. In the vertical bell-jar design the gas is distributed over the stand-up wafers, hence there is much less gas depletion and generation of few particles, but the wafer load is smaller (50 to 100 wafers per chamber). Finally, the close-spaced design, developed most recently, processes each wafer in its own separate, close-spaced chamber with the gas flowing across the wafer surface to achieve maximal uniformity.

In LPCVD no carrier gases are required, particle contamination is reduced and film uniformity and conformality are better than in conventional APCVD reactor systems. It is for these reasons that low-pressure CVD is widely used in the highly cost-competitive semiconductor industry for depositing films of insulators, amorphous and polycrystalline silicon, refractory metals, and silicides. Epitaxial growth of silicon at reduced pressure minimizes autodoping (contamination of the substrate by its dopant), a major problem in atmospheric-pressure epitaxy.

Vapor-Phase Epitaxy. Vapor-phase epitaxy (VPE)[46,47,51-55] and metal-organic chemical vapor deposition (MOCVD)[46,47,51-55] are used for growing epitaxial films of compound semiconductors in the fabrication of optoelectronic devices. Composite layers of accurately controlled thickness and dopant profile are required to produce structures of optimal design for device fabrication.

Photo-Enhanced Chemical Vapor Deposition (PHCVD). (PHCVD)[56-58] is based on activation of the reactants in the gas or vapor phase by electromagnetic radiation, usually short-wave ultraviolet radiation. Selective absorption of photonic energy by the reactant molecules or atoms initiates the process by forming reactive free-radical species that then interact to form a desired film product. Mercury vapor is usually added to the reactant gas mixture as a photosensitizer that can be activated with the radiation from a high-intensity quartz mercury resonance lamp (253.7 nm wavelength). The excited mercury atoms transfer their energy kinetically by collision to the reactants to generate free radicals. The advantages of this versatile and very promising CVD process is the low temperature (typically 150°C) needed to form films such as SiO_2 and Si_3N_4, and the greatly minimized radiation damage (compared to PECVD). The limitations at present are the unavailability of effective production equipment and the need (in most cases) for photoactivation with mercury to achieve acceptable rates of film deposition.

Laser-Induced Chemical Vapor Deposition (LCVD). (LCVD)[59-61] utilizes a laser beam for highly localized heating of the substrate that then induces film deposition by CVD surface reactions. Another mode of utilizing laser (or electron radiation) is to activate gaseous reactant atoms or molecules by their absorption of the specific wavelength of the photonic energy supplied. The resulting chemical gas phase reactons are very specific, leading to highly pure film deposits. On the other hand, the activation matching of the spectral properties with the reactant species limits the choice of reactions and hence the film deposits that can be obtained. LCVD is still in its early development stages but promises many interesting and useful applications in the future.

Ion Implantation. Recently, ion implantation[62-64] has been used to form silicon-on-insulator structures by implanting large doses of atomic or molecular oxygen ions in single-crystal silicon substrates to produce a buried oxide layer with sharp interfaces after annealing.[63] Simultaneous high-dose implantation of low energy oxygen and nitrogen ions into silicon yields very thin films of silicon oxynitride, whereas low-energy implantation of nitrogen or ammonia into silicon yields a low density silicon nitride layer.[65]

Thermal Oxidation. In the gas phase, thermal oxidation[27,28,66,67] is a chemical thin-film forming process in which the substrate itself provides the source for the metal or semiconductor constituent of the oxide. This technique is obviously much more limited than CVD, but has extremely important applications in silicon device technology where very high purity oxide films with a high quality Si/SiO_2 interface are required. Thermal oxidation of silicon surfaces produces glassy films of SiO_2 for protecting highly sensitive p-n junctions and for creating dielectric layers for MOS devices. Temperatures for this process lie in the range of about 700°C to 1200°C with either dry or moist oxygen or water vapor (steam) as the oxidant. Steam oxidation proceeds at a much faster rate than dry oxidation. The oxidation rate is a function of the oxidant partial pressure and is controlled essentially by the rate of oxidant diffusion through the growing SiO_2 layer to the SiO_2/Si interface, resulting in a decrease of the growth rate with increased oxide thickness. The process is frequently conducted in the presence of hydrochloric acid vapor or vapors of chlorine-containing organic compounds. The HCl vapor formed acts as an effective impurity getter, improving the Si/SiO_2 interface properties and stability.

Oxidation of Silicon. Silicon oxidation under high pressure[26,66,67] is of technological interest where the temperature must be minimized, such as for VLSI devices. Since the oxidation rate of silicon is approximately proportional to pressure, higher product throughput and/or decreased temperatures can be attained. The oxidant in commercial systems is H_2O, which is generated pyrogenically from H_2 and O_2. Pressures up to 10 atm are usually used at temperatures ranging from 750°C to 950°C.

Other Gas-Phase Oxidations. Gas-phase oxidation of other materials[26] is of limited technical importance. Examples include metallic tantalum films converted by thermal oxidation to tantalum pentoxide for use as antireflection coating in photovoltaic devices and as capacitor elements in microcircuits. Other metal oxides grown thermally have also been used as capacitor dielectrics in thin-film devices, to improve the bonding with glass in glass-to-metal seals and to improve corrosion resistance.

Liquid-Phase Chemical Formation

The growth of inorganic thin films from liquid phases by chemical reactions is accomplished primarily by electrochemical processes (which include anodization and electroplating), and by chemical deposition processes (which include reduction plating, electroless plating, conversion coating, and displacement deposition). A number of extensive reviews[26,68-70] of these film formation processes discuss theory and practice. Another class of film forming methods from the liquid phase is based on chemically reacting films that have been deposited by mechanical techniques.[26,69] Finally, liquid phase epitaxy[51] is still being used for growing a number of single-crystal semiconductors.

Electrolytic Anodization. In anodization, as in thermal oxidation, an oxide film is formed from the substrate. The anode reacts with negative ions from the electrolyte in solution and becomes oxidized, forming an oxide or a hydrated oxide coating on semiconductors and on a few specific metals, while hydrogen gas is evolved at the cathode. Nonporous and well-adherent oxides can be formed on aluminum, tantalum, niobium, titanium, zirconium, and silicon. The most important applications are corrosion-protective films and decorative coatings with dyes on aluminum and its alloys, and layers for electrical insulation for electrolyte capacitors on aluminum and tantalum.

Electroplating. In electroplating a metallic coating is electrodeposited on the cathode of an electrolytic cell consisting of a positive electrode (anode), a negative electrode (cathode), and an electrolyte solution (containing the metal ions) through which electric current flows. The quantitative aspects of the process are governed by Faraday's laws. Important electroplating variables include current efficiency, current density, current distribution, pH, temperature, agitation, and solution composition. Numerous metals and metal alloys have been successfully electroplated from aqueous solutions. However, the technically most useful electroplated metals are chromium, copper, nickel, silver, gold, rhodium, zinc, and a series of binary alloys including chromium/nickel composites. Electroplating is widely used in industry and can produce deposits that range from very thin films to very thick coatings (electroforming).

Chemical Reduction Plating. Chemical reduction plating is based on reduction of a metal ion in solution by a reducing agent added just before use. Reaction is homogeneous, meaning that deposition takes place everywhere in the solution, rather than on the substrate only. Silver, copper, nickel, gold, and some sulfide films are readily plated. The oldest application of the process is the silvering of glass and plastics for producing mirrors using silver nitrate solutions and one of various reducing agents, such as hydrazine.

Electroless Plating. Autocatalytic or electroless plating is a selective deposition plating process in which metal ions are reduced to a metallic coating by a reducing agent in solution. Plating takes place only on suitable catalytic surfaces, which include substrates of the same metal being plated, hence the definition autocatalysis. Electroless (or electrodeless) plating offers a number of advantages over electroplating, such as selective (patterned) deposition, but is limited to a few metals and some alloys. Nickel, nickel alloys, and copper are most widely used commercially on conductive and on sensitized insulating substrates, including plastic polymeric materials.

Electrophoretic Deposition. Electrophoretic coating is based on deposition of a film from a dispersion of colloidal particles onto a conductive substrate. The dispersion in a conductive liquid dissociates into negatively charged colloidal particles and positive ions (cations), or the reverse. On application of an electric field between the positive substrate electrode (anode), the colloidal particles migrate to the substrate, become discharged, and form a film.

Chemical or electrochemical treatments of a metal surface can produce a thin and adherent layer on that metal. Examples of such conversion coatings are black oxides on steel, copper, and aluminum. Widely used chromate conversion coatings on zinc, cadmium, silver, copper, brass, aluminum, and magnesium are formed by reaction of hexavalent chromium ions with the metal, forming protective and decorative films that consist of oxides, chromates, and the substrate metal. Phosphate conversion coatings result from treatments, especially of iron and steel, with phosphoric acid-containing salts of iron, zinc, or manganese.

Immersion Plating. Deposition of a metal film from a dissolved salt of the coating metal on a substrate by chemical displacement without external electrodes is known as displacement deposition or immersion plating. Generally, a less noble (more electronegative) metal displaces from solution any metal that is more noble, according to the electromotive force series. Actually, different localized regions on the metal surface become anodic and cathodic, resulting in thicker films in the cathodic areas. The industrial uses of this process are limited to a few applications, mainly tin coatings on copper and its alloys.

Mechanical Methods. Mechanical techniques[71] for depositing coatings from liquid media that are subsequently reacted chemically to form the inorganic thin film product are spraying, spinning, dipping and draining, flow coating, roller coating, pressure-curtain coating, brushing, and off-set printing of reagent solutions. Chemical reaction of the coating residue, often by thermal oxidation, hydrolysis, or pyrolysis (in the case of metalorganics) produces the desired solid film. Spin-on deposition of film-forming solutions is widely used in solid-state technology.

Of the deposition techniques noted, liquid spray coating is probably the most versatile mechanical coating technique and is particularly well suitable for high-speed automated mass production. Deposition of very thin films is possible by judicious selection and optimization of spray machine parameters for forming "atomized" droplets and the reagent and solvent systems used to formulate the spray liquid. An example of the capability of this refined method is the mass production spray-on deposition of organometallic alkoxy compounds, such as titanium-(IV)-isopropoxide, in an optimally formulated spray solution. Controlled pyrolysis of the deposit can form TiO_2 films of 70 nm thickness which serves as a highly effective and low-cost antireflection coating for silicon solar cells.[72]

It should be noted that spray deposition encompasses several other types of spraying processes that are based on either liquid sources, such as harmonic electrical spraying, or on dry source reactants that include flame spraying, arc plasma spraying, electric arc spraying, and detonation coating.[71]

Liquid-Phase Epitaxy. (LPE)[51] is used for the thermally-controlled overgrowth of thin single-crystalline films of compound semiconductors and magnetic garnets from the melt on a single-crystal substrate. This relatively old and simple technique has been successfully applied in the semiconductor industry for fabricating optoelectronic devices. However, compared to MBE, LPE is limited by poor uniformity and surface morphology.

Various additional film formation techniques[71] that are used industrially have not been discussed here, since they are essentially thick-film processes. These methods include powder or glass frit sedimentation and centrifugation, dipping, screen-printing, tape transfer, fluidized bed coating, and electrostatic spraying, all followed by thermal treatments for drying and fusion, or for chemical reaction of the deposit to form a coherent coating.

CRITERIA FOR THE SELECTION OF A DEPOSITION TECHNOLOGY FOR SPECIFIC APPLICATIONS

The selection of a specific technology for the deposition of thin films can be based on a variety of considerations. A multitude of thin films of different materials can be deposited for a large variety of applications; hence, no general guidelines can be given of what the most suitable deposition technology should be. In selecting an appropriate deposition technology for a specific application, several criteria have to be considered.

Thin-Film Applications

In considering the different applications of deposited thin films,[73,74] the following generic categories can be identified.

Electronic Components. The fabrication of electronic components, especially solid-state devices and microelectronic integrated circuits, have undoubtedly found the widest and most demanding applications for thin-film depositions. These films typically consist of semiconductor materials, dielectric and insulating materials, and metal or refractory metal silicide conductors.

Electronic Displays. Electronic displays are used for interfacing of electronic equipment with human operators. There are different components and device structures required, such as:

Liquid-crystal displays

Light-emitting diodes (LEDs)

Electroluminescent displays

Plasma and fluorescent displays

Electrochromic displays

The fabrication of these displays requires conductive films, transparent and conductive films, luminescent or fluorescent films as well as dielectric and insulating layers.

Optical Coatings. Optical coatings are applied for antireflection purposes, as interference filters on solar panels, as plate glass infrared solar reflectors, and for laser optics. In the fabrication of filter optics, thin films with refractive index gradients are deposited on preforms from which the optical fibers are drawn. These coatings require dielectric materials with precisely defined indices of refraction and absorption coefficients. Laser optics require metal reflective coatings which can withstand high radiation intensities without degradation. Infrared reflecting coatings are applied to filament lamps to increase the luminous flux intensity.

Magnetic Films for Data Storage. Thin films of magnetic materials have found wide commercial applications for data storage in computers and control systems. The substrates can be metal, glass or plastic polymeric materials. Thin film deposition processes for magnetic materials and for materials with a high degree of hardness are required.

Optical Data Storage Devices. Thin films are finding increasing commercial use for optical data storage devices in compact disks and computer memory applications. Processes for the deposition of organic polymer materials as storage media and as protective overcoats are required for this technology.

Antistatic Coatings. Thin films of conductive or semiconductive materials are deposited to provide protection from electrostatic discharges.

Hard Surface Coatings. Thin film coatings of carbides, silicides, nitrides, and borides are finding increased uses to improve the wear characteristics of metal surfaces for tools, bearings, and machine parts. Of particularly great current interest are films of diamond-like carbon because of this material's heat dissipation properties, electrical insulation, hardness, and resistance to high-temperature and high-energy radiation.

Material Characteristics

The desired material characteristics of the deposited films[73,74] will be in most cases the decisive factor for the selection of a preferred deposition technology. In many, if not most, instances the characteristics of a thin film can be quite different from the bulk material properties, since thin films have a large surface area to bulk volume ratio. In addition, the morphology, structure, physical and chemical characteristics of the thin film can also be quite different from those of the bulk materials. The surface and/or interface properties of the substrate to be coated can influence thin film characteristics drastically due to surface contamination, nucleation effects, surface mobility, chemical surface reactions, adsorbed gases, catalytic or inhibitory effects on film growth, surface topography, and crystallographic orientation, and stress effefts due to thermal expansion mismatch.

The major physical and chemical parameters of the thin film to be considered can be listed as follows:

Electrical: Conductivity for conductive films
  Resistivity for resistive films
  Dielectric constant
  Dielectric strength
  Dielectric loss
  Stability under bias
  Polarization
  Permittivity
  Electromigration
  Radiation hardness

Thermal: Coefficient of expansion
  Thermal conductivity

16 Thin-Film Deposition Processes and Techniques

	Temperature variation of all properties
	Stability or drift of characteristics
	Thermal fusion temperature
	Volatility and vapor pressure
Mechanical:	Intrinsic, residual, and composite stress
	Anisotropy
	Adhesion
	Hardness
	Density
	Fracture
	Ductility
	Hardness
	Elasticity
Morphology:	Crystalline or amorphous
	Structural defect density
	Conformality/step coverage
	Planarity
	Microstructure
	Surface topography
	Crystallite orientation
Optical:	Refractive index
	Absorption
	Birefringence
	Spectral characteristics
	Dispersion
Magnetic:	Saturation flux density
	Coercive force
	Permeability
Chemical:	Composition
	Impurities
	Reactivity with substrate and ambient
	Thermodynamic stability
	Etch rate
	Corrosion and erosion resistance

Toxicity

Hygroscopicity

Impurity barrier or gettering effectiveness

Carcinogenicity

Stability

Process Technology

As discussed before, a wide variety of process technologies is available for the deposition of thin films. The technologies differ to a large degree in their physical and chemical principles of operation and in the commercially available types of equipment. Each process technology has been pursued or developed because it has unique advantages over others. However, each process technology has its limitations. In order to optimize the desired film characteristics, a good understanding of the advantages and restrictions applicable to each technology is necessary.

The desired film thickness is closely related to the deposition or formation rates, since economical considerations determine, to a large degree, the selection of the most appropriate deposition technology. Thin films cover a thickness range from about 1 nm to several micrometers, or from film monolayers to thicknesses approaching bulk material characteristics.

Of increasing importance is the particle density associated with the deposited film. Particles originating from the equipment, the substrate, the environment, or from the reactant materials supplied to the deposition equipment can impose serious limitations to the utility of a deposition process or the equipment used. This is especially true for the fabrication of high-density microelectronic devices where the particle size can be equal to or exceed the minimum device dimensions. Deposition processes for very-large-scale integrated (VLSI) circuits require, or will require in the near future, particle densities of less than 0.1 per cm^2 for particles down to 0.2 μm in diameter.

All thin film deposition equipment is quite susceptible to the integrity of the processing environment. The deposition processes described in the following chapters operate over a wide range of pressures in the reaction chamber. For proper pressure or residual gas control inside the reaction chamber, the leakage of external gases has to be minimized. For molecular beam epitaxy of thin films, for example, a maximum leak rate of 10^{-12} torr-liter/cm^2-sec is required, whereas some other process technologies can easily tolerate much higher leak rates.

Considerable attention has to be given to the source materials and their delivery into the deposition reactor. They can not only be a hard-to-detect source of impurity contamination, but can also influence the uniformity of the deposited films. The source materials used for the thin film deposition can be either solid, liquid or gaseous. Special techniques for the source material delivery have been developed for each type of material and rate of delivery to the deposition reactor.

The purity of the deposited film not only depends on the purity of the source materials delivered to the reactor and the leak tightness of the system, but also to a large degree on the substrate cleaning procedure used. Deposition techniques have been developed which permit film deposition with purity levels down to 10^{12} impurity atoms per cm^3, or roughly 1 part in 10^{10}.

Of considerable importance is the stability and repeatability of the processes used, especially in the large-scale manufacturing of semiconductor devices. Many different factors can influence the deposition process, and it is extremely important to understand and control these parameters. To make a film deposition process acceptable for device manufacturing, an extensive and careful characterization of the processes and equipment is often required. Any process instabilities or uncontrolled deposition parameters should be discovered and rectified during this procedure.

The uniformity of the deposited films, both in thickness and composition, is of great concern for most deposition processes. In the manufacturing of integrated circuits, small variations in film uniformity can have a large influence on the manufacturing yield. For VLSI circuits, film uniformity deviations of less than 5% are required at present. It is expected that the uniformity requirements will become even more stringent, decreasing to a deviation limit of 1 to 2% in the near future. These requirements impose severe restrictions on the design of the deposition reactor, the delivery of the reactant materials, and the control of the process parameters.

Thin-film deposition processes for solid-state device fabrication are needed in many steps in the fabrication process sequence. It is important that compatible deposition processes are selected that do not interfere with the structures already built into the device. The process integration, which has to consider thermal effects, chemical and metallurgical compatibility as well as functional requirements and limitations, is a major consideration in successful process selection.

Frequently, the deposition processes have to offer a high degree of flexibility in meeting the demands for specific device requirements. The process selection has to be based on adjusting deposition parameters such as film thickness, uniformity and composition. During the process characterization described above, a good understanding of the sensitivity and control of the film characteristics can be obtained.

In a device fabrication process sequence, one frequently has to deposit films on a nonplanar surface. The deposited film should be uniform across all structural details of the substrate topography. For example, in VLSI structures, contact holes with micron or submicron dimensions should be uniformly coated with metal films not only inside the small contact cavities, but also on their vertical walls. This is referred to as step coverage or conformality. The different deposition processes described in the following chapters can result in quite different degrees of step coverage.

An important requirement for high-density VLSI devices is planarization of the substrate topography after film deposition. This is necessary in multilevel device fabrication processes where subsequent photolithographic pattern definition of very small geometries is required, or where deposited material step coverage is essential. The focal depth of the photolithographic equipment is on the order of one micron. To image the pattern across the entire field of view demands a highly planar surface topography.

Deposition Technologies and Applications   19

The large number of process parameters that can affect the uniformity and composition of the deposited films make *in situ* monitoring of the deposition parameters highly desirable. Many process parameters, such as reactor pressure, substrate temperature, reactant gas composition and deposition rates can be monitored in real time. However, significant improvements in process monitoring devices, real-time analytical instruments and process simulation can be expected in the next couple of years that promise to enhance the overall fabrication yield for devices manufactured by thin-film deposition processes.

Thin-Film Manufacturing Equipment

The equipment for the deposition of thin films can be classified into three basic categories:

- Thin-film deposition equipment for device research and development
- Prototype equipment for the study of new or established deposition processes
- Thin-film production equipment for device manufacturing

The prototype equipment for the development of new deposition processes can encompass a wide variety of designs and constructional details. Only the future outlook for the requirements and applications of such equipment will be discussed in the following chapters in this book.

The deposition equipment for the research and development of new device structures has to meet, in general, quite different requirements than the equipment used in the manufacturing of devices on a large scale. Research and development equipment requires a high degree of flexibility in deposition parameters, in accommodation of a variety of substrates, and in process monitoring equipment. High product throughput and a high degree of equipment automation is usually not required.

Thin-film deposition equipment used in high volume manufacturing is designed, in most cases, for a very limited range of applications. The major consideration is the cost-effectiveness of the equipment, which can be characterized as the ratio of cost per device divided by the value added to the device. The equipment cost, consisting of acquisition cost, amortization and maintenance cost is a major consideration for the selection of production-worthy deposition equipment. System throughput, expressed in the number of substrates processed per hour for a given film thickness, is a major selection criterion. Equipment reliability, characterized by equipment uptime, mean time between failures (MTBF) and mean time to repair (MTTR) is becoming increasingly a part of equipment specifications and is subject to standards for fabrication equipment. Expectations for equipment uptime, as currently required in the semiconductor industry, is in excess of 90%. The high cost of manufacturing equipment for film depositions mandates the need for ease of maintenance and repair. Undesirable deposition on reactor walls and fixtures should be minimized. Self-cleaning features using reactive gas plasma discharges are rapidly gaining in popularity.

Until recently, most deposition equipment used in thin film manufacturing has been of the batch processing type. A limited number of substrates to be coated is loaded into the deposition chamber and processed as a unit. Batch processing, although offering good process control, suffers in general from limited throughput. The time required for substrate loading, pump-down, purging, thermal equilibration, recovery, and cool-down accounts for a considerable percentage of the total batch processing time. With the availability of load-lock systems, permitting the insertion of substrates into the reactor chamber without major disturbance of the process chamber pressure and environment, a considerable improvement in product throughput and film quality can be obtained. Therefore, continuous process reactors, including advanced single-substrate reactors, are finding increasing acceptance for film deposition in device manufacturing. The continuous process reactors offer the additional advantages of incorporating process steps for pre-deposition cleaning of the substrate as well as post-deposition treatments, such as thermal annealing.

The automation of deposition equipment is making rapid progress. Automated deposition systems provide automatic loading and unloading, process sequencing, and control of variables, such as reactor pressure, gas flows, and substrate temperature. In addition, diagnostic capabilities for detecting and analyzing equipment failures, maintenance requirements, and process integrity are being incorporated. Equipment communication of process parameters, failure modes and product status with upstream host computers as well as remote process recipe generation are becoming increasingly available for modern deposition equipment.

SUMMARY AND PERSPECTIVE FOR THE FUTURE

The importance of thin-film deposition technology in modern fabrication processes has been discussed with examples from the production of semiconductor devices. Following this overview, thin-film deposition technologies have been outlined and classified into four major generic categories: (1) evaporative methods, (2) glow-discharge processes, (3) gas-phase chemical processes, and (4) liquid-phase chemical film formation techniques. Important technologies from each of these categories have been discussed with respect to the underlying principles, salient features, and typical applications. Criteria for selecting a particular thin-film deposition technology have then been described in terms of specific applications, material characteristics, and processing. Finally, thin-film manufacturing equipment has been discussed and categorized.

A variety of different thin-film deposition technologies and equipment is available from which a selection can be made. These technologies are described in the following chapters in some detail. It is possible to a large degree to tailor the deposition process to the specific needs, based on the

- Physical and chemical material characteristics,
- Specific applications,
- Advantages and limitations in process technology,
- Manufacturing technology and equipment.

The fabrication processes established for semiconductor devices during the last two decades have provided an important stimulus for the development of new thin-film materials, processes and equipment. It can be expected that this trend will continue for the decade.

The current trends[52,53,75,76] in deposition technology for the fabrication of semiconductor devices are characterized by

- Shift to larger substrate sizes, e.g. silicon wafers of 150 to 200 mm diameter,
- Automation in substrate handling and process controls,
- Reduction in particle and metal contamination,
- Improvements in equipment reliability and ease of service and maintenance,
- Lower process temperatures,
- Improvements in film uniformity,
- Reduced in-process damage (due to high voltage, radiation, particle bombardment, electrostatics, etc.).

The number of deposition steps in the fabrication sequence of integrated circuits is expected to increase with the advent of more complex circuits, as shown in Table 2.[75] Less complex devices, such as those introduced a decade ago, typically NMOS and CMOS integrated circuits, have required only three deposition steps for inorganic film deposition, whereas more advanced devices such as high-performance VLSI silicon integrated circuits, now require 8 to 11 deposition steps. With increasing demands for cost reductions in the manufacturing of integrated circuits, cost-effective, high-volume manufacturing equipment for all deposition processes will be required.

Table 2: Film Deposition for Silicon Integrated Circuit Fabrication*

Number of CVD Steps required for IC Fabrication

DEPOSITION	H-CMOS	HMOS	CMOS/NMOS	SCHOTTKY TTL	ECL
Poly-Silicon	2	1-3	1	-	1
Epitaxy	1	-	-	1	1
PSG/Si_3N_4	2-3	1-2	1	2	-
Aluminum**	1	1-2	1	2	2-3
Other Metals	1-2	1	-	1	-
Silicides	1-2	0-2	-	-	-
Total Deposition Steps	8-11	4-10	3	6	4-5

*Data from Reference 75, Courtesy of *Semiconductor International*, the Cahners Publishing Company.
**Not usually deposited by CVD.

New semiconductor device structures based on III-V semiconductor compounds for microwave, high-speed and optoelectronic applications will require improved deposition systems and higher throughput capabilities for economic device fabrication, especially molecular beam epitaxy (MBE) and metal-organic chemical vapor deposition (MOCVD) equipment. Superlattice structures with alternating films of a few atomic layers with different compositions will demand a high degree of process control.

Three-dimensional integrated structures consisting of 3 to 5 layers with active components are under development in various laboratories. These structures are likely to require highly sophisticated deposition technologies and equipment.

Considerable interest exists to enhance the survivability of integrated circuits in hostile environments, such as high-energy radiation, high operating temperatures, and polluted atmospheres. The development of protective layers and more resistant device structures will require new thin-film materials and deposition processes.

Complex high-density integrated circuits face increased limitations in interconnecting the numerous components on a chip. Optical interconnection schemes are under development which can reduce this problem considerably. This trend will lead to new technologies combining optoelectronic device technology with the existing semiconductor process technologies. Thin-film deposition techniques will most likely play an important role for the fabrication of these interconnections.

This overview of the important processes and techniques used industrially for forming thin films indicates the extremely powerful and versatile arsenal of methods that is now available to the technologist. Coatings can be prepared that can meet a very wide range of requirements for specific industrial or scientific applications. While remarkable advances in thin-film technology have been made, there are still areas in which the technology is leading science. This is particularly true for photo-induced, ion-assisted, and plasma-enhanced processes; these areas should provide fertile ground for future research.

Acknowledgments

The authors wish to thank Norman Goldsmith and George L. Schnable, both from RCA Laboratories, for critically reviewing the manuscript, and for their many helpful comments and suggestions.

REFERENCES

1. Maissel, L.I. and Glang, R., editors, *Handbook of Thin Film Technology,* McGraw-Hill, New York (1970).
2. Vossen, J.L. and Kern, W., editors, *Thin Film Processes,* Academic Press, New York (1978).
3. Bunshah, R.F., editor, *Deposition Technologies for Films and Coatings: Developments and Applications,* Noyes Publications, Park Ridge, New Jersey (1982).
4. Ghandhi, S.K., *VLSI Fabrication Principles,* John Wiley & Sons, New York (1983).

5. Glang, R., Chapter 1, pp. 1-3 to 1-130 in Reference 1.
6. Glang, R., Holmwood, R.A. and Kurtz, J.A., Chapter 2, pp. 2-1 to 2-142 in Reference 1.
7. Bunshah, R.F., Chapter 4, pp. 83-169 in Reference 3.
8. Gossard, A.C., in *Treatise on Material Science and Technology,* Vol. 24, pp. 13-66, K.N. Tu, editor, Academic Press, New York (1982).
9. Singer, P.H., *Semiconductor Internatl.,* Vol. 6 (10), pp. 73-80 (Oct. 1983), and Vol. 9 (10), pp. 42-47 (Oct. 1986).
10. Aleksandrov, L., Thin Film Science and Technology, Vol. 5, Elsevier, New York (1984).
11. Wang, K.L., *Solid State Technol.,* Vol. 29 (10), pp. 137-143 (Oct. 1985).
12. Chang, L.L. and Ploog, K., editors, *Molecular Beam Epitaxy and Heterostructures,* Martinus Nijhoff Publishers, Dordrecht/Boston/Lancaster (1985).
13. Parker, E.H., editor, *The Technology and Physics of Molecular Beam Epitaxy,* Plenum Press, New York (1985).
14. Bean, J.C., editor, Proc. First Internatl. Symp. on *Silicon Molecular Beam Epitaxy,* Proc. Vol. 85-7, The Electrochem. Soc., Inc., Pennington, New Jersey (1985).
15. Narayanamurti, V. and Gibson, J.M., *Physics Today,* Vol. 39 (1), p. S40 (Jan. 1986).
16. Knodle, W., *Research and Development,* Vol. 28 (8), pp. 73-85 (Aug. 1986).
17. Wehner, G.K. and Anderson, G.S., Chapter 3, pp. 3-1 to 3-38 in Reference 1.
18. Maissel, L., Chapter 4, pp. 4-1 to 4-44 in Reference 1.
19. Vossen, J.L. and Cuomo, J.J., Chapter II-1, pp. 12-73 in Reference 2.
20. Thornton, J.A. and Penfold, A.S., Chapter II-2, pp. 76-113 in Reference 2.
21. Fraser, D.B., Chapter II-3, pp. 115-129 in Reference 2.
22. Waits, R.K., Chapter II-4, pp. 131-173 in Reference 2.
23. Harper, J.M.E., Chapter II-5, pp. 175-206 in Reference 2.
24. Thornton, J.A., Chapter 5, pp. 170-223 in Reference 3.
25. Mattox, D.M., pp. 244-287 in Reference 3.
26. Campbell, D.S., Chapter 5, pp. 5-1 to 5-25 in Reference 1.
27. Ojha, S.M., *Phys. of Thin Films,* Vol. 12, pp. 237-296, G. Hass, M.H. Francome and J.L. Vossen, editors, Academic Press, New York (1982).
28. Ghandhi, S.K., Chapter 7, pp. 371-417 in Reference 4.
29. Hollahan, J.R. and Rosler, R.S., Chapter IV-1, pp. 335-360 in Reference 2.
30. Reinberg, A.R., *Ann. Rev. Mater. Sci.,* Vol. 9, pp. 341-372, R. Huggins, editor, Annual Reviews, Inc., New York (1979).
31. Thornton, J.A., Chapter 2, pp. 19-62 in Reference 3.
32. Bonifield, T.D., Chapter 9, pp. 365-384 in Reference 3.
33. Adams, A.C., in *VLSI Technology,* pp. 93-129, S.M. Sze, editor, McGraw-Hill, New York (1983).
34. Sherman, A., *Thin Solid Films,* Vol. 113, pp. 135-149 (1984).
35. Catherine, Y., Proc. Fifth Symp. on *Plasma Processing,* Proc. Vol. 85-1, pp. 317-344, Mathad, G.S., Schwartz, G.C. and Smolinsky, G., editors, The Electrochem. Soc. Inc., Pennington, New Jersey (1985).
36. Adams, A.C., Symp. on *Reduced Temperature Processing for VLSI,* Proc. Vol. 86-5, pp. 111-131, R. Reif and G.R. Srinivasan, editors, The Electrochem. Soc. Inc., Pennington, New Jersey (1986).

37. Nguyen, S.V., *J. Vac. Sci. Technol.*, Vol. B4 (5), pp. 1159-1167 (Sept./Oct. 1986).
38. Yasuda, H., Chapter IV-2, pp. 361-398 in Reference 2.
39. Matsuo, S. and Kiuchi, M., *Japan. J. Appl. Phys.*, Vol. 22 (4), pp. L210-L212 (April 1983).
40. Matsuo, S., 16th 1984 Internatl. Conf. Solid State Devices and Materials, Extend. Abstr., pp. 459-462, Kobe, Japan (1984).
41. Murray, J.J., *Semiconductor Internatl.*, Vol. 7 (4), pp. 130-135 (April 1984).
42. Dunn, G. and Kellogg, E.M., *Semiconductor Internatl.*, Vol. 7 (4), pp. 139-141 (April 1984). Shearer, M.H. and Cogswell, G., *ibid*, pp. 145-147.
43. Kirkpatrick, A., *Semiconductor Internatl.*, Vol. 7 (4), pp. 148-150 (April 1984).
44. Younger, P.R., *Solid-State Technol.*, Vol. 27 (11), pp. 1430-147 (Nov. 1984).
45. Appleton, B.R., editor, *Ion Beam Processes in Advanced Electronic Materials and Device Technology*, Materials Research Soc., Pittsburgh, Pennsylvania (1985).
46. Neugebauer, C.A., Chapter 8, pp. 8-3 to 8-44 in Reference 1.
47. Kern, W. and Ban, V.S., Chapter III-2, pp. 258-331 in Reference 2.
48. Kern, W. and Schnable, G.L., *IEEE Transact. on Electron Devices*, Vol. ED-26, pp. 647-657 (1979).
49. Hersee, S.D. and Duchemin, J.P., in *Ann. Rev. Mater. Sci.*, Vol. 12, pp. 65-80, R. Huggins, editor, Annual Reviews, Inc., New York (1982).
50. Blocher, J.M., Jr., Chapter 8, pp. 335-364 in Reference 3.
51. Ghandhi, S.K., Chapter 8, pp. 419-474 and Chapter 5, pp. 213-297 in Reference 4.
52. See also numerous excellent papers in Proc. Ninth Internatl. Conf. on *Chemical Vapor Deposition 1984*, Proc. Vol. 84-6, Robinson, McD, van den Brekel, C.H.J., Cullen, G.W. and Blocher, J.M., Jr., editors, The Electrochem. Soc. Inc., Pennington, New Jersey (1984).
53. Burggraaf, P., *Semiconductor Internatl.*, Vol. 9 (5), pp. 68-74 (May 1986).
54. Burggraaf, P., *Semiconductor Internatl.*, Vol. 9 (11), pp. 46-51 (Nov. 1986).
55. Mawst, L.J., Costrini, G., Emanuel, M.A., Givens, M.E., Zmudzinski, C.A. and Coleman, J.J., *Semiconductor Internatl.*, Vol. 9 (11), pp. 61-63 (Nov. 1986).
56. Peters, J.W., Gebhart, F.L. and Hall, T.C., *Solid State Technol.*, Vol. 23 (9), pp. 121-126 (Sept. 1980).
57. Mishima, Y., Hirose, M., Osaka, Y., Nagamine, K., Ashida, Y., Kitagawa, N. and Isogaya, K., *Japan J. Appl. Phys.*, Vol. 22 (1), pp. L46-L48 (Jan. 1983).
58. Numasawa, Y., Yamazaki, K. and Hamano, K., *Japan J. Appl. Phys.*, Vol. 22 (12), pp. L792-L794 (Dec. 1983).
59. Osgood, R.M. and Gilgen, H.H., *Ann. Rev. Mater. Sci.*, Vol. 15, pp. 549-576, R. Huggins, editor, Annual Reviews, Inc., New York (1985).
60. Solanski, R., Moore, C.A. and Collins, G.J., *Solid State Technology*, Vol. 28 (6), pp. 220-227 (June 1985).
61. Houle, F.A., *Appl. Phys.*, Vol. A41, pp. 315-330 (1986).

62. Ghandhi, S.K., Chapter 6, pp. 299-370 in Reference 4. See also: Ziegler, J.F. and Brown, R.L., editors, *Ion Implantation Equipment and Techniques,* North-Holland Press (1985).
63. Pinizotto, R.F., *J. Vac. Sci. Technol.,* Vol. A2 (2), pp. 597-598 (April-June 1984).
64. Wilson, S.R., *Semiconductor Silicon 1986,* Proc. Vol. 86-4, pp. 621-641, Huff, H.R., Abe, T. and Kolbesen, B., editors, The Electrochem. Soc. Inc., Pennington, New Jersey (1986).
65. Chin, T.Y. and Oldham, W.G., *J. Electrochem. Soc.,* Vol. 131 (9), pp. 2110-2115 (Sept. 1984).
66. Wolf, H.F., Chapter 4-2, pp. 342-362 in *Semiconductors,* Wiley-Interscience, New York (1971).
67. Cravin, D.R. and Stimmell, J.B., *Semiconductor Internatl.,* Vol. 4 (6), pp. 59-74 (June 1981); O'Neill, T.G., *ibid,* pp. 77-89.
68. Lowenheim, F.A., Chapter III-1, pp. 209-256 in Reference 2.
69. Schwartz, M., Chapter 10, pp. 385-453 in Reference 3.
70. Duffy, J.I., editor, *Electrodeposition Processes, Equipment & Compositions,* (Chemical Tech. Rev. No. 206), Noyes Publishing Company, Park Ridge, New Jersey (1982). See Also: *Metal Finishing '86,* 54th Guidebook - Directory Issue 1986, Vol. 84, No. 1A, Metals and Plastics Publications, Inc., Hackensack, New Jersey (1986). Durney, L.J., *Electroplating Engineering Handbook,* Fourth Ed., Van Nostrand Reinhold Co., New York (1984).
71. Vossen, J.L., Chapter I-1, pp. 3-5 in Reference 2.
72. Kern, W. and Tracy, E., *RCA Review,* Vol. 41 (2), pp. 133-180 (June 1980).
73. Ghate, P.B., Chapter 13, pp. 514-547 in Reference 3.
74. Ghandhi, S.K., Chapter 11, pp. 567-637 in Reference 4.
75. Hutcheson, G.D., *Semiconductor Internatl.,* Vol. 9 (1), pp. 46-49 (Jan. 1986).
76. Semiconductor Processing and Equipment Symposium, Technical Proc., SEMICON/EUROPA, Zurich, Switzerland, March 4-6, 1986. Semiconductor Equipment & Materials Institute, Mountain View, California (1986).

2

Silicon Epitaxy by Chemical Vapor Deposition

Martin L. Hammond

INTRODUCTION

The world epitaxy is derived from the greek "epi" - upon, and "taxis" - to arrange. Thus, epitaxial silicon deposition requires the ability to add and *arrange* silicon atoms *upon* a single crystal surface. Epitaxy is the regularly oriented growth of one crystalline substance upon another. Specific applications require controlling the crystalline perfection and the dopant concentration in the added layer.

Two different kinds of epitaxy are recognized:

Homo-epitaxy—growth in which the epitaxial layer is of the same material as the substrate.

Hetero-epitaxy—growth in which the epitaxial layer is a different material from the substrate.

Virtually all commercial silicon epitaxy is homo-epitaxy, with the exception of silicon-on-sapphire.

Epitaxial silicon layers can be created by a wide range of techniques, including evaporation, sputtering, molecular beams, and various regrowth concepts. This chapter discusses epitaxial deposition by chemical vapor deposition (CVD) in which the silicon and dopant atoms are brought to the single crystal surface by gaseous transport.[1,2]

Chemical vapor deposition is the formation of stable solids by decomposition of gaseous chemicals using heat, plasma, ultraviolet, or other energy sources, or a combination of sources. CVD is a relatively old technology. It was used to refine refractory metals in the 1800's, to produce filaments for Edison's incandescent carbon filament lamps in the early 1900's, for hard metal coatings in the 1950's, and for semiconductor material preparation beginning in the 1960's.[2a]

Silicon Epitaxy by Chemical Vapor Deposition

Commercial silicon epitaxy production, at present, is accomplished primarily by CVD using heat as the energy source for decomposing the gaseous chemicals.

With the silicon epitaxy process, radical changes in materials properties can be created over small distances within the same crystal. This capability permits the growth of lightly-doped single crystal silicon on top of heavily-doped single crystal silicon. At present, no other process technique permits this configuration of doped regions within a single crystal substrate.

Many different configurations are made possible by the CVD epitaxial deposition process. Some of the possibilities in use today include:

- n-type silicon over p-type silicon,
- p-type silicon over n-type silicon,
- Lightly-doped over heavily-doped of either type,
- Lightly-doped over heavily-doped buried layer patterns,
- Conducting silicon layers over insulating surfaces,
- Silicon layers with controlled dopant profiles,
- Silicon selectively deposited through oxide.

Applications of Silicon Epitaxy

Silicon epitaxy is required for isolation and for device performance in bipolar integrated circuits. It is also important for discrete device performance, and is becoming important for MOS integrated circuits (IC's).

For bipolar devices, epitaxy provides a wide range of device performance benefits too numerous to describe here. Among these benefits are:[1]

- higher switching speeds
- improved high voltage, linearity characteristics
- simplified isolation
- lower base resistance
- independently controlled dopant profiles
- buried layer patterns

MOS IC's have not required an epitaxy layer to create device isolation; however, as MOS IC's have become more complex, epitaxial layers can provide many device benefits.[3] MOS IC's are usually created in lightly-doped substrates. When a lightly-doped epitaxial layer is used over a heavily-doped substrate, the benefits to MOS IC's include:

- Lower diffused-line capacitance
- Better diffused-line charge retention
- Better control of spurious charge (such as alpha particles and static charge)
- Improved dynamic random access memory performance

28 Thin-Film Deposition Processes and Techniques

For complementary MOS (CMOS) IC's, the benefit is a very significant improvement in latch-up protection.

Because of its wide-ranging applications to semiconductor technology, approximately 50% of all silicon processed requires an epitaxy layer, and this percentage is expected to increase to 60-70% when epitaxy is used more extensively in MOS IC production.[4]

Silicon epitaxy is used in a wide range of thicknesses and resistivities. In commercial silicon epitaxy, layer thickness is usually expressed in micrometers (μm), commonly abbreviated microns (μ) or 10^{-4} cm. Resistivity is expressed as ohm-cm, and resistivity is related to the electrical carrier concentration by Irwins's curves.[5]

Table 1 lists typical specifications for the common silicon device structures now in production. With the trend to ever smaller feature sizes and ever higher circuit densities, there is a long range tendency toward thinner epitaxial layers; however, the values noted in Table 1 will probably be valid until well into the 1990's.

Table 1: Typical Epitaxy Specifications

Device Type	Thickness (microns)	Resistivity (ohm-cm)
Bipolar discrete devices		
High frequency	0.5-3	0.15-1.5
Power	5-100+	0.5-100+
Bipolar integrated circuits		
Digital memory	0.5-5	0.3-1.5
Microprocessor	0.5-5	0.3-1.5
Linear	3-15	2-20
MOS-on-epitaxy integrated circuits		
P/P+	4-20	10-40
(Back-sealed substrates)		
N/N+	0.5-7	1-10
BiMOS*	0.5-3	0.5-3

*BiMOS = Bipolar and MOS devices together in the same device.

Evaluation of silicon epitaxy is a major technology and detailed information for standardized techniques is available in References 5 and 6. This chapter on silicon epitaxy will address:

- Theory of silicon epitaxy by CVD,
- Silicon epitaxy process chemistry,
- Process adjustments,
- Equipment considerations for silicon epitaxy,
- Other equipment considerations,
- Defects in epitaxy layers,
- Safety,

Silicon Epitaxy by Chemical Vapor Deposition 29

- Key technical issues.

THEORY OF SILICON EPITAXY BY CVD

Successful silicon epitaxy depends upon having:

- High surface mobility for the arriving atoms,
- Numerous, equivalent growth sites,
- Commercially significant growth rates.

Production silicon epitaxy since the early 1960's has been manufactured by chemical vapor deposition in H_2 from the chlorosilanes: $SiCl_4$, $SiHCl_3$, SiH_2Cl_2, and SiH_4 in open tube, vapor transport systems. Alternate chemistries using iodides and bromides have been investigated but no overwhelming advantages have been noted. Fluorine chemistry has not been explored significantly because the Si–F bond is thermodynamically very strong and very high temperatures would be required to crack most Si–F compounds.

As illustrated in Figure 1, CVD is a heterogeneous reaction involving at least the following steps:[1d,2b,7]

Arrival
1. bulk transport of reactants into the process volume,
2. gaseous diffusion of reactants to the surface,
3. absorption of reactants onto the surface,

Surface reaction
4. surface reaction (reaction can also take place in the gas volume immediately above the surface),
5. surface diffusion,
6. crystal lattice incorporation,

Removal of reactant by-products
7. reaction by-product desorption,
8. gaseous transport of by-products,
9. bulk transport of by-products out of process volume.

The rate of chemical vapor deposition is primarily controlled by one of the following major groups of process steps:

- The rate of arrival of reactants,
- The surface reaction rate,
- The rate of removal of by-products.

For typical epitaxy process, the reaction conditions are established so that the rate of arrival of the reactants controls the growth rate. This procedure gives the best crystal quality, a feature necessary in good device performance.

The crystal quality of the epitaxial layer is controlled primarily by:

- The nature of the surface prior to epitaxial growth,

- The arrival rate relative to the surface diffusion rate,
- The nature of the lattice incorporation.

If the surface prior to deposition has contamination, such as oxides, which are not removed during heat-up and etch, or if the crystal upon which the epitaxial layer is to be grown is defective, then the epitaxial layer will have crystal defects. If the rate of arrival of reactants greatly exceeds the surface diffusion rate, then the diffusing atoms cannot move to positions of lowest energy, and again, crystal defects occur.

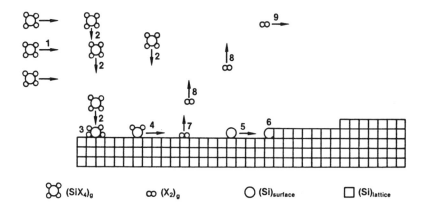

Figure 1: Heterogeneous reaction rate model illustrating arrival (1, 2, 3), surface reaction (4, 5, 6), and by-product removal (7, 8, 9).

The rate of lattice incorporation is a function of crystal orientation because the density of atomic sites is a function of which crystallographic faces are exposed. Figure 2 shows the effect of substrate orientation on growth rate for (111), (110), and (100) faces. For silicon, the (110) plane has the highest growth rate, followed by the (100) and (111) planes.[8]

For epitaxial growth on crystal faces directly on orientation, the low growth rate for the (111) planes encourages a defect called faceting or orange peel. Growth perpendicular to the (111) surface is slow, while growth on a facet not parallel to the surface is faster. The result is a shingle-like faceted surface. To prevent faceting, (111) surfaces for epitaxial growth are cut a few degrees off the (111) to cause the growth to proceed in waves across the surface. Epitaxial growth on the (100) plane does not have this problem; therefore, (100) surfaces are usually cut directly on orientation.

Surface preparation is an important factor in producing good epitaxial crystal quality. The surface requires a high quality, defect-free chemical/mechanical polish that leaves the surface polished without mechanical damage to the crystal structure. Foreign matter such as organic compounds must be removed because it will react with the crystal surface during heat-up to form undesirable silicon compounds, leading to defects.

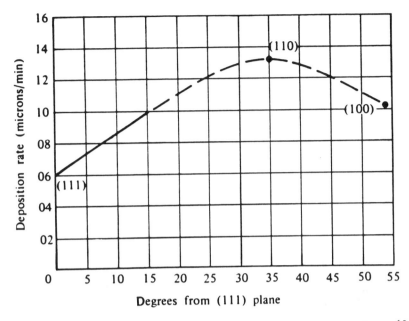

Figure 2: Effect of substrate orientation on epitaxial silicon growth rate.[1c]

SILICON EPITAXY PROCESS CHEMISTRY

The conditions for acceptable commercial silicon epitaxy are outlined in Table 2. Generally, the lower the temperature, the lower the growth rate for acceptable epitaxy quality, and the lower the tolerance for oxidizers in the process space. If silicon oxides, including SiO, are allowed to form during deposition, they interfere with surface diffusion and lattice incorporation, leading to various defects in the epitaxy layer.

Table 2: Silicon Epitaxy Growth Conditions

Chemistry	Growth Rate (microns/minute)	Temperature Range (°C)	Allowed Oxidizer (ppm)
$SiCl_4$	0.4-1.5	1150-1250	5-10
$SiHCl_3$	0.4-3.0	1100-1200	5-10
SiH_2Cl_2	0.3-2.0	1050-1150	<5
SiH_4	0.1-0.3	950-1050	<2

Epitaxy quality is especially sensitive to the presence of oxidizers during heat-up. Water vapor absorbed on cold wafer carriers can be a significant source of oxidizer in the process chamber; therefore, there is a substantial difference in surface quality between beginning the epitaxy cycle with a cold or warm wafer carrier. In a leak-free system, water vapor from a cold wafer carrier can easily be the largest source of oxidizer in the reactor.[9a]

At low concentration, water vapor etches Si surfaces by formation of volatile SiO in H_2. However, at about 1 part per million (ppm) H_2O at 900°C, water vapor begins to form SiO_2 on the Si surface, and this SiO_2 can lead to surface and crystal defects.[9b] The presence of excess H_2 will reverse this reaction somewhat; however, only a few ppm water vapor during heat-up will seriously degrade the epitaxy quality. (See Table 2 and Reference 9a.)

Epitaxial silicon can be grown from less than 1 micron thickness to more than 100 microns. Thicknesses below about 2 microns are considered very thin and careful consideration must be given to the dopant transition from a heavily doped substrate to a more lightly doped epitaxy layer. Thicknesses above about 30 microns are considered very thick and care must be taken to reduce defects that can develop during long deposition times.

Epitaxial silicon is normally doped with PH_3, AsH_3, and B_2H_6. Resistivity levels below about 0.1 ohm-cm are considered heavily doped and require proportionately much higher dopant concentrations in the gas stream to achieve the desired resistivity levels. As the dopant solid solubility is approached (nominally 0.001 ohm-cm), very high concentrations of dopant are required in the gas stream and the epitaxial layer can become a two-phase polycrystalline deposit if the solid solubility is exceeded. When very high concentrations of dopant are added to the reactor, dopant can be absorbed onto the reactor walls and be liberated into the room upon exposure to room air and water vapor. Rigorous safety procedures and adequate ventilation are required when using dopant hydrides, especially when using concentrations in the supply cylinder greater than about 100 ppm.

Resistivity levels above about 10 ohm-cm are considered very lightly doped because the intended dopant level is low enough to be influenced by the typical background dopant levels in the reactor. The undoped or intrinsic dopant level in a commercial epitaxy reactor is a function of the purity of the deposition of the purity of the deposition chemicals, the integrity of the SiC coating on the graphite wafer carrier, the level of other unwanted contamination, and the dopant history of the heated parts within the system. A clean reactor with a fresh wafer carrier and commercially available deposit chemicals should be able to demonstrate an intrinsic resistivity level of greater than 100 ohm-cm n-type on 0.005-0.010 ohm-cm Sb (n+) and on >10 ohm-cm p-type substrates. In most reactor designs, autodoping from more heavily doped As-, P-, or B-doped substrates will affect the intrinsic resistivity. The larger the area of heavily-doped silicon in the reactor, the more effect.

Gaseous HCl is used to etch the silicon surface and remove surface damage prior to epitaxial deposition. The etch rate slowly increases with increasing temperature and, at about 1150°C, a 0.1% HCl/H_2 mixture will remove silicon at 0.1-0.3 micron/minute. If the HCl etch rate is too high for a given temperature, the surface will be pitted instead of being polished. For temperatures above about 1150°C, silicon can be etched at up to 1.5 microns/minute without pitting. In the 1050°-1100°C range, an etch rate of 0.1-0.5 micron/minute will polish the surface. Below 1050°C, HCl is more likely to pit rather than polish the water surface.[10a-d] SF_6 has been considered as a Si etchant because it provides a smooth surface at etch temperatures below 1100°C.[10e] Unfortunately,

SF_6 reacts with the H_2 to form H_2S, and the resulting odor makes SF_6 unusable in a commercial environment.

Because of the quality of chemical/mechanical polishing available today,[10f] there is generally little requirement to remove silicon from the surface to achieve good epitaxy quality. A H_2 bake at 1150°C for 10 minutes will remove native oxide and provide good surfaces.[10g,10h] Nonetheless, commercial epitaxy processes often call for a light HCl etch to remove a micron or less Si just to ensure a low defect density.

When depositing with chlorosilanes, HCl is created by surface and gas phase reactions. This HCl enhances film quality by etching the high energy surface atoms during deposition. When HCl is added to SiH_4, this simultaneous etch/deposit process also occurs and the etch reaction can be treated as if it were separate from the SiH_4 deposition process.[10d]

For conditions used in commercial production, the epitaxial silicon growth rate is proportional to the concentration of silicon source gas in the process stream.[1,11a,11b] Dopant incorporation, for a given temperature, is approximately proportional to the dopant concentration; however, temperature is a primary factor. As and P exist as metallic vapor at deposition temperature;[11b] therefore, higher temperatures increase the escaping tendency of these dopants and, thereby, increase the resistivity for fixed dopant concentration in the gas stream. As is more sensitive to temperature than P. B forms complex hydrides at deposit temperatures; therefore, higher temperatures promote higher reaction rates near the wafer surface and produce lower resistivities. B doping is more sensitive to the presence of oxidizers than As or P.[11d] Oxidizers generally inhibit the incorporation of dopant.

COMMERCIAL REACTOR GEOMETRIES

The reactor chamber geometry affects the gas flow characteristics which, in turn, affect the properties of the deposited layers.

Chemical reactors can be described by one of two flow types:

- Displacement or plug flow in which the entering gas displaces the gas already present with a minimum of mixing
- Mixed flow in which the entering gas thoroughly mixes with the gas already present before exiting the reactor.

When making a gas composition change in a displacement or plug flow reactor, one volume change can produce an approximately 100% change in gas composition by displacing the volume of gas already present. When making a gas composition change in an ideal mixed flow reactor, the gas composition changes exponentially with the number of volume changes.

Because the epitaxy process involves complex intermediate chemical reactions with trace quantities of dopant making significant changes in material properties, the nature of gas flow in the reactor can have a major effect on reactor performance. In general, mixed flow reactors have very uniform wafer-to-wafer resistivity control, even if there is significant autodoping from the substrate. Displacement flow reactors can have sharper dopant transition widths,

34 Thin-Film Deposition Processes and Techniques

with autodoping effects increasing from wafer to wafer along the length of the process gas stream. High total flowrate, reduced pressure operation, and other reactor design elements will strongly influence these generalizations.

Figure 3 illustrates the three principal reactor geometries used in commercial epitaxy production.

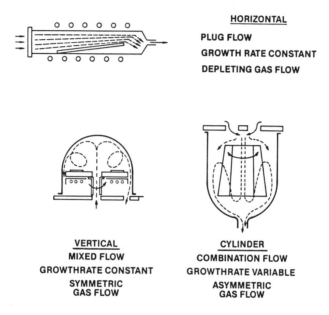

Figure 3: Principal CVD reactor geometries for silicon epitaxy.

Horizontal Reactor

The horizontal reactor is a displacement flow system in which the depletion of reactants downstream is accommodated by tilting the susceptor up to increase the local velocity, and thereby, increase the local growth rate. Overall gas velocities are in the 30-70 cm/sec range.[12]

Cylinder Reactor

The commercial cylinder reactor is a combination flow system with a substantial degree of gas mixing and a complex flow path in which the entering gas eventually displaces the depleted gas out the exhaust port.[13] Depletion is accommodated by having the wafer carrier tilt out at the bottom to increase the local velocity and by having two gas jets which give a strong downward velocity to the gas as it enters the reaction space. This downward velocity is locally 40-70 cm/sec and averages 10-20 cm/sec over the surface of the cylindrical wafer carrier. Rotation of the carrier through the high velocity/high concentration and low velocity/low concentration portions of the mainstream improves uniformity. Tangential velocities for the cylinder reactor are nominally 2-3 cm/sec.

Vertical Reactor

The vertical reactor is a mixed flow system in which fresh process gas enters the process space through a central port and mixes with the depleted gas as it flows radially inward over the wafer carrier surface. The fresh process gas stimulates a convection current which rises from the center of the carrier plate, cools at the bell jar surface, and flows back toward the carrier plate along the bell jar wall. Radial gas velocities near the carrier plate are estimated to be 5-10 cm/sec, as compared to the tangential velocities of 2-4 cm/sec where wafers are placed on the wafer carrier.[14]

New Reactor Geometry

A new commercial epitaxy reactor geometry was introduced in May 1986[15] in which a circular cluster of wafer carriers create tapered cavities, with each cavity having two wafers facing each other, as illustrated in Figure 4. Process gas enters at the outer diameter defined by the tapered cavities and flows by displacement flow toward the center of the circular cluster of cavities. Process gas depletion is accommodated by the acceleration of the gas in the tapered cavities in a manner similar to that for the horizontal reactor geometry. Because of displacement flow, each cavity is isolated from the others so that very little interaction occurs. With this design, very high intrinsic resistivities can be achieved, even with up to 50 150-mm diameter heavily-doped substrates in the reactor. Radial gas velocities of 15-30 cm/sec are typical as compared with 2-3 cm/sec tangential velocities.

Figure 4: Tapered cavity epitaxy reactor.[15]

THEORY OF CHEMICAL VAPOR DEPOSITION

An understanding of the theory of CVD is useful in developing techniques for making process adjustments in commercial production. The theory of CVD is based on chemical kinetics, fluid mechanics, chemical engineering principles, and an understanding of growth mechanisms.[1,2,12,14]

Heterogeneous CVD reactions follow the general reaction path outlined in Figure 1. For simplicity, one reaction step is considered to be rate controlling. When CVD reaction rates for a particular chemistry and reactor geometry are plotted over reciprocal temperature (Figure 5), the deposition or growth rate varies exponentially with[1] temperature, and two regimes are recognized:[16]

- The kinetic or surface reaction rate controlled regime which occurs at lower temperatures
- The diffusion or rate of arrival controlled regime which occurs at higher temperatures.

The diffusion controlled regime is also termed the mass transport controlled regime.

Below some temperature, the surface reaction rate normally controls because, for a given reactor geometry, the rate of arrival is greater than the surface reaction rate. As the temperature is increased, the overall surface reaction rate usually increases more rapidly than the overall mass transport rate, and the rate of arrival or mass transport rate becomes the limiting factor.

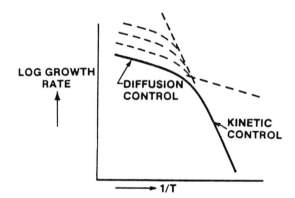

Figure 5: Typical overall reaction rate for CVD reactors versus reciprocal temperature.[16]

Growth rates vs. reciprocal temperature for the chemicals commonly used in silicon epitaxy are plotted in Figure 6.[16] The individual curves in Figure 6 can be moved vertically on the scale by changing the reactant concentration and reactor geometry; however, the shape of the curves is relatively stable. Each growth rate curve can be described by a kinetic and a diffusion controlled regime, as done in Figure 5. Note that, for a given temperature and concentration, the growth rate is lower for the more thermodynamically stable compounds.

Epitaxial silicon is normally grown in the diffusion or mass transport limited regime, at a temperature near the high temperature side of the knees of the curves in Figure 6. This temperature region is selected to be high enough to thermally decompose the selected chemical at a rate commensurate with the surface diffusion rate.

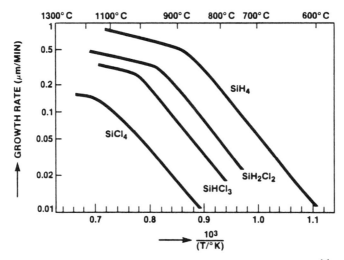

Figure 6: Temperature dependence of silicon growth rate.[16]

Bloem's data[16b] in Figure 7 demonstrate how growth rate controls the morphology of the deposited layer. In all cases, increasing the rate of arrival causes a decrease in the degree of crystalline order for the polycrystalline-to-amorphous transition in the 550°-650°C range and the single crystal-to-polycrystalline transition in the 1000°-1400°C range.

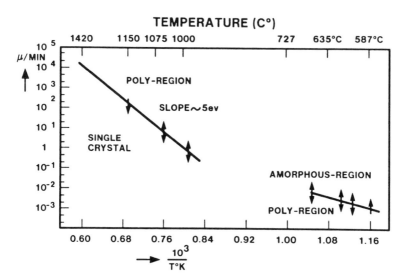

Figure 7: Relationship of growth rate, morphology, and growth temperature for silicon epitaxy.[16b]

38 Thin-Film Deposition Processes and Techniques

The concept that morphology is controlled by the rate of arrival relative to the surface diffusion rate applies on the finer scale of epitaxial crystal perfection. In general, the lower the temperature of deposition, the lower the growth rate must be to accommodate a given level of crystal perfection. It is this concept that controls the growth rate/temperature conditions used in commercial silicon epitaxy.

PROCESS ADJUSTMENTS

The important CVD reactor control parameters can be divided into two categories: reactor design variables and operator variables. Each has a direct influence on the uniformity, productivity, and quality of the epitaxial layer.

The reactor design variables are:

- Tilt angle or equivalent,
- Gas inlet geometry,
- Wafer/carrier configuration,
- Wafer/reactor wall configuration,
- Exhaust configuration.

Operator variables include:

- Gas flow rate,
- Gas composition,
- Temperature profile,
- Temperature value,
- Chemistry.

Uniformity of an epitaxy layer refers to uniformity of both thickness and resistivity. Thickness is primarily controlled by the mass transport and surface reaction rates. Resistivity is controlled by dopant incorporation which, in turn, depends primarily on local concentration of dopant and local temperature.

The most effective process adjustment strategy is:

1. establish a flat temperature profile with the desired chemistry, temperature, pressure, and growth rate;
2. mechanically adjust for thickness uniformity;
3. fine tune resistivity with local temperature adjustments.

Selections of chemistry, temperature, pressure, and growth rate are determined by the desired materials and device properties. The usual operating conditions for commercial silicon epitaxy are presented in Table 1.

The key to thickness control in a mass-transport-limited epitaxial CVD process is compensation for depletion.

Horizontal Reactor

The horizontal reactor, depicted in Figure 3, offers a simple model for describing how to compensate for depletion and the lessons learned here are applicable to other reactor geometries.

It can be useful to break the thickness profile in the direction of gas flow along the wafer carrier or susceptor into the front and back half, as illustrated in Figure 8. Each half can then be described as thick or thin in front, or thick or thin in back. The advantage is that the mechanical adjustments for thickness uniformity are also divided into those that affect the front half and those that affect the back half of the profile.[17]

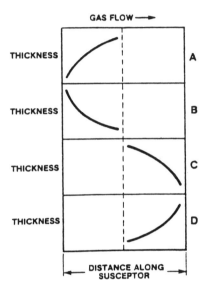

Figure 8: Thickness variation possibilities along the direction of process gas flow for horizontal reactor.[17a]

The overall concave-down profile in Figure 9 can be described as thin in front and thin in back, rather than thick in the middle. The S-shaped profile in Figure 9 can be described as thin in front and thick in back.

Figures 10 and 11 illustrate the action necessary to correct a thickness profile that is thick or thin in the front or back. The underlying principle is that local growth rate is a direct consequence of local reaction rate. Therefore local growth rate can be increased by increasing local velocity, temperature, and concentration, and vice-versa.

Cylinder Reactor

Corrective action for the cylinder reactor is presented in Figure 12. In this reactor configuration, thickness is increased in the front (top of carrier) by increasing the temperature and/or concentration of reactant. Thickness is increased in the back (bottom of carrier) by increasing the total flow rate, directing the gas jets in a more downward direction, and bringing the back pressure toward a more positive value. Thickness within a wafer from left to right is improved by balancing the two jet flows, by lowering the temperature to make the overall reaction become more surface reaction rate controlled and less dependent upon rate of arrival, and by increasing the distance between the flat wafer surface and the curved bell jar. Reducing the process pressure can cause the process gas to heat more slowly, and thereby, decrease the thickness in front and increase the thickness at the back. Thickness variation at any distance along the carrier is averaged by rotation.[17b]

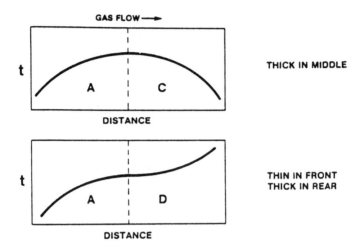

Figure 9: Thickness variation combinations along direction of process gas flow for horizontal reactor.[17a]

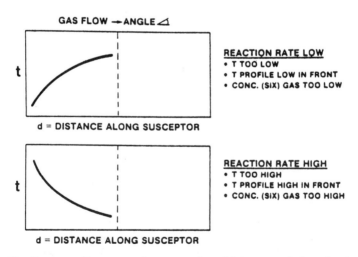

Figure 10: Process adjustments for correcting thickness variations in the front half of the horizontal reactor.[17a]

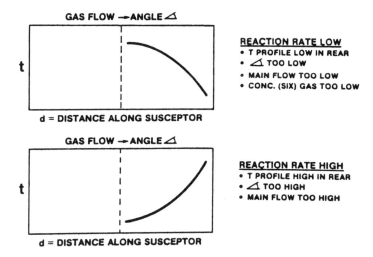

Figure 11: Process adjustments for correcting thickness variations in the back half of the horizontal reactor.[17a]

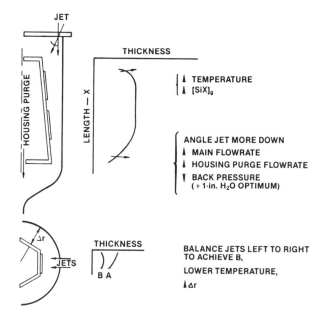

Figure 12: Process adjustments for correcting thickness variations in the cylinder reactor.[17b]

Vertical Reactor

Vertical epitaxy reactor behavior is depicted in Figure 13. Thickness variation at any radius is averaged by rotation of the susceptor. The radial variation is a function of total flow rate and tends to have an optimum value of flow rate for a particular reactor configuration. For low total flow rates, the thickness tends to be thin to the outside, and for high total flow rates, the thickness tends to be thick toward the outside.[17c]

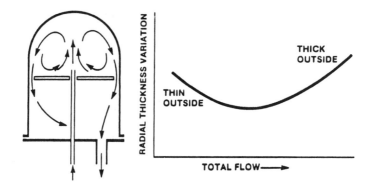

Figure 13: Effect of total flowrate on thickness variation in the vertical reactor.[18]

Assuming that the total flow rate is approximately optimum for a particular vertical reactor configuration, Figure 14 analyzes further effects on thickness uniformity in a vertical reactor. To correct thick on the outside and thin on the inside, one must reduce the reaction rate on the outside by lowering the total flow rate, decreasing the concentration of reactants, and/or lowering the temperature. To correct the opposite situation, one must make the opposite changes. When optimized, the vertical reactor should be thin on the inside and outside, and thickness uniformities of ±1-3% are possible.[33]

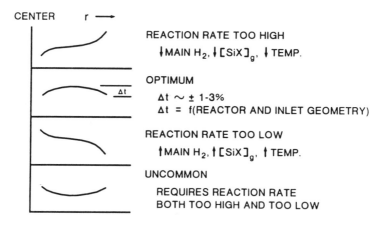

Figure 14: Process adjustments for correcting thickness variations in the vertical reactor.[18]

Being thick on the inside and outside at the same time is uncommon with a vertical reactor because that profile requires the reaction rate to be too high at the inside and outside of the carrier at the same time. Such a radical thickness profile is normally an indication of not having a flat temperature profile. Depending upon the total flowrate, temperature, and reactant concentration, the temperature can be too high on either the inside or outside of the wafer carrier to achieve this thickness profile.

Control of Variables

The operator variables of flowrate, gas composition, temperature, temperature profile, and reactant chemistry are primary control items for any reactor geometry. Table 3 lists the primary control items for reactor design variables and notes the importance of each for the three different commercial reactor geometries.

Table 3: Thickness Adjustment Variables

Variable	Vertical	Horizontal	Cylinder
tilt angle	NA	major	major, fixed
gas inlet	varies	minor	major
wafer/carrier	recessed	recessed	recessed
wafer/wall	minor	major	major
exhaust pressure	minor	minor	major

Tilt angle is not applicable to the vertical, is a major adjustment parameter for the horizontal, and is a major but fixed item in the cylinder. The gas inlet system is a major adjustment parameter for cylinder but is less important for the vertical and horizontal systems.

For all reactors, the wafers usually reside in recessed pockets because the recessed pockets prevent the wafers from moving when being unloaded; thereby minimizing generation of particles. Edge crown and crystallographic slip can also be prevented by appropriate design of the recessed pockets.[18]

Spacing between the wafer and the reactor wall is a minor parameter in vertical reactors because the wall is so far away from the wafer. In horizontal and cylinder reactors, the wafer/wall spacing is a major parameter because the distance is short and the local gas velocity is directly controlled by this spacing.

Exhaust pressure is a major adjustment parameter for the cylinder because it directly affects the local gas velocity at the bottom of the susceptor. If the exhaust pressure is too low, the gas inlet jet apparently goes directly into the exhaust region and does not circle back up into the process space. As a result, the growth rate is reduced near the bottom of the susceptor.

Adjustment parameters for the tapered cavity reactor illustrated in Figure 4 generally follow those for the horizontal reactor. Rotation of the circular cluster of tapered cavities averages the initial velocities and concentrations from multiple gas inlets.

EQUIPMENT CONSIDERATIONS FOR SILICON EPITAXY

The silicon epitaxy process requires control of:

- reaction temperature
- wafer temperature
- reactant concentrations
- process sequence
- reactor pressure
- carrier temperature gradient
- total flowrate and velocity
- exhaust effluents

Equipment design must include consideration of:

- operator safety
- process reliability
- automation
- operator convenience
- clean air load/unload
- maintenance access

A silicon epitaxy reactor is a complex assembly of subsystems including:

- power supply
- gas flow control
- cabinetry
- exhaust treatment
- reaction chamber
- wafer carrier
- automation
- gas supply

The reactor chamber geometry has already been reviewed. Here, gas flow control, power supply, temperature measurement/control, and some wafer carrier effects will be reviewed. Cabinetry, automation, exhaust treatment, gas supply and other subsystems are vital parts of an epitaxy reactor system; however, they will not be discussed in this chapter.

Gas Control System

The silicon epitaxy process requires precise control of the mainstream H_2, etching gas, silicon source gas, and dopant, while maintaining appropriate purge flows in various parts of the reactor. Provision must be made to vent the process gases without directing them through the process chamber to avoid contaminating that space with excessive dopant or unwanted silicon gas. This venting capability is often used just before depositing with a process gas to provide fresh gas to the process space.

Gas flow control is one of the most critical subsystems. Gas shutoff must be better than 10^{-6} cm^3/sec to eliminate process problems, and multiple gaseous and liquid sources must be precisely controlled. Gas compositions must be maintained with less than ±1% variation because reaction rates are generally proportional to composition. In addition, the plumbing system must safely handle noxious, corrosive, poisonous, pyrophoric, and flammable gases.[19]

Passivated 316 stainless steel is required for all plumbing components. Some consideration has been given to using more exotic metals such as Ta or Ni alloys. In general, these metals are only marginally more corrosion resistant against wet HCl than 316 stainless steel. If the plumbing system is kept clean, leak-free, and dry, there is little or no performance advantage to working with exotic metal alloys. If wet atmosphere is allowed to leak into a chlorine-containing line, corrosion products will contaminate the deposition, regardless of what alloy is being used.

Leak Testing

Leak-tight plumbing is absolutely essential for epitaxy reactor gas systems. Leak testing techniques include bubble testing with sensitive detergent solutions, He mass spectrometers, high sensitivity combustible gas detectors, and pressure decay techniques. Soap bubble testing is useful but is not very sensitive, inside of the plumbing system if large leaks are present when the detergent solution is applied. He mass spectrometers offer excellent sensitivity and highly localized detection sensitivity. Unfortunately, He mass spectrometers are complex, require considerable operator training, and are relatively expensive. Highly sensitive combustible gas detectors can provide a lower cost portable method of leak detecting; however, the He mass spectrometer provides the most sensitive leak detection technique.

Leak testing individual components and joints in a system is useful; however, pressure decay testing assures that no leaking joint has been overlooked. For a leak rate specified in atmosphere cm^3/sec, the leak rate of a system is measured by pressure decay as follows:[20]

$$\text{leak rate (atms cm}^3\text{/sec)} = \frac{\text{(pressure change) (system volume in atms cm}^3\text{)}}{\text{(absolute pressure) (time in seconds)}}$$

The purity of the H$_2$ supply must be considered along with the system leak rate for the silicon epitaxy process. Assuming that the H$_2$ supply has approximately 1 part per million (ppm) combined oxidizer and that the epitaxy process requires less than 2 ppm, then the entire gas control system and reactor chamber seal must add less than 1 ppm to the process gas. For a mainstream flowrate of 100 liters/minute operating in a displacement flow mode, the total system leak rate must be less than 2 x 10^{-3} cm^3/sec in order to maintain the required oxidizer concentration of less than 2 ppm.

Leak rates for individual components are routinely specified at less than

10^{-6} cm^3/sec; however, a collection of 100 such components can have a leak rate approaching 10^{-4} cm^3/sec. A total system leak rate of 10^{-4} cm^3/sec has been considered satisfactory; however, as process are extended to lower temperatures where less total oxidizer can be tolerated, this specification will have to be improved.

For a plumbing system with 500 cm^3 total internal volume, a pressure decay of 0.45 psig from 30 psig (45 psia) in 16 hours (57,600 seconds) corresponds to a leak rate of approximately 10^{-4} atms cm^3/sec (5 cm^3/57,600 seconds). For relatively small volumes incorporating less than 100 feet of 0.25-in tubing at 5 cm^3/ft, pressure decay is a useful leak checking procedure. For volumes exceeding 500 cm^3, the time of measurement or the resolution of pressure difference must be substantially increased to be able to measure meaningful leak rates. A more effective approach for complex systems is to divide the system into several sections, each having less than 500 cm^3 total volume. For a plumbing section with 100 cm^3 total internal volume, a pressure decay of 2.2 psig/45 psia corresponds to approximately 10^{-4} cm^3/sec.

Pressure decay testing of large process volumes is not practical because the reactor walls, typically made from quartz, will not withstand high pressures, and the system volume is too great for adequate sensitivity. Pressure decay is useful as gross leak check of the process chamber; however, more sensitive He mass spectrometers or combustible gas detectors are required to properly leak check the process chamber O-ring seals.

Gas Flow Control

Four different methods for controlling the flowrate of individual gases in a CVD epitaxy reactor are:

- Fixed or variable orifices,
- Ball-in-tube flometers,
- Mass flow controllers,
- Source controllers.

Fixed or variable orifices have the advantages of being simple, reproducible, and low cost. They do not offer a read-out of flow and fixed orifices have an easily controlled dynamic range of only about 5:1 using pressure as a variable.

Ball-in-tube flowmeters offer visual read-out and a dynamic range of 10:1 at relatively low cost; however, there are numerous elastometer seals which can leak, and the units must be mounted on a front panel for observation. The calibration of a ball-in-tube flowmeter is a function of pressure within the gauge tube, and that relationship is:

$$\frac{\text{Actual flowrate}}{\text{Calibrated flowrate}} = \sqrt{\frac{\text{Actual pressure in absolute units}}{\text{Calibrated pressure in absolute units}}}$$

Mass flow controllers based on the principle of measuring mass flow using gas heat capacity provide remote location, closed-loop control with an electronic readout, and a dynamic range of 50:1. The disadvantages of mass flow controllers are higher cost, increased complexity, and the fact that only a portion of the gas passing through the unit may actually be measured. Most mass flow controllers depend upon the performance of a by-pass orifice which can become clogged and give false readings. Nonetheless, the control, convenience, and dynamic range of mass flow controllers have led to their widespread use in gas systems.

Some chemicals, such as $SiCl_4$ and $SiHCl_3$, are liquids at room temperature with fairly high vapor pressure. These silicon source chemicals require bubblers to convert the liquid into a gas. H_2 is bubbled through the chemical and the quality of chemical transported is proportional to the efficiency (eff.) of the bubbler, the H_2 flow rate $[F(H_2)]$, and the ratio of the vapor pressure of the chemical $[P(gas)]$ divided by the total pressure (absolute units) within the bubbler; i.e., for $SiCl_4$:

$$F(SiCl_4) = \frac{[F(H_2)][P(SiCl_4)][eff.]}{\text{Total absolute pressure in the bubbler}}$$

Bubbler flowrates are often controlled by source controllers which combine an H_2 mass flow controller with a second sensor that measures the concentration of chemical in the combined flow stream.[21] For the most repeatable performance, bubblers require a constant temperature and pressure.

Bubbler efficiency is fixed primarily by the size of the bubbles and the path length the bubbles travel within the liquid. The smaller the bubble diameter, the closer it will be to saturation before it breaks through the liquid surface. To achieve a saturated flow stream, it is better to have many small diameter holes in the bubbler dip tube than one big hole. Because of the influence of path length, there is some reduction in chemical concentration as the liquid level in the bubbler is lowered. This reduction in concentration can be reduced by installing two bubblers in series. Chemical in the first bubbler is transported into the second, which will maintain an approximately constant liquid level. Two bubblers in series provide effective concentration control; however, the second bubbler can accumulate contamination and must be replaced and cleaned regularly.

Bubblers must be adequately sized for their total mass flow rate. Evaporation of the chemical absorbs energy, and if the heat loss caused by evaporation reduces the bubbler temperature by even a few degrees centigrade, there is a strong effect on vapor pressure.

It is critical that the "H_2 to Bubbler" flow line be automatically valved to turn on and off with the valve controlling the "$H_2 + SiH_xCl_y$" flow line. If the "H_2 to Bubbler" valve is not present, the bubbler will remain pressurized at line pressure and any reduction in H_2 supply pressure will cause SiH_xCl_y to flow back into and contaminate the H_2 supply line. Pressure regulators and check valves help prevent this back flow but they should not be relied upon. White powder in the H_2 supply line and haze on the epitaxial surface are strong indications that silicon source liquid has leaked back into the H_2 supply line.

Dopant Flow Control

Epitaxial silicon is normally doped with the hydrides B_2H_6, PH_3, or AsH_3.

In order to dope epitaxial layers to the 0.05–50 ohm-cm resistivity range, dopant concentrations in the process chamber are nominally 10^{-6} times the silicon source gas concentration.

Silicon gas and dopant are both added to the mainstream and the ratio of dopant to silicon source is an important factor in determining the concentration of dopant in the epitaxial layer. In practical application, the additions of dopant and silicon source to the mainstream are treated as totally separate and independent variables because that approach simplifies the control requirements. Graphs in Figures 15 and 16 show how the gas phase ratio of dopant to silicon can affect the epitaxial layer impurity or dopant concentration and resistivity.[22] Also see Reference 11.

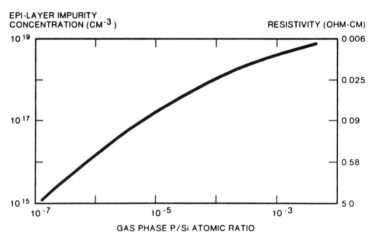

Figure 15: Epitaxy layer impurity concentration and resistivity versus gas phase phosphorus/silicon ratio.[22]

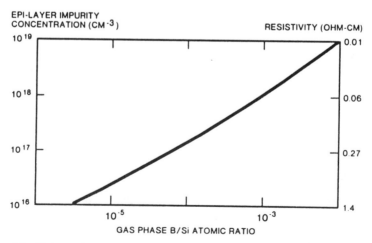

Figure 16: Epitaxy layer impurity concentration and resistivity versus gas phase boron/silicon ratio.[22]

Silicon Epitaxy by Chemical Vapor Deposition 49

The 25-250 ppm dopant supply must be diluted prior to joining the mainstream in order to be at the desired concentration relative to the silicon source gas.

Two different dopant dilution schemes are presented in Figure 17. In Figure 17a, the H_2 flow is added, unmeasured, from a controlled pressure source to the dopant flow from the "Tank" mass flow controller. The combined gas streams pass through a gas mixer and are directed to "Inject" and "Dilute" flow controllers. The Inject flow is directed to the reactor mainstream or to vent and the Dilute flow goes directly to vent.

The dilution system in Figure 17b adds a measured H_2 flow from the Dilute flow controller to the Tank flow. As before, the Inject flow goes to mainstream or vent and the unmeasured flow goes to vent through a back pressure regulator.

Both systems perform equally well; however, most commercial systems utilize the flow schematic of Figure 17a. The thought is that the upstream pressure regulator for the unmeasured H_2 flow in Figure 17a is more reliable than the back pressure regulator for the unmeasured vent flow in Figure 17b.

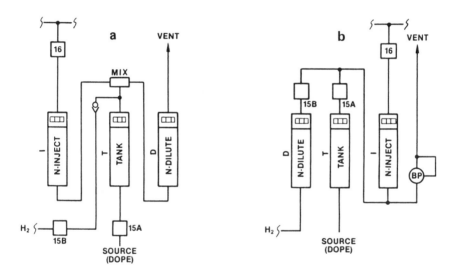

Figure 17: Dopant dilution control systems.

Mathematically, the flow of dopant F, already diluted in the gas cylinder to concentration C, passing through the Inject flow controller and into the mainstream in Figure 17a is given by:

$$F = C \frac{T}{D + I} I$$

where D, T, and I are the measured flows through the Dilute Tank, and Inject flow controllers, respectively, as each is defined in Figure 17a. It is important to note that the units of F are volume/unit time. If the Dilute flow were zero, the Inject flow values would cancel in the equation and the dopant flow F would simply be the concentration C in the gas cylinder times the tank flowrate. Similar reasoning is applied to Figure 17b, using the following equation:

$$F = C \frac{T}{D + T} I$$

In commercial epitaxy reactors, it is convenient to make the Tank and Inject flow controllers have equal full scale flowrates, with the Dilute flow controller being 10–100 times that full scale value. For systems in which the mainstream flowrate is 100–500 liters/min, the Tank and Inject flow controllers are usually sized to be 300 cm^3/min while the Dilute controller is 30 liters/min.

When the Inject and Tank flow controllers have the same flow values for all settings, as done in the control scheme described below, the equations governing Figures 17a and 17b become identical because the values for I and T become identical.

To a first approximation, the resistivity is inversely proportional to the dopant flow into the mainstream, assuming that all other parameters are constant. Thus, the dopant dilution equation should be exercised each time the epitaxy resistivity is altered.

One simple approach to dopant dilution is to keep the ratio of Tank to Dilute constant and vary only the Inject. This approach is limited to about a 10:1 ratio before the Tank/Dilute ratio must be changed.

For adjustment convenience over a wider range of values, an automatic dopant control system can be created by causing the Dilute, Inject, and Tank flow control setpoints to have related values that automatically adjust the dopant concentration.

In the automatic dopant control system shown schematically in Figure 18, the Tank and Inject flow controllers always have the same setpoint and that setpoint is always inversely proportional to the Dilute setpoint. In Figure 18, this relationship is depicted schematically by ganged slide-wide potentiometers in which the Tank and Inject setpoints move from 0 to 100% full scale while the Dilute flow controller setpoint moves from 100 to 0% full scale. These relationships can be achieved by various electronic control circuits.

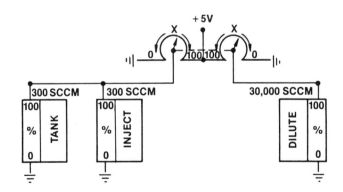

Figure 18: Automatic dopant control system.

If X represents the intermediate setpoint value for the Tank and Inject flow controllers, with a value between 0 and 1, then:

$$I = X\,I(max) = T = X\,T(max)$$
$$D = (1-X)\,D(max)$$
$$I(max) = T(max) = 300 \text{ cm}^3/\text{min}$$
$$D(max) = 30{,}000 \text{ cm}^3/\text{min}$$

and

$$F = C\,\frac{T}{D+I}\,I.$$

Therefore,

$$F = C\,\frac{X\,T(max)}{(1-X)D(max) + X\,I(max)}\,X\,I(max).$$

But: $D(max) = 100\,I(max)$;

therefore,

$$F = C\,\frac{X^2\,I(max)^2}{D(max) - X\,D(max) + \dfrac{X\,D(max)}{100}}.$$

But: $I(max)^2/D(max) = 3$;

therefore:

$$F = C\,\frac{X^2}{(1 - 0.99\,X)}\,3.$$

With this setpoint algorithm, the dopant flow F equals three times the concentration of dopant in the tank C times the function $X^2/(1 - 0.99\,X)$. The factor 3 is an arbitrary result of the choice of the full scale flowrates of the Dilute and Inject flow controllers and is usually ignored in dopant flow calculations because changes in resistivity are usually expressed in ratios of resistivity and dopant flow.

The function, $X^2/(1 - 0.99\,X)$, has the advantage of changing from 0.0026 when X = 0.05 to 15 when X = 0.95. Thus, the dopant flowrate can be changed over a range of about 6000:1 by changing X only 19:1.

The function $X^2/(1 - 0.99\,X)$ is sometimes called the dopant number and a slide rule has been created to easily ratio resistivity values with X values.[23]

OTHER EQUIPMENT CONSIDERATIONS

Heating Power Supplies

Silicon epitaxy requires temperatures in the 900°–1300°C range. Process limitations concerning gas purity have dictated batch operations with cooled elastomer seals; therefore, relatively rapid heating and cooling is necessary for commercially significant production.

The primary heating systems for silicon epitaxy are:

- Induction heating, using frequencies from 3 kHz to 400 kHz, with coils located internal and external to the process chamber,
- Radiant heating, using high intensity infrared heating lamps, located external to the process chamber,
- Radiant heating, using resistance heaters, located internal to the process chamber,
- Combinations of the above.

For the simple cold-wall reactor geometries illustrated in Figure 3, power densities of 15-30 W/cm^2 are required to achieve the temperatures and heat/cool cycles times.

Resistance and induction heating were the first techniques to be used to achieve the high power densities required for silicon epitaxy. In the case of induction heating, the wafer carrier was directly induction heated and was properly called a susceptor. This terminology has been carried over to other heating methods and wafer carriers are often referred to as susceptors, regardless of the heating method.

Each heating method has advantages and the technique used does not significantly alter the information provided in this chapter.

Effect of Pressure

Prior to 1978, virtually all silicon epitaxial deposition was done at atmospheric pressure. Process pressure does affect certain properties of epitaxy films such as pattern shift and autodoping. There are process benefits for epitaxial growth in the 20-250 torr range, and these benefits are discussed in the Key Technical Issues Section.

In the pressure range of 5-760 torr, gas flow is characterized by viscous flow. From the fluid mechanics viewpoint and *for the condition of constant mass flowrate,* there is little difference in the rate of arrival of reactants between operating at 760 torr or 5 torr absolute pressure. If the total mass flowrate is kept constant, the different effects of velocity, density, and diffusion rate balance and comments made about the effect of gas flow on deposition uniformity apply equally well for all pressures in this range. Chamber pressure, at constant total mass flowrate, can have a slight effect on how quickly the process gas is heated. A cooler gas has a lower reaction rate and this effect can be used to reduce the reaction rate at the beginning of the reaction path.

When the mean free path of a gas molecule becomes comparable to the dimensions within the process chamber, molecular diffusion is the dominant mode of mass transport. Molecular diffusion begins to be important below about 5 torr and is normally the dominant mode of mass transport below about 0.5 torr.[20]

Because epitaxy processing is usually done in the mass transport limited or rate of arrival regime, the total mass of gas passing through the reactor is kept approximately the same regardless of process pressure. The result is the number of volume changes per unit time is increased in inverse proportion to the absolute pressure. This increased number of volume changes per unit time can have a strong effect on the materials properties of the epitaxy layer, such as, pattern shift, dopant incorporation, and crystal perfection at lower temperatures.

Temperature Measurement

Temperatures for the silicon epitaxy process must be measured and controlled with a precision of a few degrees centigrade in the 900°-1300°C range, in H_2 with a silicon source chemical. In these process conditions, the only convenient temperature measurement method is radiation pyrometry. Thermocouples are generally not stable in a H_2 environment, and most thermocouple shield materials react with silicon at high temperature.

Radiation pyrometry measures the brightness or intensity of light in a narrow range of wavelengths being emitted by the target. The intensity of radiation is related to temperature by the Stefan-Boltzmann radiation law:

$$W = \epsilon \sigma T^4$$

where ϵ is the emissivity, σ the Stefan-Boltzmann constant and T the absolute temperature. If the emissivity is known, the temperature of the emitting body can be determined.[24]

Pyrometers are generally of two types:

- Disappearing filament pyrometers in which the brightness within a narrow wavelength range of a heated filament of known temperature is matched with the brightness of the target,

- Electronic pyrometers in which the intensity of light within a narrow range of wavelengths is directly measured by a photovoltaic cell or a thermopile.

The disappearing filament pyrometer operates at 0.65 micron wavelength and can provide approximately ±5°C precision. Calibration sources are available which permit absolute temperature reference.

Electronic pyrometers with silicon photocells operate in the 0.8 micron range; PbS photocells detect in the 2.2 micron range. Thermopiles are simply assemblies of thermocouples on which the radiation is focused. As such, they are broad band detectors and are easily affected by a wide range of absorptive effects. To prevent this source of imprecision, narrow band-pass optical filters are provided to minimize absorptive effects.

The emissivity of silicon ranges from 0.48 to 0.60[25] for a light wavelength of 0.65 micron, while that of SiC-coated graphite is slightly higher. Emissivities are a function of wavelength and generally increase for longer wavelengths. In the 2.2 micron range, the emissivity of silicon is approximately 0.70 for temperatures above 600°C. Below 600°C, the emissivity of Si is a strong function of temperature for wavelengths of 1-5 microns.[26]

Optical temperature measurements are further complicated by absorption effects of reactor walls, safety glass windows, and deposits on the reactor wall.

Figure 19 illustrates two configurations for optical radiation pyrometry in epitaxy reactors. In the case of the vertical reactor, an electronic radiation pyrometer is located above the bell jar and has a wide field of focus on top of the susceptor. A window is provided in the cabinetry to permit separate observation and calibration with a hand-held pyrometer. For this configuration, the pyrometer above the jar will be somewhat sensitive to load because the emissivity of rough polycrystalline silicon is slightly greater than polished single crystal. This pyrometer configuration is also sensitive to bell jar coating. In an improved version, fiber optics are inserted through the baseplate to measure the temperature of the bottom of the susceptor.

For the radiantly heated cylinder reactor in Figure 19, electronic pyrometers are located inside the wafer carrier, within a cooled and purged quartz tube. For calibration, a thermocouple can be inserted into a special wafer carrier plate. When the thermocouple is in place, the carrier cannot be rotated and the averaging effect of rotation cannot be observed. In this pyrometer configuration, deposits on the quartz tube will alter the measurement accuracy.

Because of reflected light having the same wavelength that the pyrometer is using, it is difficult to measure temperature in a radiantly heated reactor with a pyrometer aimed at the same surface heated by the radiation source. As a practical matter, pyrometers using a nominally 2.2 micron wavelength are acceptable for this application because the radiant heat lamps have very little radiation with wavelengths longer than 2 microns.

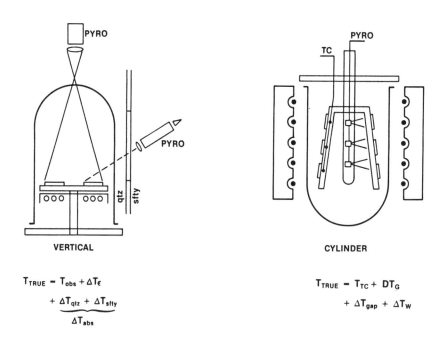

Figure 19: Optical pyrometer configurations for the vertical and cylinder reactors.

Backside Transfer

While the CVD process takes place on the face of the wafer, chemical activity also occurs between the wafer and its carrier. If silicon has been deposited on the carrier where a wafer is to be placed, and if the wafer is cooler than the carrier during process, then it is possible for silicon to be transferred from the carrier to the cooler wafer. This transfer is driven by the temperature difference, and its rate is significantly enhanced by the presence of chlorine which creates volatile silicon compounds in the space between the wafer and the carrier. If the wafer is hotter than the carrier, then chlorine can promote the transfer from the wafer to the cooler carrier.

The method of heating directly affects the relative temperature between the wafer and its carrier. As noted in Figure 20, when the wafer is located on the same side of the carrier as the radiant heat source, the wafer is hotter than its carrier and the backside of the wafer can be etched during process. For heavily doped substrates, this etching can release dopant into the process space. This backside transfer/etch process occurs more rapidly at higher temperatures because of greater temperature differences and faster chemical reaction rates.

When the wafer carrier is directly heated, as with induction heating, the wafer is usually cooler than the carrier, and backside transfer of silicon occurs. If the substrate is heavily doped, this backside transfer will partially seal the back of the wafer.

Where wafer flatness is critical, backside transfer can be undesirable. Backside transfer is reduced by depositing at lower temperatures and by operating with less chlorine in the process gas. Careful placement of a wafer into the pocket formed by the previous wafer can prevent such transfer.

In the tapered cavity reactor described in Figure 4, the wafers facing each other across the cavity are at nearly the same temperature as their carriers, and conditions can be adjusted so that neither backside transfer or backside etching occurs.

If a back-seal is required for resistivity control, it should be applied during the wafer making process. Adding oxide, nitride, or polycrystalline Si to the back of a polished wafer usually leads to unacceptable defect densities in the epitaxial layer caused by front surface damage.

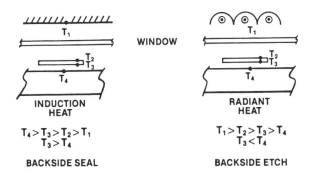

Figure 20: Wafer/carrier temperature difference as a function of heating method.

Intrinsic Resistivity

Intrinsic resistivity is the doping level from the reactor when no dopant is supplied in the gas stream during deposition. Doping silicon is an algebraically additive process; therefore, the intrinsic doping level must be substantially less than the intentional doping level to achieve resistivity control.

Unwanted dopant comes from inadequately purified wafer carriers, dopant in the silicon source gases, thermal degradation of seal o-rings, residual dopant in the wafer carrier from prior heavily doped runs, and contamination such as vacuum grease touching hot parts. Insufficiently purified graphite and vacuum grease normally contribute p-type dopant to the background while gas contamination is usually n-type.

Individual wafers can also be contaminated by such sources as sputtering from ion-implant hardware, spin-on dopant solutions, and Al scratched onto the wafer during handling.

Intrinsic resistivities should be at least ten times the intentional level in order to maintain adequate control. The intrinsic resistivity value of a clean epitaxy reactor should exceed 100 ohm-cm n-type and values of 100-1000 ohm-cm are often obtained.

Phantom p-Type Layer

Most bipolar IC's require an n-type epitaxial layer over n+ buried layer regions. Unwanted p-type contamination can create a so-called phantom p-layer in the epitaxial silicon which can be observed by groove and stain or by spreading resistance probe profiles. Sources of this unwanted p-type contamination include:

- A low p-type intrinsic resistivity caused by the contamination sources described above,
- A p-type contamination (usually B) in the substrate due to use of contaminated remelt during crystal pulling (especially true for heavily doped n-type substrates),
- A p-type contamination from the doping technique used to create the buried layer,
- A p-type contamination coming from B in a dual B/As buried layer structure,
- A p-type contamination coming from Al_2O_3 used in some H_2 purifiers.

Intrinsic resistivities and spreading resistance probe profiles should be measured and recorded on a regular basis so that the origin of phantom p-layers can be determined when they occur.

DEFECTS IN EPITAXY LAYERS

Defects in silicon epitaxy layers include:

- haze

- pits
- orange peel
- faceted growth
- edge crown
- slip dislocations
- growth dislocations
- stacking faults
- spikes and hillocks
- shallow pits

Methods for observing and measuring these defects are described in Reference 6.

Haze

Haze, as revealed by reflected light, is a fine pitting or slightly textured surface caused by oxidizer in the process space or by subsurface crystal damage in the substrate. To cure or reduce haze, eliminate all sources of oxidizer (such as air leaks), verify the purity of the H_2, extend purge times, hold reactor at 850°C during heat-up to dry it out before proceeding, and increase the deposition temperature.

Pits

Pits are localized etching of substrate defects or local inhibition of epitaxy growth, often caused by contamination left on the surface prior to heat-up. Improving the pre-epitaxy clean, extending the HCl time, increasing the HCl etch temperature, and verifying the HCl purity are steps to prevent pitting.

Orange Peel

Orange peel is a roughened surface appearance caused by having the growth rate too high for the deposition temperature. Subsurface crystal damage can also cause orange peel. Increasing the temperature and lowering the growth rate will reduce orange peel if the wafers have been correctly polished.

Faceted Growth

Faceted growth is irregular, stepped growth over the surface that is caused by growing on (111) faces cut on-orientation. Substrates should be cut 3-5 degrees off the (111) plane towards the nearest (110) plane for smooth growth. Epitaxial growth on the (100) plane does not have this problem.[1a,1b,2c,2d,27]

Edge Crown

Edge crown occurs when the growth rate at the edge of the wafer greatly exceeds that over the rest of the wafer surface. Edge crown is caused by local convection currents at the edge of the wafer and by growth enhancement due to a sharp wafer edge. Recessed wafer pockets and edge rounding will eliminate edge crown.

Etch Pits

Etch pits created by decorative etch techniques reveal crystal dislocations produced by slip, as well as other defects. Crystallographic slip can be distinguished from other dislocations because the bases of the etch pits caused by slip [triangles for (111) and squares for (100) surfaces] lay on a common line.

Slips

Crystallographic slip during epitaxial deposition is dislocation motion in response to thermal stress. Slip can be reduced by uniform heating of the wafer, and operating at lower temperatures where the wafer is stronger. Defects at the wafer edge can be a major source for dislocation generation. Rounded and defect-free edges are essential for slip-free epitaxy.

Stacking Faults

Stacking faults are geometric defects nucleated at crystallographically disturbed sites on or near the growth surface. A uniform size of stacking faults indicates that the disturbance was all at the original surface prior to epitaxy. In this case, the likely candidate is residual oxide which survived the pre-epitaxy clean and H_2 bake, or which was created in the reactor during heat-up. Stacking faults of varying size indicate that foreign matter or particles were being generated during the deposition cycle.

For stacking faults that originate at the original wafer surface, the length L of the side of the fault is related to the thickness t of the epitaxy layer as follows:[28]

- For (111) planes: $t = 0.816\ L$
- For (100) planes: $t = 0.707\ L$

Cleaning the wafer surface, eliminating sources of particulate or other foreign matter, and increasing the deposition temperature all help to reduce stacking fault density.[29]

Spikes and Hillocks

Mounds, tripyramids, hillocks, and spikes are more severe defects also caused by crystallographically disturbed sites. The greater the disturbance, the more severe the defect. The larger defects are usually the result of particle contamination such as would occur by scraping the wafer on the carrier during loading and/or particles coming from dirty surfaces such as the reactor walls.

Shallow Pits

Shallow pits or haze revealed by decorative etching alone or decorative etching following an oxidation cycle indicate the presence of heavy metal precipitates.[30] Clean, dry plumbing does not contribute measurable amounts of metal contamination; however, any corrosion in the plumbing or inside the reactor can cause metal contamination. Other sources of metal contamination include the use of metal tweezers, metal spinner chucks in pre-epi clean, or sputtering from the wall of a plasma etcher or ion implanter.

SAFETY

The epitaxy process and reactor utilize several hazardous items, including gases which are flammable, noxious, poisonous, corrosive, and pyrophoric; high voltages; and chemicals which are strongly oxidizing or flammable.

Table 4 lists some of the common gases used in the epitaxy and CVD processes, together with information about their relative hazard. A well thought-out safety plan, in combination with good ventilation is essential to safe operation.

Good safety practice requires knowledge of the hazards involved, a well-developed preventive action plan, awareness of the specific safety plan that has been developed for the area, and a trained reaction to unexpected events. Many safety plans are not well thought out, do not make available the necessary information, and do not reinforce the plan with sufficient practice.

References 6 and 31 provide further information regarding CVD safety practice.

KEY TECHNICAL ISSUES

The key technical issues for commercial silicon epitaxy are productivity/cost, uniformity/quality, autodoping/buried layer control, and new materials technology.

Productivity/Cost

The maximum reactor productivity,

$$P_{(max)} = \frac{runs}{hr} \cdot \frac{wafers}{run} \cdot \frac{area}{wafer}$$

must be derated by the factors of yield, uptime, and utilization. Yield to commercial specification is typically 90%; uptime is approximately 70-95%, and utilization is 50-70%, for a total derating of approximately 50%. Utilization of the reactor includes time lost for scheduling (3-7%), etch/cost (5-20%), test runs (3-10%), test wafers in place of production material (2-15%), and facilities downtime (3-10%).[32]

The typical epitaxy process sequence listed in Figure 21 is accomplished in approximately 60 minutes; however, the epitaxy process can be divided into steps which require the heating power supply to be on and those which do not. A system designed with two process chambers that time-share a single power supply can process up to 1.8 times the number of runs that a single chamber system can produce.[32]

To understand epitaxy cost per unit area, it is useful to include the various factors as follows:

$$\frac{cost}{area} = \frac{fixed\ cost + variable\ cost}{hour} \cdot \frac{hour}{area\ processed}$$

Fixed cost includes all amortized items such as system price, installation, initial spare parts, and facilities allocations which continue whether the reactor

60 Thin-Film Deposition Processes and Techniques

Table 4: CVD Gas Safety Summary*

Gas	Flammable % In Air	Pyrophoric % In Air Auto Ignition °C	PPM Lethal Few Min.	PPM Lethal Few Hrs.	PPM Irritant Level	PPM Approx. Odor Level	PPM 8 Hour TLV	Comments
AsH_3	Yes	?	250	6	–	1	0.05	Highly poisonous
PH_3	Yes	40-50	2,000	100	8	2	0.3	Highly poisonous
B_2H_6	0.8-88	37-52	160?	?	–	3	0.1	Highly poisonous
NH_3	15-28	650	30,000	–	25	5	50	Reacts strongly with chlorides.
N_2O	Supports Combustion	non-	?	–	100	10	5	Anesthetic. Possible nerve damage.
NO_2/N_2O_4	Supports Combustion	non-	200?	–	60	10	5	
CO_2	No	non-	?	20,000	–	–	5,000	Keep separate from reducers. Supports fierce combustion.
O_2	No	non-	non-	–	–	–	–	
SiH_4	Yes	{0.5% SiH_4/H_2, 4% SiH_4/N_2}	non-	non-	?	–	0.5	Forms fine silica dust and vigorous flame.
SiH_2Cl_2, $SiHCl_3$, $SiCl_4$	No	–	?	~8,000	~10	~1	–	Decomposes to HCl and SiO_2 in air.
H_2	4-80	585	Asphyxiant	–	non-	non-	–	Store <2000 ft³ in building.
N_2	No	–	Asphyxiant	–	non-	non-	–	
HCl	No	–	1,300	1,000	10	1	5	Noxious
HF	No	–	100(?)	?	~30	?	3	Noxious

*See Reference 31a.

is utilized or not. Variable cost only include those items directly related to system operation such as chemicals, wafers, power, labor, and maintenance.

An epitaxy reactor must be sized for the production requirement. If the system is not fully utilized, the fixed costs become very high per unit area.

For a typical reactor having a total amortized value of $1,200,000, the fixed cost per hour will range from $15 to $200 per hour, depending upon the period of depreciation, with the typical value being about $50 per hour.[32]

RUNS/HOUR

STEP	SINGLE CHAMBER SMALL LOAD	SINGLE CHAMBER LARGE LOAD	DUAL CHAMBER LARGE OR SMALL LOAD SIZE MAIN POWER OFF	DUAL CHAMBER LARGE OR SMALL LOAD SIZE MAIN POWER ON
	(MINUTES)		(MINUTES)	
UNLOAD/LOAD	9	11	9-11	-
N$_2$/H$_2$ PURGE	2	2	2	-
HEAT TO 1120°C	12	12	-	12
ETCH	2	2	-	2
CLEAR	1	1	-	1
HEAT TO 1060°C	2	2	-	2
DEPOSITION	15	15	-	15
CLEAR	1	1	-	15
H$_2$ COOL	6	12	6-12	-
H$_2$/N$_2$ PURGE	2	2	2	-
TOTAL	53	60	19-27	33
RUNS/HR	1.1	1.0		1.8

Figure 21: Typical epitaxial silicon process sequence for single and dual chamber silicon epitaxy reactors.

Variable (operating) costs are summarized in Figure 22. When maintenance labor is considered, the hourly cost of repair/maintenance is about equal to the operating cost; therefore, it is convenient to use $50-$75/hour as a constant variable cost, whether the system is in an operational or maintenance mode.

Combined fixed and variable costs are $100-$125/hour for a typical reactor in 1986 dollars. Other factory costs add $25-$50/hour. With a maximum productivity of 10-20 wafers/hour and a 50% derating factor, the manufacturing cost of silicon epitaxy is approximately $20 per 125-150 mm diameter wafer. Substrate cost, including yield loss and test wafers, brings the manufacturing cost of an epitaxy wafer to the $40-$50 range. Non-manufacturing costs, plant utilization, and profit further increases the market price of a 125-150 mm diameter epitaxy wafer to $75-$100, in 1986 US dollars.

Epitaxy cost is expected to be reduced during the late 1980's as more productive reactors are put into service and better utilization is achieved through centralized operations, especially where buried layer patterns are not required.[4,15]

62 Thin-Film Deposition Processes and Techniques

ITEM	TYPICAL COST/HOUR
PROCESS GAS	$12 to $16
POWER	$ 6 to $12
CONSUMABLES	$ 8 to $12
TEST WAFERS	$ 5 to $10
LOADED LABOR RATE	$19 to $25
TOTAL (EXCLUDING MAINTENANCE)	$50 to $75
LOADED MAINTENANCE LABOR RATE	$50 to $75

Figure 22: Typical epitaxy reactor operating costs.[32]

Uniformity/Quality

Uniformity and quality are the most important technological aspects of silicon epitaxy because device performance often depends directly upon these factors.

Thickness uniformity of ±2-4% and resistivity uniformity of ±4-10% are available in commercial systems.[33] Such capability is more than adequate for most device applications.

Statistical techniques can be used to characterize variations in film properties. Average and standard deviation are well known concepts; however, results are often expressed as ±xy % without specifying the definitions, as was done in the paragraph above.

The uniformities noted above are for 90% of all points measured. Another specification frequently used is the total variation from maximum to minimum value, expressed as ±xy% of the average. The maximum-minimum variation is easily calculated from the formula:

$$\pm \text{ variation \%} = \frac{\max - \min}{\max + \min} \frac{\%}{100}$$

The term "90% of all data points" can be directly related to the standard deviation. Assuming there is a normal distribution of data, "90% of all data points" would be equal to 1.64 standard deviations. The maximum-minimum variation is not directly related to statistical calculations; however, three standard deviations will normally include 99.6% of the data points.

Epitaxy quality includes crystal and surface defects, as well as metallic contamination. Causes and cures for these defects are discussed in an earlier section.

Buried Layer Pattern Transfer[27]

Buried layer pattern transfer is vital to the construction of bipolar integrated circuits. The buried layer (actually a pattern of heavily-doped regions in

Silicon Epitaxy by Chemical Vapor Deposition

the original substrate surface) are created in the wafer surface before epitaxial growth to provide low resistance paths for the collectors of the bipolar transistors. If the subsequent patterns are not aligned directly above these buried layer patterns, the transistors will not operate to specification.

Buried layer patterns in the wafer surface are created because the oxidation rate is higher in areas where the dopant concentration is higher. When the oxide is removed after buried layer diffusion, the higher oxidation rate leaves depressions in the wafer surface over each buried layer region. These depressions are used to align subsequent masks in the creation of an IC.

Because of different growth rates in different crystallographic directions, the buried layer patterns can be shifted relative to the region of high doping, and the pattern can be distorted or washed out. Figure 23 describes the various possibilities of symmetric and unsymmetric distortion with and without size change.[27a,27b]

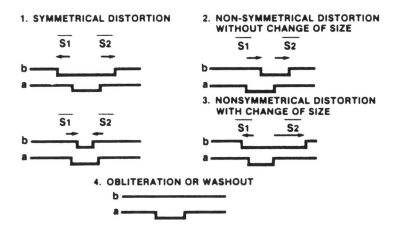

Figure 23: Possibilities for epitaxy buried layer pattern shift/distortion.[27a,27b]

Pattern distortion is a change in size of the original pattern dimensions, often accompanied by sidewall faceting. Patterns can enlarge, shrink, facet and virtually disappear, depending upon the deposition conditions and orientation of the wafer. Pattern distortion occurs:

- With symmetric pattern shift on off-orientation (111) surfaces, when the off-orientation is in the direction of the nearest (110) plane,
- With unsymmetric pattern shift on off-orientation (100) surfaces,
- Without pattern shift on on-orientation (100) surfaces.

Pattern distortion is reduced by:

- Decreasing the chlorine concentration in the depositing gas,

64 Thin-Film Deposition Processes and Techniques

- Increasing the growth temperature,
- Decreasing the growth pressure.

Pattern shift is the motion of the surface pattern relative to the heavily doped buried layer region. The pattern shift ratio is the amount of shift divided by the thickness of the epitaxy layer. Pattern shift must be controlled to permit accurate registry of subsequent masks with the actual buried layer pattern. A zero pattern shift ratio is desirable but it is not achieved in commercial production except by reduced pressure deposition. A pattern shift ratios of 1:1 to 1.5:1 are commonly utilized.

When a wafer is cut from a (111) oriented crystal, it is cut in a specific off-orientation direction, as depicted in Figure 24.[27f] This orientation off the (111) plane toward the nearest (110) plane eliminates orange peel surface growth and provides for symmetric pattern shift during epitaxy growth. It is convenient to note that the symmetry of a dislocation etch pit reflects the symmetry of the off-orientation cut, as shown in Figure 24. This observation provides a useful test for incorrect off-orientation wafers.

For atmospheric pressure deposition, pattern shift is minimized by:

- Orienting (111) plane 3-5 degrees toward the nearest (110) plane,
- Orienting the (100) plane parallel to the wafer surface within ± 0.15 degrees,
- Reducing the growth rate,
- Increasing the growth temperature,
- Reducing the growth pressure,
- Reducing the chlorine content of the process gas.

Fortunately, the conditions for reducing pattern shift are essentially the same as those for reducing pattern distortion, except that pattern distortion can be significant for on-orientation (100) surfaces.

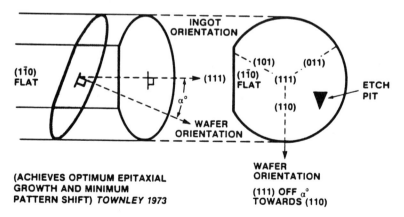

Figure 24: Off-orientation (111) wafer cut from (111) crystal ingot.[27f]

SiH_4 offers the best results in terms of reduced pattern shift and distortion; however, the SiH_4 growth rate is normally less than 0.25 micron/minute because of bell jar coating and gas phase reactions.

For epitaxy layers thicker than about 1.5 microns, the best compromise is considered to be SiH_2Cl_2 operating at reduced pressure with growth rates in the 0.3-0.5 micron/minute range. For epitaxy layers less than 1.5 microns, SiH_4 offers the advantage of lower temperature and the disadvantage of being more sensitive to oxidizer leaks.

For lower growth pressures, pattern shift occurs to a lesser degree, becomes zero, and then becomes "negative," in that the pattern shift is in the direction opposite that for atmospheric pressure. The effect of pattern shift on pressure is illustrated in Figure 25[34] where "negative" pattern shift was found for pressures below 100 torr at 0.3 micron/minute using SiH_2Cl_2 at 1080°C in a cylinder reactor. Because of the "negative" pattern shift concept, process changes that "reduce" pattern shift must be considered in the algebraic sense. That is, if "negative" pattern shift is present, process changes that would "increase" pattern shift would actually cause the "negative" shift to be closer to zero. Process changes that would "decrease" pattern shift will actually cause "negative" pattern shift to become more "negative" or further from zero.

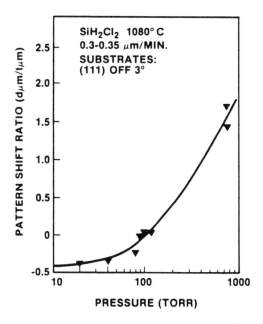

Figure 25: Effect of pressure on pattern shift ratio.[34]

Pattern shift ratios versus pressure are provided in Figure 26 for a radiantly heated cylinder and an induction heated vertical reactor.[35] For pattern shift ratios above zero, decreasing the pressure and growth rate brings the shift closer to zero. For pattern shift ratios below zero, increasing the pressure and growth rate bring the shift closer to zero.

Growth temperature has the strongest effect on pattern shift. A change of only 20-30°C has a major impact on the pattern shift ratio. Decreasing the deposition temperature causes the pattern shift ratio to increase in absolute terms; i.e., for "negative" pattern shift, decreasing the deposition temperature brings the pattern shift closer to zero.

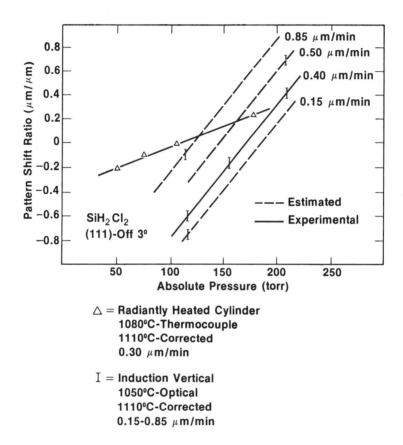

Figure 26: Pattern shift ratio versus pressure for various growth rates in vertical and cylinder reactors.[35]

Autodoping[36]

Autodoping is the presence of unwanted dopant contributed to the epitaxy layer by the wafers themselves. The intrinsic doping level is unwanted dopant contributed by the reactor parts. Two kinds of autodoping are recognized:

- Macro-autodoping in which dopant from the wafer surfaces (front and back) contribute dopant generally to all the growing layers.
- Micro-autodoping in which dopant from one location on a wafer migrates to another location on the same wafer.

Autodoping increases with increasing vapor pressure and increasing diffusion rates of the dopant. Sb causes the least autodoping, followed by As, B and P. Macro-autodoping can be reduced by many techniques, including:

- Sealing the back of the wafer with CVD oxide, nitride, or polycrystalline silicon during the wafer preparation steps before epitaxy (most effective),
- Operating the reactor so that backside transfer of silicon occurs to seal the back of the wafer during HCl etch (effective but requires etch/coat time),
- Using a two-step process in which a thin, undoped cap layer is deposited first, the system is then purged with H_2 and the desired epitaxy layer is grown (more effective for plug flow reactors where the gas composition can be rapidly changed),
- Using a high/low temperature sequence in which the surface of the wafer is depleted of dopant by a high temperature bake, followed by a lower temperature epitaxy growth step (most effective for As),
- Decreasing the gas residence time by increasing the total flow rate and/or reducing the reactor pressure or volume,
- Operating at reduced pressure where the escaping tendency of the dopant on the wafer surface and the number of volume changes per minute is greatly increased (most effective for As),
- Reducing the concentration of dopant on the wafer surface by using ion implantation instead of dopant diffusion to dope the wafer surface prior to epitaxy growth (effective for all dopants).

Micro-doping is depicted schematically in Figure 27. Here, the n+ buried layer encroaches into the epitaxy layer by a combination of solid state diffusion and vapor transport during growth. Both vertical and lateral autodoping occur. Vertical autodoping is dopant moving vertically into the growing layer; lateral autodoping is movement to the sides of the buried layer region.

Micro-autodoping is reduced by the same techniques as macro-autodoping. Backside sealing is usually not required because IC processing is normally structured to avoid backside doping.

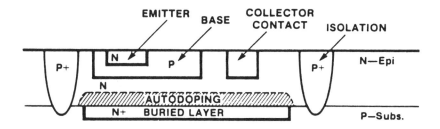

Figure 27: Micro-autodoping in a buried layer bipolar device structure.

68 Thin-Film Deposition Processes and Techniques

Pre-epitaxy oxidation can have a strong effect on autodoping. B is preferentially absorbed by a growing thermal oxide on silicon and the concentration near the single crystal surface is reduced.[37] The opposite is true for As and P. This effect of the oxide segregation coefficient for different dopants and different pre-epitaxy oxidation conditions is summarized in Figure 28.[38] The number of arrows pointing up indicates an increasing degree of autodoping; arrows pointing down indicate a decreasing degree. If possible, pre-epitaxy doping for As and P should be accomplished with higher temperature, dry oxidation rather than lower temperature, wet oxidation. The opposite is true for B.

	HIGH TEMP	LOW TEMP	WET OXID	DRY OXID
As	↓	↑↑	↑↑↑	↑
P	↓	↑↑	↑↑↑	↑
B	↓	↑	↓↓	↓

Figure 28: Effect of pre-epitaxy drive-in oxidation on autodoping.

Vertical and lateral autodoping are measured by spreading resistance probe (SRP) resistivity profile techniques as illustrated in Figure 29.[34] When the SRP profile is taken through the epitaxy layer and into the buried layer region, the vertical autodoping profile is measured. When the profile is taken to the side of the buried layer region, one slice through the lateral autodoping profile is measured.

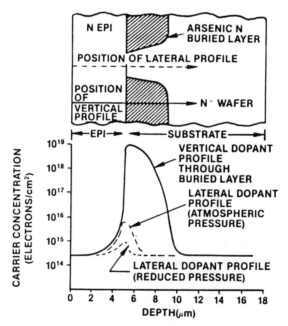

Figure 29: Vertical and lateral autodoping measurement using spreading resistance probe resistivity profiles.[34]

The SRP profile measures many different contributions to autodoping and these contributions are summarized in Figure 30.

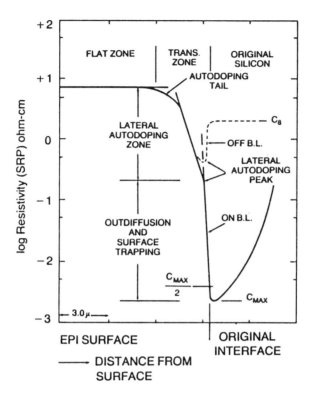

Figure 30: Spreading resistance probe resistivity profile on and off buried layer region (150 torr, $SiCl_4$, 1060°C optical).

Dopant Transitions. As silicon devices become faster and smaller, the epitaxy layer becomes thinner, and autodoping effects on transition width and circuit density become more important.

Figure 31 is an SRP resistivity profile through an epitaxy layer over a heavily doped As buried layer.[35] The epitaxy layer was grown in a vertical reactor at 1110°C in SiH_2Cl_2 at 0.5 micron/minute with a reactor pressure of 150 torr. The buried layer minimum resistivity is 0.0027 ohm-cm and the controlled dopant region has a resistivity of 0.27 ohm-cm. If the autodoping tail is ignored, a straight line can be drawn using semi-log paper through the resistivity points and a "transition width" can be defined as the distance required for this straight line to traverse two orders of magnitude in resistivity. For consistency, this straight line should pass through the resistivity value corresponding to one-half the maximum dopant concentration (the location of the original wafer surface). This unusual definition of transition width permits an analysis of the effect of pressure on the initial slope of the vertical autodoping profile.

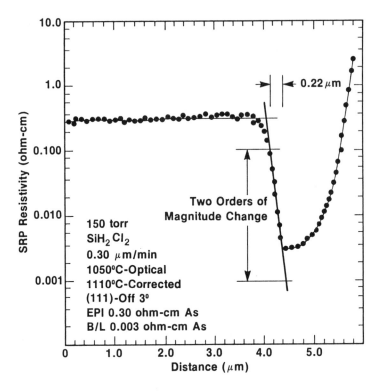

Figure 31: Spreading resistance probe resistivity profile over buried layer region illustrating one definition of transition width (distance required for a two order of magnitude resistivity change).[35]

Figure 32 is a plot of the "transition width" as defined above versus reactor pressure in the 50-250 torr region for a vertical reactor operating under the deposition conditions described above.[35] Data for both B and As are presented. For pressures above about 250 torr, the transition width decreases fairly slowly with decreasing pressure. In the 100-200 torr range, the transition width decreases rapidly with decreasing pressure and then becomes essentially constant as the pressure is decreased below about 100 torr.

The data in Figure 32 are explained as follows. Above about 200 torr, the escaping tendency for either B or As is not strongly affected by pressure, and the volume changes per unit time are increased by only 3.8 times compared to 760 torr for the same total flowrate. From 200 torr to 100 torr, the number of volume changes per unit time doubles to 7.6 per unit time and the lower pressure encourages dopant to escape from all surfaces. As the pressure is reduced to below 100 torr, the autodoping by gas transport is essentially eliminated and only solid state diffusion remains as the mechanism for dopant encroachment into the epitaxy layer. Solid state diffusion is virtually unaffected by reduced pressure; therefore, the transition width no longer decreases with decreasing pressure.

The radiantly heated cylinder reactor has been reported to show increased

B autodoping for process pressures below 100 torr,[39] in contrast with the data presented here.[35]

Device structures can have both B and As buried layer patterns. The data presented in Figure 32 support the idea[39a] that the 100-150 torr pressure range is optimum for minimizing both B and As autodoping.

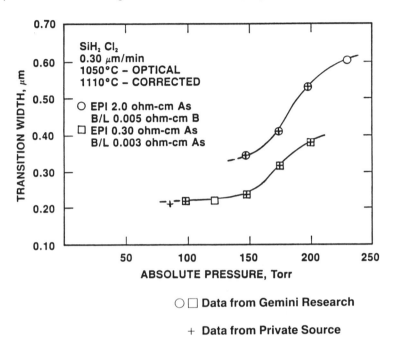

Figure 32: Transition width (as defined in Figure 31) versus deposition pressure.[35]

For some bipolar IC's, B buried layers are used to reduce the amount of diffusion time required to create the p+ isolation regions. The concern is that B autodoping will cause a compensation resistivity peak above the As buried regions and adversely affect device performance. Figure 33 includes a schematic cross section through an epitaxy layer grown in a vertical reactor over both B and As buried layers. Figure 33 illustrates how the SRP resistivity profiles and groove/stain cross section would be affected if there were or were not significant autodoping.[35]

Figure 34 contains an SRP resistivity profile through an epitaxy layer grown in a vertical reactor for a real device structure using both B and As buried layers. Figure 35 presents a grooved and stained cross section through the same structure represented in Figure 34. Clearly, for the conditions cited in Figure 34, there is little or no lateral autodoping effect of the B above the As buried layer region.[35]

Autodoping is complex and all the process conditions are closely related. Key issues must be resolved by direct experiment on each device structure using the autodoping reduction techniques described above.

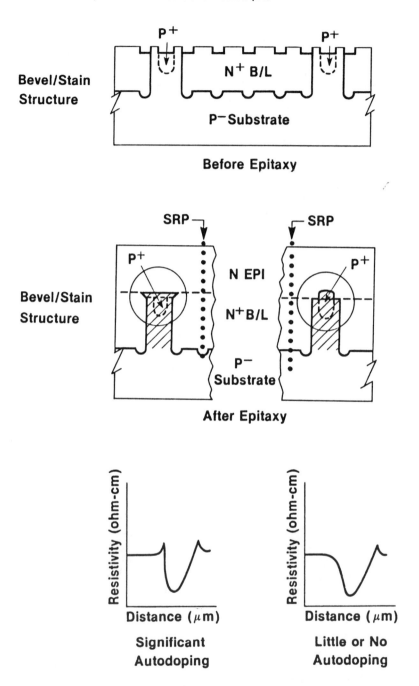

Figure 33: Schematic diagrams illustrating possible effects of autodoping in an As-doped epitaxy layer grown over both B and As buried layer patterns.[35]

Silicon Epitaxy by Chemical Vapor Deposition 73

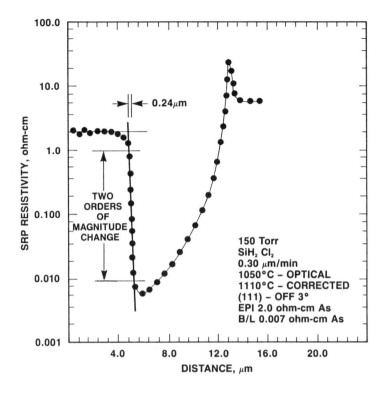

Figure 34: Actual spreading resistance probe resistivity profile over As buried layer region in a device structure with both B and As buried layer patterns.[35]

Figure 35: Grooved and stained cross section through same device structure in Figure 34.[35]

NEW MATERIALS TECHNOLOGY FOR SILICON EPITAXY

The epitaxy process has been well-defined since the mid-1970's in terms of temperature and pressure ranges, growth rates, and chemical choices. Special techniques which were developed in that time frame but are only now becoming more widely used include:

Silicon-on-sapphire,

Selective epitaxy,

Epitaxial lateral overgrowth,

Epitaxial trench refill,

Low temperature epitaxy.

Silicon-on-sapphire provides single crystal (100) silicon grown on top of single crystal sapphire substrates cut and polished on the (1102) plane where there is a geometrical relationship between the two different crystal structures. Successful growth of silicon-on-sapphire requires:[40]

A complete absence of chlorine which attacks the sapphire at high temperature and liberates Al, a p-type dopant,

An oxidizer level below 1 ppm in the reactor,

A deposition temperature of 900-920°C in SiH_4,

A well-polished sapphire substrate,

A high initial growth rate, followed by one somewhat slower after the growth mechanism becomes silicon on silicon rather than silicon on sapphire.

Selective epitaxy is the growth of single crystal silicon in windows etched into thermally grown oxide on the wafer surface. Selective epitaxy with and without controlled polycrystalline silicon being grown on top of the oxide is possible.[35,41]

Selective epitaxy *without* polycrystalline silicon overgrowth on the oxide is accomplished by depositing with conditions that do not promote nucleation. These conditions include depositing:

With approximately a 3:1 ratio of chlorine to silicon, created by adding HCl to the deposit gas,

At a slightly higher temperature to enhance the etching rate of the chlorine,

At a reduced pressure where escape of volatile silicon chlorides is enhanced.

A growth temperature of 1000°C using SiH_2Cl_2 with a 3/1 HCl/SiH_2Cl_2 ratio at 100 torr is currently recommended for selective epitaxy without polycrystalline overgrowth.[35]

Epitaxial lateral overgrowth is another variation on selective epitaxy in which the selectively grown epitaxy is permitted to grow up out of the windows and over the oxide surface. The single crystal layer on top of the oxide can be etched to disconnect it from the single substrate, thereby creating high quality single crystal islands on top of thermally grown oxide. Such structures have considerable promise for three-dimensional device structures.[41]

Selective epitaxy *with* polycrystalline silicon overgrowth utilizes deposition conditions that promote nucleation, such as:

- The absence of chlorine,
- The use of SiH_4,
- The use of lower temperatures and higher pressures.

Selective epitaxy *with* polycrystalline silicon overgrowth can be accomplished with SiH_4 at 975°C and 760 torr deposition pressure.[35,42]

In device applications, the polycrystalline silicon overgrowth provides a place to create heavily doped contacts to the single crystal area without affecting the more lightly doped device regions.

LOW TEMPERATURE EPITAXY

Silicon epitaxy by CVD at lower temperatures is now of significant interest.[43,44] Silicon epitaxy has been successfully demonstrated:

- In the 800°-1000°C range in a conventional reactor using SiH_4 or SiH_2Cl_2 after first initiating growth at a higher temperature,[45]
- In the 800°-1000°C range using SiH_4 and substituting He for H_2 as the main carrier gas,[46]
- In the 850°-1000°C range in a conventional reactor using SiH_4 with wafers that were etched in HF just prior to placing in the reactor,[47]
- In the 850°-900°C range in a conventional reactor using SiH_2Cl_2 at reduced pressures of 10-20 torr,[41b]
- In the 800°-900°C range using photodissociation of silane and other silicon compounds,[48]
- In the 750°-950°C range in a cold wall reactor after removing the native oxide with a plasma etch,[49]
- In the 700°-900°C range by operating at very low pressure in a load-locked, hot wall reactor.[50]

All these low temperature epitaxy processes have limited application because they require very low growth rates to achieve even partially satisfactory crystal quality.[16b,16c,51] As the required epitaxy layer thicknesses go below 0.6 micron, such techniques must be further developed and characterized for commercial production.[44]

CONCLUSIONS

Silicon epitaxy is a vital and growing semiconductor process technology that is becoming more important as device geometries shrink and MOS IC's begin to depend upon its unique advantages.

Fluid mechanics and other chemical engineering principles can be used to understand the process adjustments. The technology is somewhat complex but understandable and reproducible in a production environment.

Epitaxy quality is very sensitive to pre-epitaxy crystal surface preparation and to the deposition conditions. Crystal quality is especially sensitive to the presence of oxidizers during heat-up and deposit. Commercial equipment is progressing to meet the industry's needs in terms of productivity and quality, and CVD will continue to be the dominant silicon epitaxy technology through the century.

REFERENCES

1a. Berry, B.M., *Fundamentals of Silicon Integrated Circuit Technology,* Vol. 1, R.M. Burger and R.P. Donovan, eds., Prentice Hall (1967).
1b. Li, C.H., *Phys. Stat. Sol.,* Vol. 15, p. 419 (1966)
1c. Gise, P.E. and Blanchard, R., *Semiconductor and Integrated Circuit Fabrication Techniques,* Reston (1979).
1d. Wolf, S. and Tauber, R.N., *Silicon Processing for the VLSI Era,* Vol. 1 - Process Technology, Lattice Press (1986).
1e. Grove, A.S., *Physics and Technology of Semiconductor Devices,* Wiley (1967).
1f. Sze, S.M., *Semiconductor Devices and Technology,* Wiley (1985).
2a. *Proceedings of International Symposium on Chemical Vapor Deposition,* held alternate years since 1968. See Electrochemical Society, Pennington, N.J. for most recent sponsorship and early year issues.
2b. Vossen, J.L. and Kern, W., *Thin Film Processes,* Academic Press (1978).
2c. Hammond, M.L., *Solid State Tech.,* Vol. 21 (11), p. 68 (1978).
2d. Bollen, L.J.M., *Acta Electronica,* Vol. 21, p. 185 (1978).
2e. Gupta, D.C., *Solid State Tech.,* Vol. 15 (10), p. 33 (1971).
3a. White, L.S., et al, *Silicon Processing, ASTM STP 804,* D.C. Gupta, ed., American Society for Testing Materials, p. 190 (1983).
3b. Borland, J.O. and Deacon, T., *Solid State Tech.,* Vol. 27 (8), p. 123 (1984).
3c. Kopp, R.J., *Information Services Seminar,* Semiconductor Equipment and Materials Institute, Mountain View, CA (1986 and 1987).
3d. Troutman, R.R., *Latchup in CMOS Technology: The Problem and its Cure,* Kluwer Academic Press (1986).
4a. Hammond, M.L., *Semiconductor International,* Vol. 6 (10), p. 58 (1983).
4b. Burggraaf, P.S., *ibid.,* p. 44.
5. *Quick Reference Manual for Silicon Integrated Circuit Technology,* W.E. Beadle, et al, eds., Wiley (1985).
6. *Book of SEMI Standards,* Vols. 1-4, published annually by Semiconductor Equipment and Materials Institute, Inc., Mountain View, CA.

7a. Bloem, J. and Giling, L.J., *VLSI Electronics,* N.G. Einspruch and H. Huff, eds., Vol. 12, Academic Press (1985).
7b. Nishizawa, J. and Saito, M., *Proc. of the VIIIth International Conf. on Chemical Vapor Disposition,* Electrochemical Society, p. 317 (1981).
8a. Tung, S.K., *J. Electrochem. Soc.,* Vol. 112, p. 436 (1965).
8b. p. 35 of Reference 1a.
9a. Roberge, R.P., et al, *Process and Equipment Reliability in the Fab Environment,* Semiconductor Equipment and Materials Institute, Mountain View, CA (1986); also published in Semiconductor International, Vol. 10 (1), p. 177 (1987).
9b. Ghidini, G. and Smith, F.W., *J. Electrochem. Soc.,* Vol. 131, p. 2924 (1984).
10a. p. 36 in Reference 1a.
10b. Medernach, J.W. and Wells, V.A., *Emerging Semiconductor Technology, ASTM 960,* D.C. Gupta and P.H. Langer, eds., American Society for Testing Materials (1986).
10c. Shephard, W.H., *J. Electrochem. Soc.,* Vol. 112, p. 968 (1965).
10d. Bloem, J., *J. Electrochem. Soc.,* Vol. 117, p. 1397 (1970).
10e. Rai-Choudhury, P., *J. Electrochem. Soc.,* Vol. 118, p. 266 (1971).
10f. Rai-Choudhury, P., *J. Electrochem. Soc.,* Vol. 118, p. 1183 (1971).
10g. Silvestri, V., et al, *J. Electrochem. Soc.,* Vol. 131, p. 877 (1984).
10h. Srinivasen, G.R., U.S. Patent 4,153,486 (1979).
11a. Bloem, J., *J. Electrochem. Soc.,* Vol. 118, p. 1838 (1971).
11b. Swanson, T.B. and Tucker, R.N., *J. Electrochem. Soc.,* Vol. 116, p. 1271 (1969).
11c. Shephard, W.H., *J. Electrochem. Soc.,* Vol. 115, p. 541 (1968).
11d. Bloem, J., et al, *J. Electrochem. Soc.,* Vol. 121, p. 1354 (1968).
12a. Severin, P.J. and Everstyn, F.C., *J. Electrochem. Soc.,* Vol. 122, p. 962 (1975).
12b. Everstyn, F.C., et al, *J. Electrochem. Soc.,* Vol. 117, p. 925 (1970).
12c. Kern, W. and Ban, V.S., in Reference 2b, p. 271.
12d. Gillis, J. and Hammond, M.L., Abs. 151 in *Extended Abstracts 84-1,* The Electrochemical Society (1984).
13. Corboy, J.F. and Pagliaro, Jr., F., *RCA Rev.,* Vol. 44, p. 231 (1983).
14. Manke, C.W. and Donaghey, L.F., *Proc. of the VIth International Conf. on Chemical Vapor Deposition,* L.F. Donaghey, et al, eds., p. 151 (1971).
15a. Burggraaf, P., *Semiconductor International,* Vol. 9 (5), p. 68 (1986).
15b. *Ibid,* p. 250.
15c. *TETRON ONE Product Brochure,* Gemini Research, Inc., Fremont, CA.
16a. Everstyn, F.C., *Philips Research Rept.,* Vol. 19, p. 45 (1974).
16b. Bloem, J., *J. Cryst. Growth,* Vol. 50, p. 581 (1980).
16c. Bloem, J., *Semiconductor Silicon 1973,* H.R. Huff and H.R. Burgess, eds., The Electrochemical Society, p. 180 (1973).
17a. Hammond, M.L., *Solid State Tech.,* Vol. 22 (12), p. 61 (1979).
17b. *Series AMC 7800 Instruction Manual,* Applied Materials, Inc., Santa Clara, CA.
17c. *Gemini-2 Operations Manual,* Gemini Research, Inc., Fremont, CA.
18a. Robinson, McD., et al, *J. Electrochem. Soc.,* Vol. 129, p. 2858 (1982).

18b. Patents pending, Gemini Research, Fremont, CA.
19. Thomas, R.C., *Solid State Tech.,* Vol. 27 (9), p. 153 (1985).
20. Roth, A., *Vacuum Technology,* 2nd Ed., North-Holland Pub. Co. (1982).
21. *Tylan Source Controller Product Brochure,* Tylan Corp., Torrance, CA.
22. p. 40 of Reference 1a.
23. *Dopant Slide Rule,* Gemini Research, Inc., Fremont, CA.
24a. *Introduction to Radiation Thermometry,* Tech. Note TN 102, Ircon, Inc.
24b. Baker, H.D., et al, *Temperature Measurement in Engineering,* Omega Press, Division of Omega Engineering, Stamford, CT (1975).
25. pp. 2-61 of Reference 5.
26a. Sato, T., Japan *J. Appl. Phys.,* Vol. 6 (3), p. 339 (1967).
26b. Runyan, W.R., *Silicon Semiconductor Technology,* McGraw-Hill (1965).
26c. Wolf, H.F., *Semiconductors,* Wiley (1971).
27a. Drum, C.M. and Clark, C.A., *J. Electrochem. Soc.,* Vol. 117, p. 1401 (1970).
27b. Drum, C.M. and Clark, C.A., *J. Electrochem. Soc.,* Vol. 115, p. 664 (1968).
27c. Dixit, A., Abst. 209 in *Extended Abstracts, 78-1,* The Electrochemical Society (1978).
27d. Weeks, S.P., *Solid State Tech.,* Vol. 24 (11), p. 111 (1981).
27e. Boydston, M.R., et al, in *Silicon Processing, ASTM 804,* American Society for Testing Materials, p. 174 (1983).
27f. Townley, D.O., *Solid State Tech.,* Vol. 16 (1), p. 43 (1973).
27g. Corboy, J.F., et al, *Proc. of the IXth International Conf. on Chemical Vapor Deposition,* p. 434, McD. Robinson et al, eds., The Electrochemical Society (1984).
27h. Ogirima, M., et al, *J. Electrochem. Soc.,* Vol. 124, p. 903 (1977).
27i. Krullman, E. and Engl, W.L., *IEEE Trans. Elec. Devices,* Vol. ED-29, p. 491 (1982).
28. Runyan, W.R., *Semiconductor Measurements and Instrumentation,* Mc Graw-Hill (1975).
29. Bansal, I.K., *Solid State Tech.,* Vol. 29 (7), p. 75 (1986).
30. Pearce, C.W. and McMahon, R.G., *J. Vac. Sci. Tech.,* Vol. 14 (1), p. 40 (1977).
31a. Hammond, M.L., *Solid State Tech.,* Vol. 23 (12), p. 104 (1980).
31b. Murray, C., *Semiconductor International,* Vol. 9 (8), p. 60 (1986).
32. Hammond, M.L., et al, *Microelectronics Manufacturing and Testing,* Vol. 7 (10), p. 80 (1984).
33. *Gemini-2 Product Brochure,* Gemini Research, Inc., Fremont, CA.
34. Herring, R.B., *Solid State Tech.,* Vol. 22 (11), p. 75 (1979).
35. Fisher, S.M., et al, *Solid State Tech.,* Vol. 29 (1), p. 109 (1986).
36a. Srinivasen, G.R., *J. Electrochem. Soc.,* Vol. 127, p. 1334 (1980).
36b. Srinivasen, G.R., *J. Electrochem. Soc.,* Vol. 125, p. 146 (1978).
36c. Srinivasen, G.R., *Solid State Tech.,* Vol. 24 (11), p. 101 (1981).
36d. Srinivasen, G.R., *Silicon Processing, ASTM STP 804,* D.C. Gupta, ed., American Society for Testing Materials, p. 151 (1983).
36e. Wong, M. and Reif, R., *J. Solid State Circuits,* Vol. SC-20 (1), p. 3 (1985).
36f. *Ibid.,* p. 9.
36g. Bozler, C.O., *J. Electrochem. Soc.,* Vol. 122, p. 1705 (1975).
37a. Grove, A.S., et al, *J. Appl. Phys.,* Vol. 35, p. 2629 (1964).
37b. p. 229 of Reference 1b.

38. Ackermann, G.K. and Zybur, H., Abs. 522, *Extended Abstracts, 84-2,* The Electrochemical Society (1984).
39a. Kulkarni, S.B. and Kozul, A.A., Abs. 540, *Extended Abstracts, 80-2,* The Electrochemical Society (1980).
39b. Graef, M.W.M. and Leunissen, B.J.H., Abs. 456, *Extended Abstracts, 84-2,* The Electrochemical Society (1984).
40a. McMahon, R.A., et al, *Solid State Tech.,* Vol. 28 (6), p. 208 (1985).
40b. Cullen, G.W., et al, *J. Cryst. Growth,* Vol. 56, p. 287 (1982).
41a. Jastrzebski, L., *Solid State Tech.,* Vol. 27 (9), p. 239 (1984).
41b. Borland, J.D. and Drowley, C.I., *Solid State Tech.,* Vol. 28 (8), p. 141 (1985).
41c. Liaw, H.M., et al, *Solid State Tech.,* Vol. 27 (5), p. 135 (1985).
41d. Zinng, R.D., et al, *J. Electrochem. Soc.,* Vol. 133, p. 1274 (1986).
41e. Ishitani, A., et al, *Microelectronic Engineering,* Vol. 4, p. 3 (1986).
42. p. 155 in Reference 1b.
43. Su, S.C., *Solid State Tech.,* Vol. 25, p. 72 (1981).
44. Fisher, S.M., et al, *Emerging Semiconductor Technology, ASTM STP 960,* D.C. Gupta and P.H. Langer, eds., American Society for Testing Materials (1986).
45. Nakanuma, S., *IEEE Trans. Elec. Devices,* Vol. ED-13, p. 578 (1966).
46. Richman, D., et al, *RCA Rev.,* Vol. 41, p. 613 (1970).
47. Fok, T.Y., et al, Abs. 498 RNP, Abstracts of Recent Newspapers, *J. Electrochem. Soc.,* Vol. 130, p. 441C (1983).
48a. Frieser, R.C., *J. Electrochem. Soc.,* Vol. 115, p. 401 (1968).
48b. Kumagawa, M., et al, *Japan, J. Appl. Phys.,* Vol. 7, p. 1332 (1968).
48c. Yamazaki, T., et al, *1984 Symposium for VLSI Technology,* Sept. 1984, IEEE and Japan Society for Appl. Phys. IEEE Catalogue #84, CH 2061-0, pp. 56-77 (1984).
49a. Burger, W.R. and Reif, R., Abs. 267, *Extended Abstracts, 85-2,* The Electro Chemical Society (1985).
49b. Donahue, T.J. and Reif, R., *Semiconductor International,* Vol. 8 (8), p. 142 (1985).
49c. Reif, R., *Emerging Semiconductor Technology, ASTM STP 960,* D.C. Gupta and P.H. Langer, eds., American Society for Testing Materials (1986).
49d. Townsend, W.G. and Uddin, M.E., *Solid State Electronics,* Vol. 16, p. 39 (1973).
49e. Suzuki, S. and Itoh, T., *J. Appl. Phys.,* Vol. 54, p. 1466 (1983).
49f. Donohue, T.J. and Reif, R., *J. Appl. Phys.,* Vol. 57, p. 2757 (1985).
50. Myerson, B.S., et al, Abs. 266, *Extended Abstracts, 85-2,* The Electrochemical Society (1985).
51. Gupta, D.C., *J. Electrochem. Soc.,* Vol. 116, p. 670 (1969).

3

Low Pressure Chemical Vapor Deposition

Ronald C. Rossi

INTRODUCTION

The deposition of thin films for semiconductor device manufacture by chemical vapor deposition at atmospheric pressure (APCVD) was a widely accepted process in 1976 when equipment for low-pressure chemical vapor deposition (LPCVD) was introduced into the marketplace. At that time, the 3-inch wafer was the predominant wafer size used in production with some residual presence of smaller wafers and the 100 mm wafer just being introduced into advanced lines. In the next few years, the LPCVD process became the preferred method for chemical vapor deposition of thin films. The transformation to a new technology that required massive capital expenditure for new equipment took place at a rapid rate throughout the industry. The reasons for this rapid change were: (1) a superior film quality, (2) a greatly reduced processing cost, and (3) greatly increased throughput per unit of capital investment. Improved film quality also means increased yields and decreased unit costs in an industry that was becoming increasingly competitive.

A comparative cost analysis for thin film deposition processing by typical atmospheric CVD reactors versus LPCVD reactors was made by Hammond and Gieske.[1] Their study projected a savings of 98-99% on process gases, 98-99% on electricity, 80-90% on labor, and 85-90% in capital investment for overall cost savings of 90% for 3-inch wafers and 94% for 100 mm wafers. These estimates did not include the increased yield of LPCVD films from a reduction of defects, better step coverage, and finer surface texture that further reduced per-wafer costs. Subsequently, more conservative and possibly more realistic cost savings of 80%-plus have been reported[2] —significant nevertheless.

If this comparison were made on 125 mm or 150 mm wafers, the total cost savings could well be expected to approach 98-99%, since the cost per-wafer for APCVD processing increases roughly proportional to the surface area of the wafer, whereas for LPCVD processing, costs are nearly independent of wafer area. Moreover, it was largely this near-independence of wafer cost on wafer size made possible by LPCVD that provided the impetus for the continuing demand for equipment for processing larger diameter wafers.[3]

While cost may be the primary motivating force for promoting technology change, the greatly improved film quality was a significant additional factor. Compared to thin films deposited by APCVD, LPCVD films have better thickness uniformity, fewer pinholes, better conformality, fewer particulate contaminants, and finer surface texture.[4] All of these aspects of film quality improvement can be directly traced to either the reduction in gas pressure or to the design of the process reactor. The higher pressure of the APCVD process promotes gas-phase reactions which produce particulates that can readily fall on horizontally positioned wafers so as to create defects. In LPCVD, low pressure inhibits the gas-phase reaction and the hot walls promote an adherent film identical to that which forms on vertically positioned wafers. However, it is primarily the lower deposition rates (typically 20% for most films[2]) of LPCVD that is responsible for the reduced pinhole density and a surface texture that is approximately 70% finer than APCVD films.[4] Thickness uniformity is both a function of process pressure and reactor design, and will be discussed at more length in a subsequent section.

EQUIPMENT

Reactors for LPCVD processing can be divided into four basic categories. The first, and until very recently, the only type is the horizontal tube reactor. The second type is the vertical tube reactor which operates in much the same manner as the horizontal tube reactor. The third type uses a bell jar, and the fourth type is a single wafer reactor. A summation of the features of each type is given in Table 1. All LPCVD reactors are batch-operated with batch size typically as large as 100 wafers for horizontal tube reactors and as few as 1 in the single wafer reactor. However, in the latter case, the semi-automatic loading feature simulates a near-continuous operation. Since most of the wafer production is done in horizontal tube reactors, the discussion will concentrate on that reactor type.

Early LPCVD reactors were modified diffusion furnaces. By outfitting the furnaces with appropriate reactant gas controls and attaching a vacuum pump to the quartz liner, a diffusion furnace was converted to an LPCVD reactor. Moreover, some vendors provided retrofit kits that did just that.[1] Even now, the majority of LPCVD processing is done by using a horizontal quartz diffusion tube as the reactor vessel. Although the basic reactor design has not significantly changed over the years, modern equipment is far advanced in control technology and overall ancillary equipment.

A schematic of a typical production LPCVD system is shown in Figure 1. The various functional elements of the reactor are shown except for the load station. Early equipment used manual operation, whereas it is now possible to obtain completely automated systems with robotic wafer loading and remote operational control. While most production equipment now in use still requires some manual operation, they all are equipped to provide process control to meet the repeatability requirements for satisfactory product yield.

Table 1: LPCVD Reactor Characteristics

Reactor Type	Throughput wafers/hr. (mm)	Maximum wafer size	Uniformity %	Processes available
Horizontal Tube	100	150	±2-6	All except W, WSi_2, Al
Vertical Tube	100	200	±2-6	All except W, WSi_2, Al
Bell Jar	50-100	150	±3-6	Hot wall – all; Cold wall – W, WSi_2
Single Wafer	70-80	200-250	±1-3	All probable*

* Processes still under development, no apparent limitations

Low Pressure Chemical Vapor Deposition 83

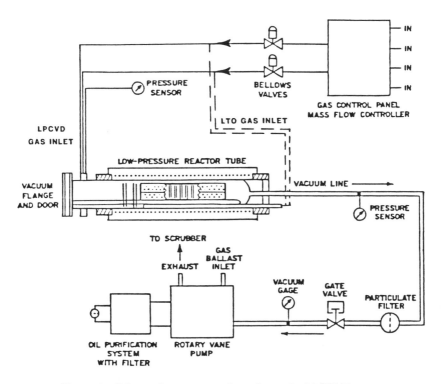

Figure 1: Schematic representation of a typical LPCVD system.

Horizontal Reactor

The heart of the reactor is a horizontal quartz tube that provides a concentric cross-section to densely spaced wafers standing on edge that allows as many as 200 wafers to be processed at one time. The tube is concentrically heated with a multiple zone heater (usually 3 zones, coil-wound) that allows precise temperature control. In the scale-up to 150 mm wafers, a 5-zone heater yields superior temperature control, typically ±1°C or less from a set point position.

Wafers are loaded into quartz cassettes each containing typically 25 wafers. Originally, they were manually transported into the furnace on a paddle that slid along the bottom of the tube. A later development used automatic loading systems and a wheeled paddle. Both of these methods generated a great number of particulate contaminants by abrading the deposit from the inner wall of the reactor vessel. The most common method now in use is a cantilever-suspended wafer support made from rigid ceramic support members (aluminum oxide and/or SiC are most often used). The load station where wafers are loaded into the furnace is usually designed as a laminar flow station so as to avoid particulates that might adhere to the wafer surfaces while exposed to the ambient. In the present state of technology, a particle level less than $0.5/in^2$ is desired. The most recent development has converted the load station into a closed cabinet wherein a robot transfers the wafers from plastic cassettes to quartz cassettes and back again after completion of the LPCVD process. The primary advantage of these robotic systems is the elimination of the human-borne contaminants that occur during manual loading operations.

Gas Control Systems. Gas control was originally accomplished with rotameters, but the higher accuracy and automatic capabilities of mass flow controllers has virtually eliminated the rotameter from the process room. In-line filters upstream of the mass flow controllers are used to eliminate particulates originating from the gas source or gas lines.

Some interest in in-line, point-of-use filters has developed to further reduce particulates at the point in which they enter the reaction chamber. There exists some evidence[5] suggesting a source of particulates between the in-line filter upstream of the gas controls and the process tube. A possible source of these particulates is the gas lines that carry reactive gases (silane, dichlorosilane) on which forms a very fine white film on the inner walls when they are exposed to air, for example during quartzware cleaning. Subsequently, nitrogen purging at high flow rates may dislodge this film from the gas line and deposit particulates on wafer surfaces. A step in alleviating this problem may be the installation of an on/off valve close to the reactor tube inlet to be used when the line would be exposed to air.

Vacuum Systems. The evolution of the low-pressure reactor was a consequence of joining the CVD technology with vacuum technology. Over the years, possibly fewer changes have taken place in the vacuum system than in any other functional element in the LPCVD system. The exhaust manifold connecting the process tube and vacuum pump contains a particle trap, a gate valve, pressure indicator, and pressure sensors. The particle trap serves several purposes. It protects the vacuum pump from abrasive particles originating from heavy deposits on the reactor or from debris that might damage the pump should the quartz tube implode. Additionally, the particle trap reduces oil backstreaming into the process tube. Without a filter, pump oil would severely contaminate the tube whenever the tube pressure dropped into the molecular flow range (typically $\leqslant 100$ mTorr). However, the high surface area of the filter provides a trap for oil so that a vacuum system can be operated for considerable time below 100 m Torr without tube contamination.

Vacuum pumps have long reached a level of technological maturity and very few changes have been made except for the trend toward larger pumps as wafer size has increased. Vane pumps are used almost exclusively and are constructed from materials resistant to chemical attack, primarily acids. Newer pumps are now available that can indicate the condition of the oil and indicate when an oil change is due. By the nature of LPCVD processes, pump oil experiences considerable abuse; therefore, oil recirculatory filtration for both acid neutralization and particulates is widely used. Further reduction in oil abuse has been made through the development of highly refined hydrocarbon oils that are less easily broken down by heat and reduction with exhaust gases. Additionally, synthetic oils, specifically perfluoropolyether, are also available that are suitably non-reactive in oxidizing conditions.

Some processes require high gas velocity to gain uniformity along the length of the load and use a roots-type booster pump installed upstream of the vane pump. Exhaust from the vane pump is generally passed through an oil demister and then exhausted through a scrubber typically installed on the building roof.

Process Control Systems. The early LPCVD equipment used essentially manual controls. A continuous evolution has taken place through analog to digital control, and has incorporated the complete interlocking of all subsystems. Time, temperature, vacuum, and gas flow control are the primary subsystems that provide control in LPCVD processing. Process recipes can be stored in a number of ways, called forth in some cases from a central console or from individual computer control modules dedicated to each tube. The number of control variations or alternatives among the various vendors are numerous. Many levels of control are available from a simple time sequencer to a fully automatic, interlocked controller that handles all equipment functions.

Vertical Reactors

The vertical tube reactor differs from the horizontal reactor basically only in the orientation of the reaction chamber. A major advantage of this reactor is a significant reduction in the clean room footprint. This type of reactor is manufactured by several companies; the major difference between them is the wafer-loading operation. In one, wafers are loaded in a horizontal position on a quartz-loading cassette that is raised into the furnace by an elevator mechanism. In the other model, wafers also loaded in the horizontal position on a quartz cassette are lowered into the furnace. Gas flow patterns are the same as for the horizontal tube reactor so that there exists no limitations on the processes that can be performed in this reactor type. Performance criteria are also the same as for horizontal reactors. A major drawback is price that is as much as 50% greater than a horizontal tube. However, for facilities with little or no growth space, this reactor could be a bargain. Moreover, they are available as single tubes whereas horizontal tube reactors are sold as four or three tube units. This reactor type represents no technological advantages; it provides an alternative for facilities with space limitations.

Bell Jar Reactors

The discussion thus far has centered around tube reactors that have evolved from diffusion furnaces. There also exists several reactors of novel design that need to be mentioned.[6] One is a cold-wall barrel reactor that is capable of processing tungsten silicide or tungsten films on a batch of eight wafers. The core of the reactor is a hollow turret shaped like a drum that supports wafers in a near vertical position and rotates slowly to compensate for orientation variations. Gas enters the chamber through eight individual mixing ports located on the outer circumference of the process chamber. Temperature to 500°C is provided by a stationary quartz lamp array located within the turret. When compared with tube reactors, the low throughput and limited processes restrict its wide acceptability. However, it performs on processes for which hot--wall tube reactors have not mastered successfully, and some believe they may never master.

Another new design uses concentric quartz bell jars surrounded by a heating system that maintains an isothermal reaction chamber in the space between the bell jars. Wafers are loaded into special enclosed cassettes in a densely packed vertical position. Load size is 50 or 100 wafers, depending on process, and wafer sizes to 150 mm can be processed. Gas is introduced through injectors juxtapositioned over the wafers in such a manner that only virgin gases flow across the wafer surfaces, and then are exhausted through the vacuum ports located in the tray on which the cassettes are placed. A schematic of this design is shown in Figure 2.

Although the throughput of this reactor is less than that of tube reactors, this system has experienced a great deal of interest because of its reputed low particulate generation. Moreover, it is capable of performing all processes being performed by tube reactors. While some believe this is the LPCVD reactor of the future, its higher operating cost has restricted its more universal acceptance.

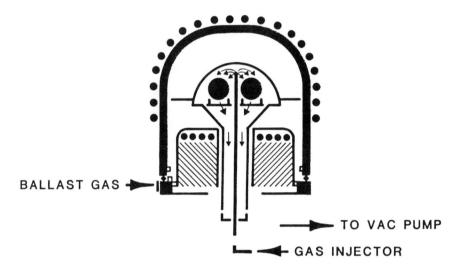

Figure 2: Schematic of bell jar reactor. (Anicon, Inc.)

Single Wafer Reactors

The single wafer reactor represents not only a novel design, but also a unique production concept. This is the newest approach to LPCVD processing, and was developed simultaneously by several manufacturers. Besides single wafer operation, these reactors have in common such features as cassette-to-cassette loading, no quartzware, high deposition rates, and exceptionally high uniformity. Although batch size is but a single wafer, these reactors are definitely a production tool with throughputs of 70–80 wafers per hour.

Additional features are cold-wall operation that should allow these reactors to deposit refractory metals, and size capability of 200 mm, and in one case to 250 mm. Although the throughput decreases as wafer size increases, the advantage normally gained by larger wafers is not totally lost. Whereas not all manufacturers claim a full spectrum of standard processes (all can deposit low temperature oxides and doped glasses), there appears to be no technical inhibitor to prevent development of all processes on these reactors. Acceptance of these reactors in production facilities would represent a significant change in wafer processing.

PRINCIPLES OF LOW-PRESSURE CVD

The basic parameters that control all CVD processes are: (1) the rate of mass transfer of reactant gases from the ambient gas stream to the wafer surface, and (2) the rate of reaction of those gases at the wafer surface. These are sequential processes and the observed deposition rate G is:

$$(1) \qquad G = A \frac{1}{1/F + 1/R}$$

where A is a geometric constant, F is the mass transfer rate and R is the surface reaction rate. Under the conditions of APCVD processing, these two rates are approximately equal. Therefore, in order to obtain uniform deposition, it is necessary to assure both rate conditions are met. The rate of surface reaction is dependent upon reactant gas concentration and temperature; the rate of mass transfer is also dependent upon reactant gas concentration as well as gas diffusion across a boundary layer. The common requirement to maintain a uniform reactant gas concentration over the surface of the wafer has been accomplished in APCVD through reactor design and appropriate flow velocity.

Reactors for APCVD processing are designed so that the main gas stream flows over the surface of the wafer. With this design, maintaining good thickness uniformities is difficult, and it does not lend itself to high throughput, especially for large wafer sizes.

In contrast, LPCVD processing depends on the transition from a primarily diffusion-controlled process to a surface reaction-controlled process. The surface reaction rate at a given temperature is:

$$(2) \qquad R = k_1 C_s$$

where k_1 is the chemical reaction constant and C_s is the concentration of the reactant at the surface. C_s is proportional to the partial pressure of the reactant gas and is minimally affected by low-pressure processing because the partial pressure of the reactant gas is similar in both APCVD and LPCVD. Therefore, surface reaction rates or deposition rates are generally not strongly changed by processing at low pressure.

However, mass transfer rates are strongly affected by a change in total pressure. The mass transfer rate is:

$$(3) \qquad F = (D/d) \Delta C$$

where D is gaseous diffusivity, d is the boundary layer thickness and ΔC is the reactant gas concentration gradient between the surface of the wafer and the ambient gas stream. The diffusivity is inversely proportional to total pressure, P, i.e., $D = k_2/P$, where k_2 is a constant. The boundary layer thickness is proportional to the inverse square root of the Reynolds number and can be described by:

$$(4) \qquad d = k_3 (k_4 \, v \, P)^{-1/2}$$

where k_3 and k_4 are constants and v is the gas flow velocity. Typical LPCVD process pressures are less than 10^{-3} atmosphere, therefore diffusivity is increased by a factor of 10^3 relative to APCVD processing. At low pressure, gas velocities are generally 10–100 times greater than in atmospheric processes; therefore, d (Equation 4) is 3 to 10 times larger in low-pressure processing. The value of F (Equation 3) is therefore increased more than 10^2 compared to atmospheric processing with all other parameters remaining nearly constant. In the resulting expression for deposition rate (Equation 1), the term containing the mass transfer rate becomes small and ineffective, and deposition rate is solely dependent upon surface reaction rates.

In physical terms, this analysis states that at low pressure, gas diffusion becomes so great even though its diffusion length increases so as to always provide sufficient gas at the surface of the wafer to accommodate the surface reaction rate. In practice, it has been found that wafers stacked only 4.76 mm (3/16-inch) apart provide sufficient spacing to allow reactant gases to penetrate to the center of the wafers without depleting the gas concentration from edge to center. This experience has been demonstrated in 150 mm wafers as well as in 3-inch and 100 mm wafers for which the spacing was first established. It is seen here that the high throughput of the LPCVD processing is the direct consequence of the reactor design changes made possible by the use of low pressure. (A more detailed discussion of LPCVD process principles is reviewed in Reference 7 from an unpublished presentation by R.S. Rosler.)

LPCVD PROCESSING

As in most industrial process technologies, the commercial utilization of a process precedes the understanding of the process. However, the continuing demand for large sized wafers requires an understanding of process mechanisms. Whereas in some processes this scale-up technology is straightforward, in others, minor and sometimes major design changes are necessary. The governing parameter in each case is the controlling step for each process and this is usually determined through the study of the kinetics of the process. While kinetic data has been collected by many investigators on LPCVD processes, a clear understanding has not been gained on any of these processes as reported in the literature.

LPCVD processes can be divided into two major categories: (1) those that deposit a film through direct pyrolysis, and (2) those for which two or more components must react to form a deposit. Examples of the first category among the more commonly used processes is the deposition of polysilicon through the pyrolysis of silane or dichlorosilane and the deposition of silicon dioxide by the decomposition of tetraethylorthosilicate (TEOS). In the second category are the deposition of silicon nitride by the reaction of dichlorosilane and ammonia, and the deposition of silicon dioxide through the reaction of dichlorosilane and nitrous oxide.

An additional subcategory exists for both types when the observed deposition rate is controlled by the action of an intermediate phase; that is, a species that forms from the primary reactants either as a gaseous phase or by a primary surface reaction from which subsequent reaction species desorb. The formation of Low Temperature Oxide (LTO) of silicon dioxide from silane and oxygen is a prime example of this latter group. Also, in-situ doped polysilicon using phosphine as the dopant source appears to be an additional example. Notice that these two examples belong to different major categories.

The formation of a film by chemical vapor deposition consists of a sequence of process steps beginning with the diffusion of reactant gases through the boundary layer, followed by a reaction at the surface. By operating at low pressure, it is possible to eliminate the diffusion process as the controlling step. Therefore, one of several possible reactions on the surface must be considered as shown in Figure 3. These include adsorption, direct reaction or partial reaction with the surface or with an adsorbed species on the surface, diffusion along the surface with possible partial reaction at energy traps, and finally a terminating reaction at a growth site. When two or more species are involved, each species may be undergoing different reactions simultaneously. In the study of such systems, it is necessary to determine the species responsible for process control before a clear understanding can be gained.

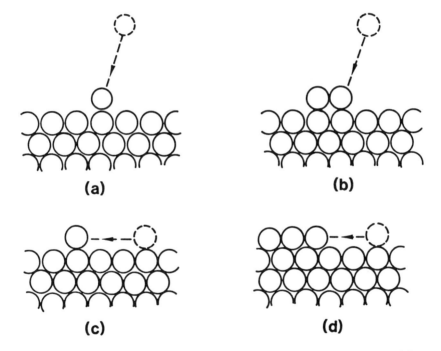

Figure 3: Possible surface reactions: (a) adsorption, (b) surface reaction, (c) surface diffusion, (d) growth reaction.

In some processes, for example Low Pressure Low Temperature Oxide (LPLTO), it was necessary to deduce from practical design the process steps responsible for control of the process. Nevertheless, in all cases it was found that the more understood about a process, the easier it was to further its development, whether for optimization or for scale-up.

In the following sections, the process mechanisms of several major LPCVD processes deduced from published data are described with the aid of the author's own experience using commercially manufactured production equipment.

Polysilicon

The LPCVD process most often investigated has been the decomposition of silane to form a film of polysilicon. In spite of the number of reports on this process, little additional insight into mechanistic understanding of film growth has emerged since the classic work of Rosler.[8] From that work, it was learned that the process was dependent on the partial pressure of silane and the activation energy was 36.8 kcal/mol; both observations are in agreement with the prior work of Joyce and Bradley[9] under atmospheric conditions. This value is also in good agreement with the value of 35 kcal/mol found by Eversteyn[10] for the growth of silicon films in one atmosphere of hydrogen from silane and from each member of the chlorosilane family (i.e., SiH_2Cl_2, $SiHCl_3$, and $SiCl_4$). These data suggest that there is possibly no difference in deposition mechanism between low and atmospheric pressure processing, and no difference between silane and chlorosilane. In a compilation of available data, Bryant[11] found a value of 33 kcal/mol to represent the activation energy in atmospheric hydrogen and 30 kcal/mol in an atmosphere of an inert gas. These values are in agreement with Eversteyn and may point out that the presence (or absence) of a specific gaseous environment affects the activation energy for growth of the silicon film.

Among the investigators of low pressure polysilicon growth, reported values of activation energy have varied between 32 and 40 kcal/mol.[8,11-14] Investigators have offered several possibilities for the controlling mechanism, but since the values do not coincide with any known independently measured decomposition/growth process using silane, there remains no clear consensus among the investigators.

From two independent and diverse methods, the energy of dissociation of silane by the reaction:

(5) $$SiH_4 \rightarrow SiH_2 + H_2$$

has been determined to be 52 kcal/mol (51.7 kcal/mol[15] and 52.7 kcal/mol[16]). However, when silane cracking on the surface of silicon was studied through the release of hydrogen, either through a change in pressure[17] or by mass spectrometry,[18-20] the activation energy for the decomposition was found to fall in the range 20 to 17 kcal/mol.

Alternatively, the possibility of surface diffusion of an adatom has been considered by several investigators[17-19,21] with activation energies of 2.4, 20, 36, and 58 kcal/mol. Since these values are determined primarily through calculation or by difference rather than by direct measurement, they are subject to considerable error. Moreover, the low values were calculated from higher temperature considerations and may not be applicable to polysilicon deposition.

Data in addition to activation energy are necessary to reveal details of the process mechanism. The following discussion presents some observations made by the various investigators on which it is possible to speculate on the controlling process steps.

In a modulated molecular beam experiment, Farnaam and Olander[20] were able to conclude that a silane molecule does not adsorb when it strikes a silicon surface, instead it either scatters or cracks. They propose that upon hitting the surface, silane reacts with a silicon adatom to form two SiH_2 molecules adsorbed on the silicon surface. These SiH_2 molecules are free to diffuse to one of two types of reaction sites whereupon hydrogen is released with an activation energy of 17 kcal/mol. This model can be supported by several other observations: The possibility that two SiH_2 molecules are also free to react to form a dimer which could readily desorb is supported by the observation of a mass 60 gaseous species made by Hitchman and Kane.[22] If we credit the relatively high amplitude of this mass species to desorption, it correlates with the observation of Farrow[18] in that the majority of silicon adatoms are desorbed from the growth surface.

Additionally, Hitchman[14] evaluated the large entropy change that takes place when silane strikes the surface of silicon and found that even the adsorption and complete immobilization of a SiH_4 molecule provides insufficient entropy change. The observed value of approximately –85 entropy units is more in line with the model of Farnaam and Olander where two adsorbed species are involved. By using this model, we can then perceive two process steps that might be controlling: (1) the reaction of silane with a Si adatom as defined in the Farnaam-Olander model, or (2) the diffusion of the SiH_2 adsorbed species to a growth site.

In the first case, the energy for the reaction,

(6) $$SiH_4(g) + Si(a) \rightarrow 2SiH_2(a)$$

should be the difference between the thermal decomposition and the adsorbed decomposition (52 kcal/mol less 17–20 kcal/mol). This value 32–35 kcal/mol agrees perfectly and maybe coincidentally with the many observed values for activation energy for polysilicon growth.

The alternative controlling process is surface diffusion of adsorbed SiH_2 to a growth site. While evidence for this process is less convincing, several observations would support it. In the presence of hydrogen, a readily adsorbed layer of H_2 forms on the silicon surface such that it poses competition to SiH_2 for low energy adsorption sites. Therefore, polysilicon growth controlled by surface diffusion in the absence of hydrogen would be expected to require less energy than when hydrogen forces SiH_2 to occupy higher energy transitional sites. This difference in energy values was found by Bryant.[11]

Noorbatcha, et al,[21] in discussing their calculated value for the activation energy for diffusion of Si adatoms on silicon believed that a value close to 5 kcal/mol most represents the correct value of diffusion at high temperatures. However, at lower temperatures the surface is characterized by ridges, steps, and kinks that could significantly increase the activation energy to values typically observed in pyrolysis measurements.

In spite of the lack of a definitive understanding of the process mechanisms, experience has revealed that when the process is run at constant temperature along the length of the process tube, the deposition on wafers in a downstream position was always thinner than those in an upstream position. Moreover, the severity of this nonuniformity became greater as temperature increased. This behavior is the consequence of depletion of silane along the length of the tube.

However, at a temperature of 619°C, the depletion effect was nearly nonexistent. Cottrin, et al[23] for APCVD and Kuiper, et al[24] for LPCVD modeled the deposition of polysilicon from silane and showed the effect of depletion under various conditions of pressure, silane partial pressure and gas velocity. Their results revealed that high flow rates favored uniform growth rate along the length of the load and becomes more effective as the partial pressure of silane increases. Therefore, at 619°C where reaction rate is relatively slow, the depletion effect can be overridden with high flow rates and high partial pressure of silane.

There is a desire in practice to keep the grain size as small as possible, and this is best done by processing at low temperature. A temperature of 620°C has been found to be an appropriate process temperature in the field.[25] In addition, a large booster pump is normally used to attain high gas velocities and in turn, this allows the use of high silane flows to increase partial pressure. The importance of the partial pressure of silane to the deposition rate is shown in Figure 4. A linear relationship exists at pressures greater than 100 mTorr. Below this pressure, the surface is reactant-starved, and a square-root dependence is observed.

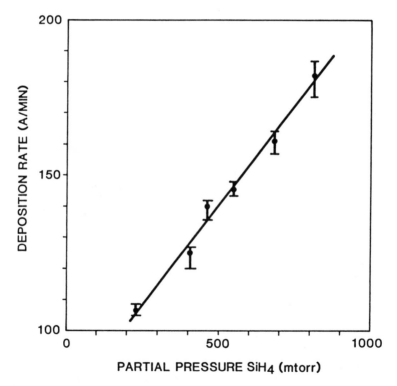

Figure 4: Polysilicon deposition rate dependence on SiH_4 partial pressure.

Nitrogen is often used as a diluent and for some conditions of processing, the results are superior to pure silane.[26] When high gas velocities are used, the presence of nitrogen in the gas stream has been shown to have no effect on the deposition rate (Figure 5) and creates a tendency to decrease uniformity within wafer. At the same time, however, uniformity along the length of the load is usually improved.

The success of the polysilicon deposition process at constant temperature is dependent on the rapid transfer of reaction gases down the reaction tube. Essential to this high gas velocity is a vacuum pumping system with sufficient capacity to provide high gas throughput. The concept is to move sufficient volume of silane along the length of the load so that depletion effects are overcompensated. Typical reactant gas utilization under these process conditions is about 25%, thus, a large quantity of unreacted silane must be properly handled and disposed in order to make this process operate safely.

The most critical parameter in the operation of this process is temperature control. The precision to which the temperature is maintained has direct bearing on the precision in thickness uniformity from front to rear along the load. Because of the relatively high activation energy, a temperature deviation of 1°C will create a 2.5% difference in thickness.

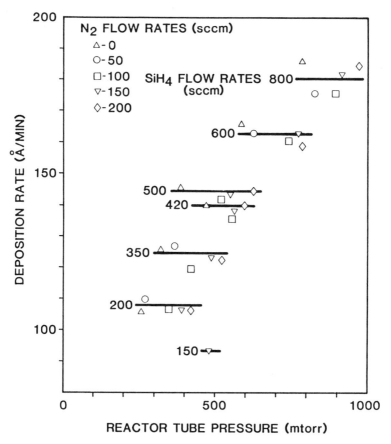

Figure 5: Dependence of polysilicon deposition rate on reactor tube pressure at various SiH_4 and N_2 flow rates.

However, this difference is often much greater because of an effect that reverses downstream the consequence of temperature deviations upstream and is shown in Figure 6. For example, if there should be a +1°C temperature deviation in an upstream position, there will be not only a 2.5% increase in thickness at that position, but it will create a 2.5% decrease in thickness downstream as a consequence of relative depletion. A negative temperature deviation of course reverses this behavior. Therefore, in order to assure acceptable wafer-to-wafer thickness uniformity, it is necessary to maintain a ±0.5°C from constant temperature at all points along the load. Many equipment manufacturers have found that this requirement was very difficult with large wafer sizes when 3-zone furnace heaters were used, and many are now using 5-zone heaters for this process in order to gain better control along the length of the reaction tube.

When constant temperature is maintained, polysilicon deposition is a straightforward process that behaves well. Typical process conditions might use silane in the flow range 200-400 sccm, which should give process tube pressure nominally between 250 and 450 mTorr and deposition rates of 100-130 A/min. The process is highly sensitive to system leaks that could introduce air (moisture) into the silane supply and/or reactor. In this case, haze is usually the first indicator with the strongest effect on wafers upstream relative to the gas flow.

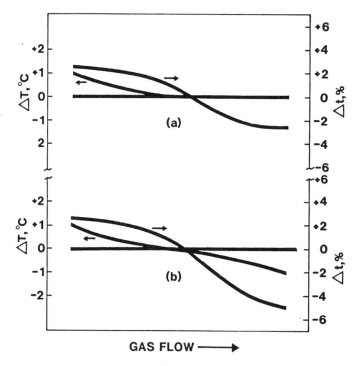

Figure 6: Effect of temperature deviation on thickness uniformity: (a) upstream temperature deviation creates thickness deviation both upstream and downstream, (b) upstream and downstream temperature deviation creates a magnified downstream effect.

Silicon Nitride

Among the commonly used LPCVD processes in semiconductor manufacture, the deposition of silicon nitride films from the reaction of ammonia (NH_3) and dichlorosilane (DCS, $SiCl_2H_2$) is one of the easiest to control, and this may account for a dearth of information on the process. Process temperature and partial pressure of $SiCl_2H_2$ determine deposition rate, whereas temperature gradient and reactant gas ratios are parameters used for process control. Other parameters that affect process performance are: gas velocity in the reactor, diameter ratio between wafer and tube, and within-wafer deposition uniformity dependence on process pressure.

Silicon nitride has been deposited between 750°C and 900°C, and between 250 and 850 mTorr reactor pressure. The relationship between reactor pressure and deposition rate is shown in Figure 7 for a variety of experimental conditions presented in Table 2. These data indicate that deposition rate is linearly increased by an increase in process pressure. The slopes of the lines in Figure 7 which represent the change in growth rate with a change in process pressure are inversely dependent upon the $NH_3/SiCl_2H_2$ ratio as shown in Figure 8.

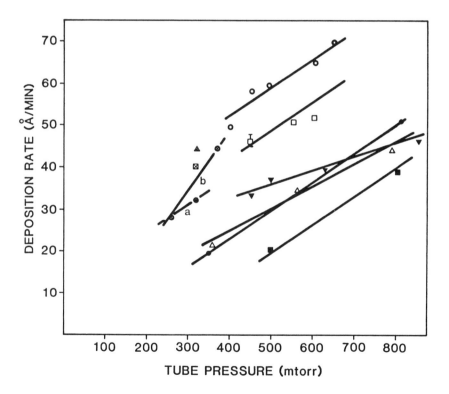

Figure 7: Relationship between Si_3N_4 deposition rate and reactor tube pressure for parameters as given in Table 2. (Lines a and b refer to slopes presented in Figure 8.) See Table 2 for key to symbols.

Table 2: Experimental Processing Parameters
(See Figure 7)

Symbol Figs. 7 & 9	Wafer size (mm)	Tube size (mm)	Flow Rates (sccm) DCS	Flow Rates (sccm) NH₃	Ratio NH₃/DCS	Wafer Space (mm)	Process Temp. °C	Ref.
□	150	203/211	125	375	3.0	4.8	790	(31)
○	125	203/211	125	375	3.0	4.8	790	(31)
●	125	170/176	125	375	3.0	4.8	790	(31)
■	100	170/176	125	375	3.0	4.8	790	(31)
⊗	76.2	135/141	27	93	3.4	4.8	800	(28)
⊠	76.2	135/141	35	85	2.4	4.8	800	(28)
◁	76.2	135/141	50	70	1.4	4.8	800	(28)
▶	76.2	135/141	15	150	10.0	4.8	752	(8)
△	50.8	74/80	30	150	5.0	3.2	750	(27)

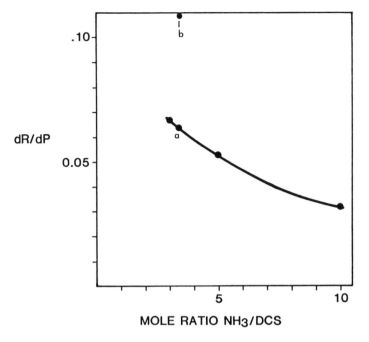

Figure 8: Effect of the reactant gas ratio on the slope of Figure 7.

When the deposition rate R is plotted against the partial pressure of $SiCl_2H_2$, P_{DCS} as shown in Figure 9, a set of linear relationships are revealed that are independent of the NH_3 content and can be expressed as:

(7) $$R = K P_{DCS} + R_o$$

where K is the reaction rate constant and R_o is the intercept value extrapolated to zero partial pressure. The significance of R_o is not known and may be dependent on reactor design. The simple linear relationship of Figure 9 infers that the controlling process is dependent only on the partial pressure of $SiCl_2H_2$, and that interaction with ammonia is not involved with the controlling mechanism of the process.

Szendro and Marton[27] plotted the logarithm of the deposition rates at constant partial pressure of $SiCl_2H_2$ as a function of reciprocal temperature and from the slope of this line, an apparent activation energy of 38.4 kcal/mol was calculated. This value is nearly identical with that observed by a number of investigators for the growth of silicon from silane. The near similarity between these values suggests that the identical process mechanism may control the deposition of both polysilicon from SiH_4 and silicon nitride from $SiCl_2H_2$.

By analogy, with the growth of silicon from silane, we may conclude that the controlling step is either the simultaneous adsorption and decomposition of dichlorosilane (Equation 6) or the diffusion of the adsorbed specified across the surface of the growing film.

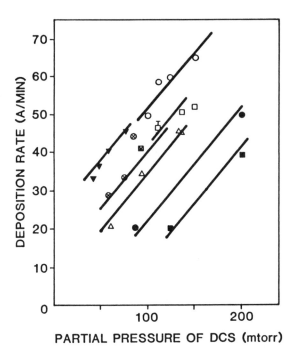

Figure 9: Dependence of Si_3N_4 deposition rate on the partial pressure of $SiCl_2H_2$. See Table 2 for key to symbols.

The overall chemical reactions have been suggested for the process:

(8) $\quad 3\ SiCl_2H_2 + 4NH_3 \rightarrow Si_3N_4 + 6HCl + 6H_2$

(9) $\quad 3\ SiCl_2H_2 + 10NH_3 \rightarrow Si_3N_4 + 6NH_4Cl + 6H_2$

Since both HCl and NH_4Cl are found in the exhaust gases, it might be that both reactions are taking place. Therefore, an alternative reaction might be operative which includes both reaction products:

(10) $\quad 3\ SiCl_2H_2 + 7NH_3 \rightarrow Si_3N_4 + 3NH_4Cl + 3HCl + 6H_2$

Kinetic data are not capable of discerning independent steps other than rate-controlling steps. Brown and Kamins[28] investigated the effect of reactant gas ratios NH_3/DCS of 3.4 (Equation 9), 2.4 (Equation 10), and 1.4 (Equation 8). They found the index of refraction changed from 2.014 for a ratio of 3.4 to 2.030 for a ratio of 1.4 suggesting a minor increase in silicon content for the lower ratio. From thermodynamic considerations, Spear and Wang[29] calculated that the boundary between Si_3N_4 and the two-phase region containing excess silicon exists at a NH_3/DCS ratio of 0.67. On the basis of thermodynamic equilibrium, therefore, a ratio even as low as 1.4 is far enough from the two-phase boundary that excess silicon should not be expected. From the apparent increase in silicon content, we may conclude that the reaction is kinetics controlled and is not in thermodynamic equilibrium.

Experience with silicon nitride film deposition in a tube reactor has generally proceeded without difficulty. By selecting reactant ratios and center-position temperature, the relative rate of depletion of $SiCl_2H_2$ is established and compensation is made by imposing a temperature gradient along the length of the tube. At lower processing temperatures, reaction rates are slower, depletion is less, and the required processing temperature gradient is less. Also, at higher $NH_3/SiCl_2H_2$ ratios, relative depletion is less, and the required temperature gradient is less. The converse is also true in each case. An example of these relationships is presented schematically in Figure 10.

The process pressure is not a factor except as it affects the partial pressure of $SiCl_2H_2$. If process pressure is increased by increasing reactant gases (no change in ratio) or by throttling the pump, either with a valve or by loading the pump with nitrogen,[30] the partial pressure of $SiCl_2H_2$ is increased with an increase in growth rate and with no change in depletion rate. If process pressure is increased by introducing nitrogen with the reactant gases, the partial pressure of $SiCl_2H_2$ will be unaffected causing no change in growth rate. However, pressure increase by this method creates a loss of in-wafer uniformity.

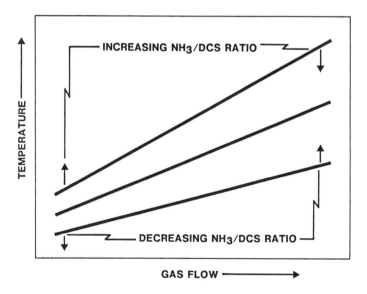

Figure 10: Effect of center point temperature and reactant gas ratios on temperature gradient.

The effect of process pressure on in-wafer uniformity was observed by Rosler[8] and by Rossi,[31] and these results are shown in Figure 11. In both experiences, the nonuniformity in thickness increased as the process pressure increased. The reason for the large disparity between the two sets of data, especially at high pressure, is not clear other than the lower partial pressure of $SiCl_2H_2$ used by Rosler and the decrease in gaseous diffusivity at higher pressure may have contributed to an enhanced nonuniformity in gas distribution throughout the wafer space in that work.

The silicon nitride process is one of few LPCVD processes that operate well at narrow wafer spacing (2.38 mm, 3/32 inch) which allows 50-wafer cassettes to be used successfully, even with 150 mm wafers. The primary difficulty with the process is the very high concentration of particulates (NH_4Cl) that accumulates in the pump oil. However, the use of acid neutralizing and particle filters in a high capacity recirculating oil system can reduce the problem so that only routine maintenance is required.

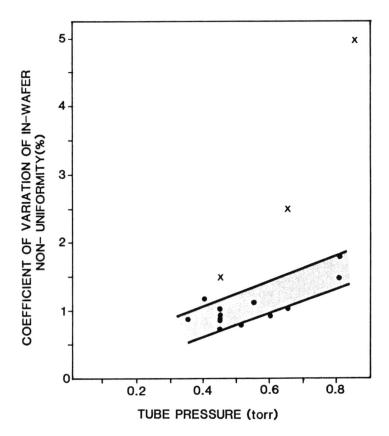

Figure 11: Variation of in-wafer uniformity on reactor tube pressure in the Si_3N_4 process (x, Rosler; ●, Rossi).

Low-Temperature Oxide (LTO)

The reaction of silane and oxide at low temperature has been one of the more difficult processes to control at low pressure and has required special hardware in order to attain deposition thickness uniformity. Moreover, most

LPLTO hardware is designed for a specific condition of gas flow rates, gas ratios, and gas velocities, therefore, the system cannot be made to work properly under a wide range of processing conditions as is the case with most other processes.

Typically, LTO films are processed between 420°-430°C and in some special conditions ranging as low as 390°C and as high as 450°C. Gas flow rates and gas velocity are interrelated so that these parameters are dictated by the pumping throughput. The gas ratio in this temperature range is nominally 1.30-1.35:1, $O_2:SiH_4$ and varies with process temperature. The deposition process is complicated and poorly understood. The following discussion attempts to shed some light on this subject.

The reaction of silane and oxygen was studied by Emeleus and Stewart[32] at low pressure and under explosive conditions. Their work showed that the oxidation of silane occurs by a chain-branching, free-radical mechanism. Subsequently, Tobin, et al,[33] showed that at standard LPLTO conditions, the primary reaction is:

(11) $$SiH_4 + O_2 \rightarrow SiO_2 + 2H_2$$

This reaction indicates no pressure change occurring as a consequence of the reaction.

A secondary reaction:

(12) $$SiH_4 + 2O_2 \rightarrow SiO_2 + 2H_2O$$

was suggested by mass spectrometric data whereby water represented 15-20% of the overall reaction product. This reaction indicates a decrease in process pressure.

When silane is introduced into a reaction tube containing oxygen, no increase in pressure is experienced when the $O_2:SiH_4$ ratio lies between 5:1 and 1.5:1; the relative decrease in pressure is equivalent to the expected pressure rise due to silane. Although (Equation 12) would indicate some reduction in pressure, it cannot account for the observed behavior.

In the earliest reported work on LPLTO, Rosler[8] reported the need for a modification to quartzware in order to achieve acceptable uniformity. The prevalent experience with LPLTO is a need to contain wafers within a smaller diameter tube; this effect is commonly experienced in chain reactions[34] where an intermediate homogenous species is formed and removed at a neutral surface. Many possibilities exist for an intermediate species in LPLTO.

In the presence of oxygen the following are but a few of the possible reactions that can take place:

(13) $$SiH_2(ad) + O(ad) \rightarrow SiH_2O(g)$$

(14) $$SiH_4(g) + O_2(g) \rightarrow SiH_2O(g) + H_2O(g)$$

(15) $$SiH_2(g) + H_2O(g) \rightarrow SiH_4O(g)$$

Wiberg[35] reported easy polymerization of siloxanes to form $(SiH_3)_2O$ and $(SiOH)_2O$ in hydrogen/oxygen environments. The formation of these compound phases as intermediate species either by surface reaction or gas-phase reaction is a possible explanation for the pressure effect. Also, the formation of H_2O by either reaction (12) or (14) might be strongly understated by the observation of Tobin, et al, as a consequence of reaction (15). The formation of H_2O and its subsequent reaction may make a major contribution to the deposition of the oxide film.

The mechanism of deposition is often discerned from activation energies. For LTO, the value of apparent activation energy appears to be dependent upon pressure (or partial pressure). In an atmospheric system, Cobianu and Pavelscu[36] report an activation energy of 7.7 kcal/mol. Skouson and Schuegraf[37] found a value of 8.03 kcal/mol at 425 mTorr and 10.8 kcal/mol at 345 mTorr. The measured value of the activation energy is typical of an adsorption process; the decrease in enthalpy values with an increase in pressure is typical of the adsorption process at low concentrations of the adsorbed species.[38] The heat of adsorption generally decreases with increasing surface coverage of adsorbed species and this coverage increases as partial pressure increases. At low pressures used in LPLTO, the surface coverage by adsorption might be expected to be considerably less than unity.

Several investigators have observed that a maximum occurs in the deposition rate as a function of the oxygen/silane ratio. Goldsmith and Kern[39] in an atmospheric process have found this maximum at an $O_2:SiH_4$ ratio of 3:1 at 325°C, 23:1 at 475°C, and 60:1 at 515°C. Cobianu and Pavelescu[36] also found this maximum at a ratio of 10:1 at 350°C. At low pressure, Tobin, et al[33] found this maximum at a ratio of 1:1 at 360°C; Learn[40] at 400°C observed this maximum at a ratio slightly above 1:1, and this author observed it at a ratio of 1.45:1 at 450°C. While these data might suggest that the reactant ratio for maximum deposition rate at typical process temperatures decreases from a nominal value of 20:1 to a value of 1:1 as a function of pressure decrease, the following argument is presented to suggest that the design of the reactor may be an integral factor in the observed ratio at maximum deposition rate.

Atmospheric LTO processes use a flat-surface wafer carrier on which wafers are placed in a horizontal position and reactant gases pass across its surface. Low pressure processes use vertically-held wafers in a dense arrangement and gases pass between wafers. The schematic illustrations in Figure 12 is representative of most LPLTO design configurations. The first wafers upstream will experience a reactant gas ratio as metered, e.g. 1.5:1, as delivered by the first matched set of injector holes. The wafers adjacent to the second set of matched holes will experience the same 1.5:1 ratio plus 0.5 of the oxygen in excess from the first set of holes. The wafers downstream will experience an increasing concentration of oxygen from the excess delivered upstream so that at the 100th wafer the actual reactant gas ratio experienced is 51:1. It is apparent, therefore, that the average reactant gas ratio at maximum deposition rate in low pressure is nominally consistent with the maximum value Goldsmith and Kern[39] found at atmospheric pressure and 450°C. Therefore, the difference in the observed behavior is design-dependent and not a difference in process mechanism.

The postulated identification of an adsorption process by the measured activation energy can be further supported by the deposition rate dependence on flow rates. When a process is dependent upon interaction between two species, the reaction rate is proportional to the interaction probability of the

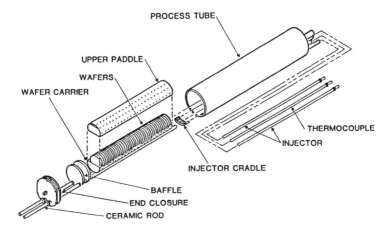

Figure 12: Schematic representation of LTO cantilever/quartzware. (Tylan Corp.)

two species. If both species are adsorbed, the probability of interaction is the product of their fractional surface coverage. Therefore, the reaction rate is defined:

$$R = K_r \theta_s \theta_o \tag{16}$$

where K_r is the reaction rate constant and θ_i is the fractional surface coverage of each species. The fractional surface coverage is dependent upon the partial pressure of the species by the relationships:

$$\theta_s = \frac{k_s P_s}{(1 + k_s P_s + k_o P_o)} \tag{17}$$

and:

$$\theta_o = \frac{k_o P_o}{(1 + k_s P_s + k_o P_o)} \tag{18}$$

where k_i are the equilibrium constants for adsorption of each gas and P_i are their partial pressures. The denominator arises from the competing probability for adsorption sites between the two species on all available surface sites.

The reaction rate can then be expressed in terms of partial pressures:

$$R_1 = \frac{K_r k_o P_o k_s P_s}{(1 + k_s P_s + k_o P_o)^2} \tag{19}$$

Alternatively, if the reaction is dependent upon the interaction of a gaseous species with an adsorbed species, the interaction probability is the product of the partial pressure of the gaseous species and the fractional coverage of the adsorbed species, and the reaction rate becomes:

$$(20) \quad R_2 = \frac{K_r k_o P_o P_s}{(1 + k_o P_o)}$$

where oxygen is assumed to be the adsorbed species.

When deposition rate is plotted as a function of $SiH_4:O_2$ ratio as in Figure 13, we do not find a maximum, but a rising curve that appears as if it would peak at the 1:1 stoichiometric ratio. The maximum in $O_2:SiH_4$ at a value of about 20:1 indicates a need for oxygen in excess of stoichiometry and that further addition of oxygen interferes with the reaction. From the point of view of silane (Figure 13) through this same range of chemical ratios, the reaction rate increases as long as sufficient oxygen exists to provide the reaction. These data suggest, therefore, that the interference with the reaction with excess oxygen (relative to the maximum) is related to the restriction of adsorption sites, i.e., $k_o P_o$ in the denominator of Equations (19 and 20), and adsorbed silane does not interfere, i.e., $k_s P_s \ll 1$ in Equation (19) or not existent as in Equation (20).

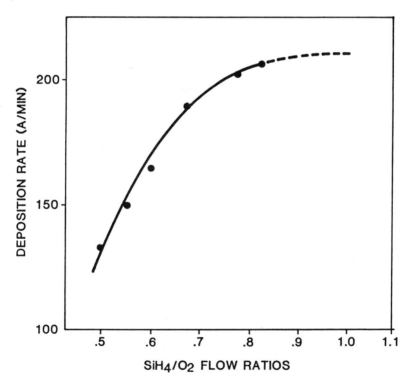

Figure 13: Deposition ratio as a function of the SiH_4/O_2 flow rate ratio.

To test this hypothesis, the deposition rate has been plotted against silane flow at two constant O_2/SiH_4 ratios in Figure 14 and the tangents to the curve are compared with the slopes defined by Equation (19) or Equation (20). Additionally, we can assume that $k_s P_s \ll 1$ so that the reaction rates are:

(19a) $$R_1 = \frac{K_r k_o P_o k_s P_s}{(1 + k_o P_o)^2}$$

(20a) $$R_2 = \frac{K_r k_o P_o P_s}{(1 + k_o P_o)}$$

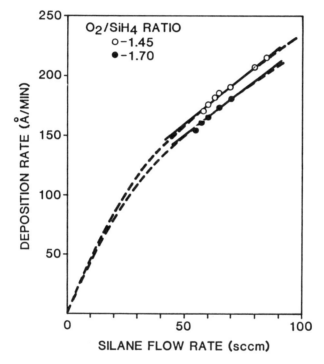

Figure 14: LTO deposition rate as a function of silane flow rate at constant O_2/SiH_4 ratios.

By comparing the ratio of the tangents at the same silane concentration for the two reactant ratios, we can determine k_o by either Equation (19a) or Equation (20a). For convenience in the calculation, we place P_s equal to unity at the point of tangency and determine $k_o = 1.560$ by Equation (19a) or $k_o = -2.166$ by Equation (20a). Since a negative value of k_o has no meaning, we are justified in assuming Equation (19) to be correct and silane, or siloxane adsorption as the controlling mechanism. The precision in determining k_o by these calculations further indicates that $k_s < 0.001$. These values also indicate a relatively high concentration of adsorbed oxygen and a very low concentration of adsorbed silane. Since interference with adsorption is found only with oxygen and not with silane or an intermediate species, we may assume that the lifetime of silane or a siloxane as an adsorbed species is sufficiently short as to lead directly to reaction and growth of the oxide film.

The LPLTO process has had a reputation for being difficult to control in production environments. This condition is often the consequence of a lack of understanding of the relationship between process chemistry and design. Since the LPLTO process is design compensated for chemical changes occurring in the process, the gas flows, gas ratios, and system pressure are interrelated. In a laboratory environment, the process was demonstrated on 100-150 mm wafers to give an average nonuniformity within wafer and wafer-to-wafer less than 0.5% calculated as a coefficient of variation. While this performance may not be expected in production environments, nonuniformity parameters of 2-3% should be normal in a well-balanced, well-designed system.

The process is very hard on pumps because SiO_2 particles are carried into the pump oil. A recirculation oil system with particulate filters is necessary to avoid severe pump damage.

Other LPCVD Processes

Tetraethylorthosilicate (TEOS). Among LPCVD processes for which interest is rising is the deposition of silicon dioxide by the pyrolysis of TEOS because of the superior step coverage offered by the deposited film. This is a process similar in all respects to the pyrolysis of silane to deposit polysilicon. When a temperature of 665°C is used, a flat temperature profile is possible with a mechanical pump. At 675°C, a roots pump is necessary, but a flat profile is still possible. At temperatures to 750°C, both a roots pump and a temperature gradient are necessary; the higher the temperature, the steeper the gradient. Adams and Capio[41] suggested a heterogeneous dissociation of TEOS at the surface and obtained an activation energy of 45 kcal/mol. Deposition rates are a function of the partial pressure of TEOS and deposition temperature. Attempts to dope the film by co-pyrolysis of $POCl_3$ or trimethylphosphate were unsuccessful because of thickness and dopant nonuniformity created by the presence of phosphorus on the wafer surface.

Diacetoxyditertiarybutoxysilane (DADBS). An alternative to TEOS with similarly superior step coverage was reported by Smolinsky[42] from the pyrolysis of DADBS. Uniformly thick deposits with perfectly conformal coatings were obtained between 450°C and 500°C. Successful doping with trimethylphosphate was reported and the resulting intrinsic stress was less than one-half that for undoped oxide. The ability to dope this oxide may generate some interest toward DADBS and away from TEOS. Unfortunately, the process temperatures are just out of range precluding deposition over aluminum.

Phosphorus-Doped Silicon. In-situ phosphorus-doped polysilicon films are processed under conditions identical to undoped polysilicon except deposition rates are reduced to about 25% and wafer carriers similar to those shown in Figure 9 are necessary to attain in-wafer uniformity. Meyerson and Yu[42] have shown that phosphorus acts as a poison on the surface by occupying low energy adsorption sites. The activation energy has been shown to increase as the phosphorus content increases[13] which supports a surface diffusion model. It is proposed that as a consequence of the presence of phosphorus on the surface, the desorption of Si_2H_4 increases relative to undoped conditions, and the growth of the film is then the consequence of the redeposition of this dimer and the deposition of Si_2H_6 that forms by insertion in the gaseous phase.

When phosphorus is introduced with silane at a concentration of 0.7%, sufficient phosphorus is incorporated into the film to saturate the polysilicon when annealed at 1000°C. Resistivities as low as 5×10^{-4} ohm-cm can be obtained. Annealing at higher temperatures drives phosphorus out of the silicon, slow cool-down rates reprecipitates phosphorus on grain boundaries; both will produce higher resistivities.

Doped LPLTO. Phosphorus-doped LPLTO is more commonly used than undoped or boron-doped LPLTO. Doping is done by substituting phosphine for silane so that the oxygen/hydride ratio remains unchanged. Doping levels of 3 to 9 wt % are common; doping concentration of phosphine is nearly proportional to its concentration in the glass film. Phosphine causes no processing problems and either increases or decreases slightly the deposition rate relative to undoped LPLTO depending on phosphorous concentration. Diborane (B_2H_6) normally used for B-doped LTO in atmospheric processes has not been used successfully in hot-wall tube reactors. Diborane is thermally unstable at deposition temperatures and decomposes in the injectors of horizontal tube reactors. Therefore, BCl_3 has been used but boron content beyond 3.5-4 wt % have caused difficulties because the simultaneous incorporation of chloride ions has produced problems with delamination of the glass from the underlying silicon surface.[44]

The jointly doped BPSG (boron phosphorus silicon glass) has been under recent intensive development because of its reflow characteristics. Reflow temperatures as low as 850°C have been obtained by using maximum safe concentrations of B (<4%) and P (<10%) with subsequent reflow in steam. Means are now under development for using diborane in horizontal tube reactors which will allow higher concentrations of boron in the glass and further reduce reflow temperature. The presence of boron doping in LPLTO creates no additional processing problems except the quartzware requires more frequent cleaning to prevent particulate problems on the wafers.

Tungsten. Tungsten films deposited on silicon at low temperature have the advantage of growing selectively on exposed silicon and not on oxide. The process is difficult to run because of problems associated with the presence of very low levels of oxygen that may originate from the native oxide on the silicon surface, leaks, impurities in process gases and desorption of oxygen from quartzware. The process is best operated in two steps:[45] (1) a reduction of WF_6 by the silicon substrate in an inert atmosphere followed by, (2) reduction of WF_6 by H_2. The process is controlled by the dissociation of H_2 adsorbed on the silicon surface with an activation energy of 16 kcal/mol.[46] Additional problems arise from the chemical attack of quartzware by WF_6 which releases oxygen over extended use. Also, selective deposition in excess of about 3000 Å is lost because the exposed SiO_2 on the wafer surface is attacked and growth of tungsten is then initiated. Although several equipment manufacturers now offer this process, the selective tungsten process and equipment are not yet production-worthy.

Tungsten Silicide. Tungsten silicide is a process similar to that for tungsten except silane is introduced instead of hydrogen to reduce WF_6.[47] Fewer problems exist with this process than tungsten because silane captures the low levels of oxygen and renders them ineffectual. WSi_2 films are not selective and for this reason they have not received as much attention. The process is not commercially offered for hot-wall tube reactors at this time; however, the process is offered in several of the cold-wall reactors described in a previous section.

Semi-Insulating Polysilicon (SIPOS). SIPOS can be deposited in an LPCVD reactor by using silane and nitrous oxide (N_2O). The process is similar to P-doped polysilicon in that the wafers require a shroud like that used for LPLTO. The process mechanism is also similar except that it is oxygen that poisons the silicon surface.[48] However, in contrast to phosphorus doping, oxygen reduces the apparent activation energy. In both cases, deposition rate is reduced relative to undoped polysilicon, but the effect is not as severe with N_2O. The increase in the N_2O/SiH_4 ratio was shown to produce a concommittant increase in the atomic mass unit 60 (Si_2H_4) intensity as observed by mass spectrometry.[22] This suggests an increased desorption of this species. The extent of oxidation that occurs in the SIPOS film was found to correlate nearly proportional to the N_2O/SiH_4 ratio.[49]

Aluminum and Aluminum Silicon Alloys. Aluminum films and aluminum silicon alloys have been deposited through pyrolytic decomposition of tri-isobutyl aluminum (TIBA). Pure aluminum films behave much as tungsten in quartz systems in that they easily passivate by low levels of oxygen. Intermittent introduction of silane alleviates this problem. Alternatively, Cooke, et al[50] alloyed aluminum with silicon to 0.8% wt by a post-deposition heat treatment in a silane pressure of 1 Torr. Deposition can be performed between 250°C and 300°C at rates between 100 and 300 Å/sec. Satisfactory films require a catalytically active layer over silicon oxide substrates. Although this process is not now being offered commercially, the advantages of a CVD aluminum film is expected to promote interest in this development for production.

Boron Nitride. Continuing interest for boron nitride films exists for applications as a passivation layer, a diffusion source for boron, a sodium barrier, and a heat dissipation coating. A process for depositing high quality boron nitride films in a horizontal reactor was described by Adams and Capio.[51] They reacted diborane and ammonia at temperatures between 250°C and 600°C at 500 mTorr on a load of 110 wafers. They found the reaction rate to be dependent upon the one-half power of the diborane partial pressure, and an apparent activation energy of 24 kcal/mol. In a subsequent work, Adams[52] pyrolyzed borazine ($B_3N_3H_6$) to form boron nitride films at temperatures between 300°C and 650°C. Only above 550°C was film quality comparable to those originating from diborane and ammonia. Below 550°C, film quality was poor and highly reactive with atmospheric moisture. At this time, the major use of boron nitride films is in the manufacture of masks used in x-ray lithography. Description of this technology has been given by Maydan et al[53] and recently reviewed by Shimkunas.[54]

SUMMARY

It was the intent of this chapter to introduce the reader to the performance that led to acceptance of LPCVD processes, a general discussion of the equipment used and its development, and the principles that have made LPCVD a process superior to others in many cases. A lengthy and detailed discussion of three processes now widely used has been given with specific emphasis being placed on process mechanisms. A more general description of processing parameters has also been given and included problems or difficulties the process presents. Additionally, less widely used processes and processes under development were also presented in a more cursory manner. It is hoped that the reader was provided not only an overview of LPCVD technology, but also an introduction to the science upon which LPCVD technology depends.

REFERENCES

1. Hammond, M.L. and Gieske, R.J., *Tempress Microelectronics,* Tech. Note (501) (1976).
2. Kern, W. and Rosler, R.S., *J. Vac. Sci. Technol.,* Vol. 14 (5), p. 1082 (1977).
3. Iida, S., *JST News,* Vol. 2 (5), p. 29 (1983).
4. Markstein, H.W., *EP&P,* p. 31 (May 1977).
5. Duffin, R., "Point-of-Use Microfiltration of Process Gases in Integrated Circuit Production – Impact on Device Quality and Yield," Presented at Microelectronics Technical Symposium on Process Gas Contamination Control Technology, May 1984.
6. Singer, P.H., *Semiconductor Internat.,* p. 72 (May 1984).
7. Kern, W. and Schnable, G.L., *Trans. Electron Devices,* Vol. ED-26 (4) (1979).
8. Rosler, R.S., *Solid State Tech.,* Vol. 20, pp. 63-70 (April 1977).
9. Joyce, B.A. and Bradley, R.R., *J. Electrochem. Soc.,* Vol. 110 (12), p. 1235 (1963).
10. Eversteyn, F.C., *Philips Res. Report,* Vol. 29, p. 45 (1974).
11. Bryant, W.A., *Thin Solid Films,* Vol. 60, pp. 19-25 (1979).
12. Harbeke, G., Krausbauer, L., Steigmeier, E.F., Widmer, A.E., Kappert, H.F. and Neugebauer, G., *J. Electrochem. Soc.,* Vol. 131 (3), p. 675 (1984).
13. Kurokawa, H., *J. Electrochem. Soc.,* Vol. 129 (11), p. 2620 (1982).
14. Hitchman, M.L., *Chemical Vapor Deposition – Seventh International Conf.,* T.O. Sedgwick and H. Lydtin, eds., Electrochemical Society, Princeton, NJ, p. 59 (1979).
15. Hogness, T.R., Wilson, T.L. and Johnson, W.C., *J. Am. Chem. Soc.,* Vol. 58, p. 108 (1936).
16. Newman, C.G., O'Neal, H.E., Ring, M.A., Leska, F. and Shipley, N., *International J. of Chem. Kinetics,* Vol. 11, p. 1167 (1979).
17. Henderson, R.C. and Helm, R.F., *Surface Science,* Vol. 30, p. 310 (1972).
18. Farrow, R.F.C., *J. Electrochem. Soc.,* Vol. 121 (7), p. 899 (1974).
19. Joyce, B.A., Bradley, R.R. and Booker, G.R., *Phil. Mag.,* Vol. 15, p. 1167 (1967).
20. Farnaam, M.K. and Olander, D.R., *Surface Science,* Vol. 145, p. 390 (1984).
21. Noorbatcha, I., Raff, L.M. and Thompson, D.L., *J. Chem. Phys.,* Vol. 82 (3), p. 1543 (1985).

22. Hitchman, M.L. and Kane, J., *J. Crystal Growth,* Vol. 55, p. 485 (1981).
23. Cottrin, M.E., Kee, R.J. and Miller, J.A., *J. Electrochem. Soc.,* Vol. 131, p. 425 (1984).
24. Kuiper, A.E.T., vanden Brekel, C.J.H., de Groot, J. and Veltkamp, G.W., *Ibid,* Vol. 129, p. 2288 (1982).
25. Bloem, J. and Beers, A.M., *Thin Solid Films,* Vol. 124, p. 93 (1985).
26. Walker, M.L. and Miller, N.E., *Semicond. Internation.* (May 1984).
27. Szendro, I. and Marton, E., *J. Electrochem. Soc.,* Vol. 128 (3), p. 708 (1981).
28. Brown, W.A. and Kamins, T.I., *Solid State Tech.,* Vol. 7, p. 51 (1979).
29. Spear, K.E. and Wang, M.S., *Ibid,* Vol. 7, p. 63 (1980).
30. de Fraiteur, M. and Goldman, J., *Semicond. International* (May 1984).
31. Rossi, R.C., *Tylan Report TB0106 984,* Tylan Corp., Carson, CA.
32. Emeleus, H.J. and Stewart, K., *J. Chem. Soc. (London),* Part I, p. 1182 (1935); Part II, p. 677 (1936).
33. Tobin, P.J., Price, J.B. and Campbell, L.M., *J. Electrochem. Soc.,* Vol. 127 (10), p. 2222 (1980).
34. Dainton, F.S., *Chain Reactions: An Introduction,* 2nd ed., p. 71, Methuen Press, London (1966).
35. Wiberg, E., *Anorganische Chemie,* p. 71, Walter DeGruyter & Co., Berlin (1951).
36. Cobianu, C. and Pavelescu, C., *J. Electrochem. Soc.,* Vol. 130 (9), p. 1888 (1983).
37. Skouson, G. and Schuegraf, K.K., unpublished work.
38. Adamson, A.W., *Physical Chemistry of Surfaces,* p. 152, Interscience Publs., John Wiley & Sons, Easton, PA (1963).
39. Goldsmith, N. and Kern, W., "The Deposition of Vitreous Silicon Dioxide Films from Silane," *RCA Review:* Vol. 37 (3), p. 153 (1967).
40. Learn, A.J., *J. Electrochem. Soc.,* Vol. 132 (2), p. 390 (1985).
41. Adams, A.C. and Capio, C.D., *J. Electrochem. Soc.,* Vol. 126 (6), p. 1042 (1979).
42. Meyerson, B.S. and Yu, M.L., *Ibid,* Vol. 131 (10), p. 2366 (1984).
43. Smolinsky, G., The Low Pressure Chemical Vapor Deposition of Silicon Dioxide Films in the Temperature Range 450° to 600°C From a New Source: Diacetoxyditertiarybutoxysilane, presented at VLSI Conference, San Diego (1986).
44. Foster, T., Hoeye, G. and Goldman, J., *Ibid,* Vol. 132 (3), p. 505 (1985).
45. Blewer, R.S. and Wells, V.A., *Proceedings of the IEEE VLSI Multilevel Interconnection Conference,* p. 153 (June 1984).
46. Broadbent, E.K. and Ramiller, C.L., *J. Electrochem. Soc.,* Vol. 131 (6), p. 1427 (1984).
47. Brors, D.L., Fair, J.A. and Monnig, K., *Semicond. International,* p. 82 (May 1984).
48. Keim, E.G. and Van Silfhout, A., *Surface Science,* Vol. 152/153, p. 1096 (1985).
49. Hitchman, M.L. and Widmer, A.E., *J. Cryst. Growth,* Vol. 55, p. 501 (1981).
50. Cooke, M.J., Heinecke, R.A., Stern, R.C. and Maes, J.W.C., *Solid State Tech.,* Vol. 12, p. 62 (1982).

51. Adams, A.C. and Capio, C.D., *J. Electrochem. Soc.,* Vol. 127, p. 399 (1980).
52. Adams, A.C., *Ibid.,* Vol. 128, p. 1018 (1981).
53. Maydan, D., Coquin, G.A., Levinstein, H.J., Sinha, A.K. and Wang, D.N.K., *J. Vac. Sci. Technol.,* Vol. 16, p. 1959 (1979).
54. Shimkunas, A.R., *Solid State Tech.,* p. 192 (Sept. 1984).

4

Plasma-Assisted Chemical Vapor Deposition

V.S. Nguyen

INTRODUCTION

The promotion of a chemical reaction by an electrical discharge through gases has been known for over a century.[1] In 1933, Robertson and Chapp were the first to observe that material can be removed from the wall of a glass tube if it is subjected to a high frequency discharge, excited through an external electrode.[2] During the next 30 years, only a few follow-up investigations of their work was reported.[3-6] In 1962, Anderson, et al.,[7] showed that one can apply radio frequency (RF) voltage to the inside of the tube. He also suggested that the method could be used to generate reactive species for deposition of thin films.

The first work specifically aimed at plasma deposition for microelectronic uses appeared in 1963,[8] but ten more years passed before commercial batch processing equipment was introduced.[9-10]

Prior to the last decade, the plasma process had little commercial success because of the difficulty in controlling it and the complexity of the phenomena involved. The productivity and yield were too low to compete with other methods. In the chemical industry, this situation persists. In recent years, the material requirements in new technologies, such as microelectronics, optics, and solar energy research have changed. Discharge-promoted deposition technology has become increasingly important. On the other hand, efficient use of material and exact identification of the product are often neglected.

Plasma-enhanced or plasma-assisted chemical vapor deposition (PECVD) of silicon insulator dielectrics and semiconducting materials, such as SiO_2, Si_3N_4, amorphous and polycrystalline silicon are the most important commercial use of glow discharge films formation for microelectronic and photovoltaic applications. Glow discharge has been used in the deposition of amorphous hydrogenated silicon (a-Si:H) by decomposition of mono or higher silane (or halosilane).

This is an important electronic material that has been extensively used for xerography, thin film field effect transistors (FETs), solar cells, and other microelectronic applications. Silicon nitride, oxide and oxynitride films which are formed from the glow discharge decomposition of silicon, nitrogen, and/or oxygen containing gases comprise the greatest use of plasma deposition processing in the semiconductor industry. These low temperature PECVD films were found to be suitable materials for the final passivation layer of integrated circuits,[11] for mutlilayer resist lithography,[12,13] and for inter-level dielectrics in multi-level metallization structures.[14] In telecommunications, optical fibers and integrated optical structures[15] have been fabricated by plasma-assisted deposition processes. Organic and organosilicon thin films deposited by radio frequency (RF) glow discharge have also been used as optic and biocompatible coatings, thin film capacitors, laser light guides in optoelectronic devices, reverse osmosis membranes, and other microelectronic applications.[16-27] Other inorganic materials including oxides and nitrides such as GaAs, BN, BCNH, TiN, and TiO_2 have been prepared by PECVD processing[28-33] for various applications.

In this review, we will discuss some general fundamental aspects and recent advances in plasma-assisted, inorganic, thin film deposition technology for microelectronic applications. We will concentrate on plasma assisted deposition of silicon dielectrics and semiconducting materials that are most widely or likely used in the semiconductor industry. Many in-depth and comprehensive reviews on basic phenomena, theory, reaction mechanisms, and other aspects of plasma-assisted processing[34-39] and plasma polymerization[25-40] in various applications have been previously published and will not be covered in this review.

GENERAL PRINCIPLES

Nature of Plasma

The primary role of plasma is to produce chemically active species that subsequently react via conventional pathways. A key factor is that substitution of electron kinetic energy for thermal energy avoids excessive heating and consequent degradation of substrates. The plasma used for semiconductor applications is produced by the application of a high frequency electric field across a body of gas. Typical operating parameters for plasma deposition are presented in Table 1. The scope of plasma reaction conditions[41] is depicted in Figure 1. Of all the regions in Figure 1, the types of plasma of particular interest for use in the semiconductor industry are glow discharge and low pressure plasma. In these regions, the electron density ranges from 10^9 to $10^{12}/cm^3$ and electron energies vary from 1-20 eV. One important characteristic of a glow discharge plasma is the electron temperature, which is typically 30 to 1000 times greater than the average gas molecule temperature (10^4-$10^{5°}$K vs. 25-300°K). Reactant gases enter the reaction zone at room temperature (25°C) and may be heated to 100°C through 200°C, depending on reactor temperature. The translational and rotational states of the excited or reactive fragments may correspond to 100°C through 700°C temperatures. Also, vibrational and electronic excited states are at thousands of degrees centigrade and electron energies are equivalent to $10^{4°}$C to $10^{5°}$C.

Table 1: Typical Operating Parameters

Pressure (torr)	0.1 – 2
Mean Free Path (cm)	≤ 0.01
Power Density (W-cm²)	≤ 0.5
Substrate Temperature (°C)	200 – 350
Applied Potential (V)	150 – 500

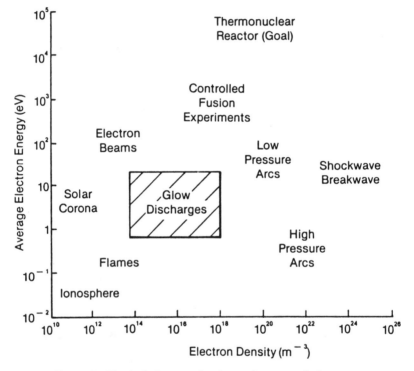

Figure 1: Typical electron density and energy of plasmas.

When the plasma process first starts, energy from the electric field is coupled into the gas almost entirely via the kinetic energy of a few free electrons. The electrons acquire energy rapidly from the field and lose it slowly to elastic collisions. Soon, the electrons are capable of ionizing or dissociating gas molecules and, thus, produce secondary electrons by electron impact reactions [see equations (1) through (5) in Figure 2]. Typical primary electron impact reactions of silane (SiH_4) molecules are shown in Table 2. The process avalanches and the discharge begins. During the plasma process, electrons are lost to the electrodes and walls by the attachment or recombination reactions cited in Figure 3.

Excitation:
(Rotational, Vibrational, and Electronic)

$$e^- + A_2 \longrightarrow A_2^* + e^- \quad (1)$$

Dissociative Attachment:

$$e^- + A_2 \longrightarrow A^- + A^+ + e^- \quad (2)$$

Dissociation:

$$e^- + A_2 \longrightarrow 2A^\cdot + e^- \quad (3)$$

Ionization:

$$e^- + A_2 \longrightarrow A_2^+ + 2e^- \quad (4)$$

Dissociative Ionization:

$$e^- + A_2 \longrightarrow A^+ + A + 2e^- \quad (5)$$

Figure 2: Electron–impact reactions.

Table 2: Primary Electron Impact Reactions of Silane

Reactant	Reaction Products	Enthalphy of Formation (eV)
$e^- + SiH_4 \longrightarrow$	$SiH_2 + H_2 + e^-$	2.2
	$SiH_3 + H + e^-$	4.0
	$Si + 2H_2 + e^-$	4.2
	$SiH + H_2 + H + e^-$	5.7
	$SiH^* + H_2 + H + e^-$	8.9
	$Si^* + 2H_2 + e^-$	9.5
	$SiH_2^+ + 2H_2 + 2e^-$	11.9
	$SiH_3^+ + H + 2e^-$	12.3
	$Si^+ + 2H_2 + 2e^-$	13.6
	$SiH^+ + H_2 + H + 2e^-$	15.3

(* = electronic excited state)

a

Penning Dissociation:

$$M^* + A_2 \longrightarrow 2A^\cdot + M$$

Penning Ionization:
$$M^* + A_2 \longrightarrow A_2^+ + M + e^-$$

Ion – Ion Recombination:

$$M^- + A_2^+ \longrightarrow A_2 + M \text{ or } M^- + A_2^+ \longrightarrow 2A^\cdot + M$$

Electron – Ion Recombination:

$$e^- + A_2 \longrightarrow 2A^\cdot$$
$$e^- + A_2^+ + M. \longrightarrow A_2 + M$$

Charge Transfer:

$$M^+ + A_2 \longrightarrow A_2^+ + M$$

$$M^- + A_2 \longrightarrow A_2^- + M$$

b

Collisional Detachment:

$$M^* + A_2^- \longrightarrow A_2 + M + e^-$$

Associative Detachment:

$$A^- + A \longrightarrow A_2 + e^-$$

Atom Recombination:

$$2A + M \longrightarrow A_2 + M$$

Atom Abstraction:

$$A + BC \longrightarrow AB + C$$

Atom Addition:

$$A + BC + M \longrightarrow ABC + M$$

M = Inert Gas or Substrate
A,B,C, = Reactant Gases

Figure 3: Inelastic collision between heavy particles.

The supply of electrons is maintained by various secondary electron reactions and positive ion impact on the electrodes. In a stable-plasma process, the number of electrons generated and lost should be the same. Stability is a function of the plasma pressure. When the pressure of the system is less than 0.1 torr, the mean free paths of electrons and gas molecules are too large, thus decreasing the collision probability and lowering the dissociation and ionization of gas molecules. The results are a lower deposition rate and plasma instability. When the pressure of the system is greater than 5 torr, molecular collisions become too frequent, especially in a parallel-plate reactor. Plasma instabilities occur and filamentary discharges may appear. These phenomena tend to promote homogeneous nucleation; thus decreasing uniformity in thickness and in composition of the deposited material.[37] The latter is associated with variation in the film's refractive index. During the deposition process, the plasma is in a totally non-equilibrium state both energetically and thermally.

Reaction Kinetics in Plasma

Elementary reactions in a plasma have been discussed by several authors.[34-37,41] Figures 2-4 list major reactions that may occur during the deposition process. Details about the kinetics of these reactions have been discussed in a book by Kondratiev.[42]

S is a Solid Surface in Contact with the Plasma

Atom Recombination:

$S - A + A \longrightarrow S + A_2$

Metastable De-excitation:

$S + M^* \longrightarrow S + M$

Atom Abstraction:

$S - B + A \longrightarrow S + AB$

Sputtering:

$S - B + M^+ \longrightarrow S^+ + B + M$

Surface Contact Ionization:

$S + B^* \longrightarrow B^+ + e^- + S$

Figure 4: Heterogeneous reactions.

Collisions between electrons and gas molecules in plasma space can be characterized as either elastic or inelastic, according to whether or not the internal energies of the colliding bodies are maintained. In the former case, only a small amount of energy is transferred, while the latter case involves a much larger energy loss and *the excitation of internal modes* (electronic, vibrational or translations) *of target molecules*. Examples of such collisions are listed in Figures 2 and 3. For the electron impact reactions in Figure 2, the kinetic rate is given by:

$$r = K n_e N_i$$

where

n_e = Electron density

N_i = Reactant concentration

K = Rate coefficient

The rate coefficient can be expressed as:

$$K = c \int_0^\infty E^{1/2} \delta(E) f_e(E) dE$$

E = energy of colliding electron

c = constant

$\delta(E)$ = reaction cross section

$f_e(E)$ = electron energy distribution function

The general form of $f_e(E)$ is a non-Maxwellian distribution with an unsymmetrical tail. When a Maxwellian distribution is assumed, the calculated rate constant results are generally unreliable.[43,44] Parameters such as electron temperature (T_e), density (n_e) and cross sections $\delta(E)$ can be measured with very limited accuracy. Furthermore, the large number of species known and unknown and their simultaneous balance equations (Figures 2-4) in the plasma make the reaction kinetics and consequently the mechanism of most deposition glow discharges essentially unknown. A simple and typical energy distribution curve for a low-powered Ar glow discharge reactor with pressure ~1 torr is shown in Figure 5. The average electron energy distribution is about 1 to 6 eV depending on operating condition. The higher energy electrons are important because they can lead to the formation of ions, free radicals, and metastable species. These products are the primary precursors to the reactions in Figures 3 and 4. In Figure 5, note the influence of plasma power is not simple; changing power will change the energy distribution of the average, especially at the tails where most of the initial reaction processes occur, even when the average energy is the same. Thus, plasmas with the same average electron energy may involve markedly different reactions. For silane deposition glow discharge at 13.56 MHz, recent studies[46] have shown that the average electron energy ranged from 1.6 to 2.5 eV at 0.05–0.15 torr. In general, the species produced by elastic electron-impact reactions will interact more with each other and with gas molecules, sustaining the glow discharge, to yield a variety of ionic and free radical species. The rate of such inelastic reactions, listed in Figure 3, is given by:

$$r = k N_i N_j$$

k = Rate coefficient

N_i, N_j = Concentration of species i and j

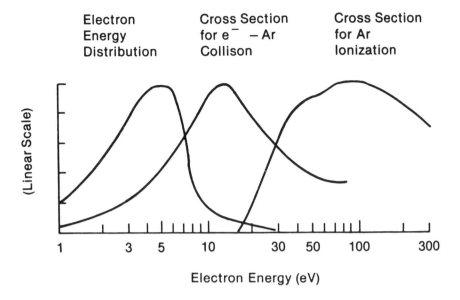

Figure 5: Electron energy distribution of typical 1 torr argon low-power glow discharge.

Rate coefficient (k) is a function of the molecular characteristics of each reactant and of the gas and ion temperatures. Inevitably, these two types of reaction processes involve the synergism of the physical and chemical aspects of the discharge, and they are far from being well understood.

In many plasma deposition processes, inert carrier gases such as argon are normally used as carrier or diluent gases. At low pressure, inert gas can absorb electron energy from the glow discharge and be excited to metastable states (Figure 2, equation 1, where A_2 = inert gases such as Ar, Kr, Xe, and He). Our recent studies[47] indicate that the metastable inert gas atoms can then transfer their energy to other reactant gases via inelastic collision (Penning effect) as shown below:

$$e^- + A \rightarrow A^* + e^-$$
$$\left. \begin{array}{l} A^* + B_2 \rightarrow A + 2B\cdot \\ A^* + B_2 \rightarrow A + B_2^+ \end{array} \right\} \text{Penning reactions}$$

where

A = inert gas, A^* = inert gas at metastable state

B_2^+, B_2 = ion and ground state reactant gases

B = radical by-product of B_2

For light, inert carrier gases with high mobility and metastable energy levels such as helium (Table 3), the Penning reaction is more efficient than other inert gases. Thus, more ions and radical reactant gases (B_2^+ and B·) are uniformly generated throughout the glow discharge. As a result, the deposited films' thickness and refractive index will be more uniform throughout the radius of reactor chamber. Recent experiments using metastable Ar ($^3P_2$) to induce the dissociation of silane further confirmed the significant effect of inert gases in plasma deposition.[48]

Table 3: Energy Levels of Inert Gases

Inert Gas	Dissociation Energy (eV)	Metastable Energy (eV)	Ionization Energy (eV)
He	——	19.8	24.53
Ne	——	16.6	21.56
Ar	——	11.5	15.76
Kr	——	9.9	14.0
Xe	——	8.32	12.13

The species produced in plasma can also interact with the substrate surface by a wide range of heterogeneous reactions, as shown in Figure 3. The kinetics of these heterogeneous reactions depend upon several surface parameters such as geometry composition, potential, and temperature. The chemistry that takes place between solid surfaces and atoms, free radicals, and ions has been discussed by Winter, Carter, and Kondratiev.[36,42,49] During the deposition process, the substrate surface is constantly bombarded by electrons and ions. The final deposited film properties are thus strongly dependent on the kind of plasma-surface interaction. Depending on the kind of species in the plasma and the manner of its interaction with the surface, various mechanisms, such as sputtering, RIE, or deposition may occur.

Deposition Mechanism

The properties of plasma deposited films are strongly dependent on process parameters, i.e., the deposition mechanism. Thus, it is necessary to understand this mechanism in detail in order to produce films with desirable properties. In general, the deposition mechanism of PECVD processes can be divided into four major steps:[50]

(1) The primary reactions between electrons and reactant gases in the plasma to form a mixture of ions and free radical reactive species.

(2) The transport of reactive species from the plasma to the substrate surface in parallel with various secondary inelastic and elastic reactions, e.g., ion-radicals, photon-molecules, etc. Steps 1 and 2 occur in plasma glow discharge and sheath regions, and can be classified as radical and ion generation steps in plasma.[51]

(3) The reaction or absorption of reactive species (radical absorption and ion incorporation) with or onto the substrate surface.

(4) The rearrangement processes where reactive species or their reaction products incorporate into the growing film or re-emit from the surface back into the gas phase. Steps 3 and 4 involve various heterogeneous reactions and interactions between ions and radicals with the surface in the sheath region.

A qualitative description of the reaction mechanism in a typical parallel plate plasma deposition reactor is shown in Figure 6. Steps 3 and 4 generally have a critical effect on the final film properties. So far, relatively few experiments have been performed to understand the transport and surface reactions in the plasma deposition process. Of all the plasma assisted processes, the deposition of amorphous and microcrystalline silicon is the most studied process and where detailed insight into the mechanism has been achieved.[34] Veprek et al.[39,52] have found that the decomposition of silane (SiH_4), in gas phase under conditions far from partial chemical equilibrium, is due predominantly to electron impact fragmentation to produce SiH_2 and H_2 (first equation, Table 2). This data is consistent with the low enthalpy formation of SiH_2 and H_2 from SiH_4. In gas phase, the SiH_2 radical undergoes the fast insertion reaction with SiH_4 to form Si_2H_6 as shown:

$$SiH_2 + SiH_4 \rightarrow Si_2H_6$$

This leads to the formation of hydrogenated silicon nuclei in gas phase and results in poor quality silicon films with large hydrogen content. For good quality silicon films by glow discharge deposition, gas phase nucleation reactions must be suppressed and the decomposition of silane at the film surface has to be the dominant reactions. In a separate study of the amorphous Si:H glow discharge deposition process, Longeway, et al.,[53] showed that SiH_3 radicals were the dominant species involved in mass transport to the substrate at 0.5 torr pressure. However, this finding has been disputed by Veprek in his recent review.[39] Perrin[54] found that the fraction of radical reactive species incorporated in surface decreased substantially from 0.83 to 0.74 as the substrate temperature was increased from 155° to 335°C. He concluded that this decrease was due to the reemission of radical species from the substrate surface as the temperature was increased. In a separate experiment, Matsuda et al.,[55] showed that for silane glow discharge, the species responsible for the deposition reaction with silane has an average lifetime of ~1.5 msec. In more recent work on the step coverage of a trench by poly Si formed by glow discharge processing, Knight[50] showed that deposition on trench sidewall is rate limited by a surface reaction involving an adsorbed radical whose rate is determined by surface temperature T_s. Overall, many questions regarding plasma deposition mechanism of amorphous and microcrystalline Si remain unanswered. Capitelli[56] showed that substantially more work is needed before the mechanisms of these processes are understood.

For silicon nitride, H. Dun et al.[51] have investigated the silicon nitride glow discharge deposition mechanism by studying the effects or process parameters such as substrate temperature, RF power, reactant gas ratios, and total pressure on the film's composition and bonding. They proposed a three-step deposition mechanism, namely, radical generation, radical adsorption, and adsorbed atom rearrangement to explain the process. Dun also suggested that an ion incorporation mechanism which would explain some change in the film's physical properties, may also exist. These proposed mechanisms are only based on the variation of film properties without any spectroscopic analysis data about the plasma state or any conclusive evidence about the existence of various proposed species (radical, atom, ions). Claassen et al.[57,58] have also studied the effect of deposition parameters on properties of plasma silicon nitride films. They proposed a simplified model of Si-H and N-H insertion mechanism, together with ion bombardment of the film surface to explain the variation in film physical properties.

In recent optical emission spectroscopic analysis of silicon nitride and oxynitride glow discharges,[59] radicals such as SiH_x and NH_y ($x = 1-3$, $y = 1-2$) were found to be much more abundant than their ionic counterparts at normal plasma deposition pressure (~1 torr). This indicates, once again, that radical species are dominant under normal plasma deposition glow discharge conditions, and probably have much more influence on deposited film properties. In general, plasma deposition mechanisms may be divided into two groups, radical and ionic, depending on the type of species that interact with the solid surface during the plasma process.

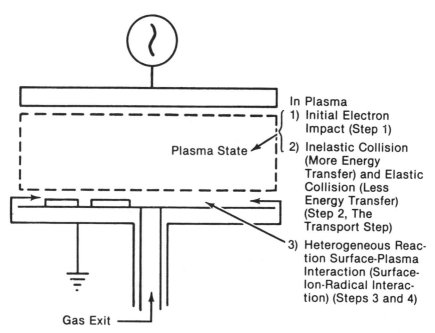

Figure 6: Reaction kinetic steps in a typical parallel plate plasma reactor system.

Radical Mechanism. During the plasma deposition process, the generation rate and lifetime for neutral radicals are usually greater than they are for ions. These two effects make the radical concentration higher than that for ions. Hence, it is believed that neutral radicals were the major deposition agents under most deposition conditions. After being generated in a glow discharge and adsorbed on (i.e., bonding with) the substrate surfaces, the adsorbed radicals have to diffuse into a stable site to become a part of growing films. Surface diffusion of such adsorbed atoms is much slower, compared with diffusion in the CVD process, at normal plasma process temperatures ($\leqslant 300°C$).

At these low temperatures, surface diffusion and rearrangement of reactive species on substrate surfaces become dominant and strongly affect the film's depth profile composition of weakly bonded and high diffusivity species such as hydrogen.[57,60] In regions where surface temperatures become too low for reactive species to react, to rearrange its bonding, and to diffuse to stable bonding sites, the film's depth composition uniformity is poor. Thus, low temperature PECVD films normally contain more trapped radicals (defects) and have a low density compared with high temperature PECVD or CVD films. In general, the radical mechanism is dominant at low power and low flow rates. Since this mechanism is less selective in bonding with the substrate surface, the deposition rate is generally higher than the deposition rate of films deposited by ion-assisted mechanisms.

Films deposited under conditions where the radical mechanism is dominant will be more porous and contain more hydrogen.[57,58,61] In addition, films deposited by a dominant radical-mechanism process should result in good step coverage, but have poor thermal and electrical stability. The instabilities are due to a large amount of weakly bonded hydrogen that breaks easily under thermal or electrical stress and thus creates cracking or more dangling bonds in the film structure.

Ionic Mechanism. Coincidental with the large number of radicals generated during the plasma process, a small number of electron and ionic species are constantly bombarding the surface. Some ionic species react (bond) with the surface to become a part of the growing film, and others bounce off after neutralizing.[62] Depending on the energy of the bombarding species and the surface state of the substrate, either deposition, densification, or sputtering may occur. Various ionic species exist during the plasma process with varying energies, so that all three reactions may happen at the same time. Because ionic species carry a charge, their attachment to the surface will be preferential. The number of sites suitable for this kind of bonding formation is limited; thus, the deposition rate is lower. The constant bombardment of the surface by electrons and ions also speeds up the rearrangement of adsorbed atoms on the substrate surface. The result is a denser film (due to tighter and more long-range order interaction bonding) with higher compressive stress and a greater scratch resistance. Recent studies on plasma deposition of microcrystalline silicon[63,64] and silicon nitride films[57,58] showed that ion bombardment will affect the bonding structure, crystallite size and hydrogen concentration in deposited films.

In general, both the radical and ionic mechanisms happen concurrently during the deposition process. Depending on the process parameters, one mechanism may dominate the other. For example, a more ionic mechanism, i.e., ion bombardment, is observed with high powers, low pressures plasma and low RF frequencies. It is necessary to understand the parameter set that results in one mechanism becoming more important and what the net effect is on film properties. The result of this work should yield a deterministic model for optimizing new processes. Again, it is quite difficult to determine the details of the reaction between reactive species and the substrate surface in plasma with present analysis technology; however, there are several plasma diagnostic techniques that will help determine plasma properties. These properties [for example, the existence of certain types of ionic and (or) radical species], may be used as evidence that certain types of mechanisms are happening. Discussion of these plasma diagnostic techniques are beyond the scope of this chapter. For those who are interested, a recently published comprehensive report on plasma diagnostics and endpoint detection[65] will provide more in-depth details.

THE DEPOSITED FILMS

Literature on the formation of plasma deposited films dates back more than 20 years.[8] Since that time, many types of organic and inorganic films have been deposited by plasma assisted processing. This discussion will be limited to certain silicon dielectrics, semiconductors, and conducting films that are presently considered important to or promising for microelectronic fabrication processing.

Silicon Nitride

In 1965, Sterling and Swan[66,67] reported that plasma deposited, low temperature silicon nitride films are suitable for passivating silicon circuit devices. The report generated interest in using PECVD silicon nitride for microelectronic applications today.[11] This is due to the excellent passivation properties of the silicon nitride layer against the diffusion of alkaline ions and moisture. In addition, PECVD silicon nitride has a low deposition temperature (300°-400°C). PECVD silicon nitride layers have also been used as the interlevel dielectric in multilevel metallization structures[67] in multilayer resist systems,[12,13] and in encapsulation layers for GaAs devices.[68,69] Normally, silicon nitride is formed by reacting silane (or halide silane gases) and ammonia (and/or nitrogen) in the glow discharge. The physical properties and uniformity of the deposited films are found to be dependent on the type of reactant gas, flow rate, RF frequency, power density, substrate temperature, and various other factors.[47,51,60,70,71]

These dependencies are generally specific to the configuration and dimension of the reactor used. In many deposition processes where highly reactive silane gas is used as the reactant, inert carrier gas such as argon was normally used to reduce the risk of spontaneous explosion. For silicon nitride film deposition, our studies[47] show that helium inert carrier gas tended to enhance the film thickness and refractive index uniformity as compared to other inert carrier gases (Ar, Kr, Xe) under the same deposition conditions. This is due to the better thermal conductivity of the helium plasma and higher Penning reaction efficiencies between helium and the reactant gas in the plasma. These properties will suppress runaway temperatures during the deposition process and enhance the uniform distribution of reactive species across the diameter of the plasma reactor. As a result, the deposited film thickness and refractive index uniformity was significantly improved. Recent studies[73,74] of plasma silicon nitride deposition using helium carrier gas has further confirmed this uniformity enhancement.

One of the important advantages of plasma deposited silicon nitride is its low intrinsic stress as compared to thermal CVD films (10^9 vs. 10^{10} dynes/cm^2). In general, the intrinsic stress of plasma deposited silicon nitride is affected by the preparation conditions, and consequently, its bonding structure and composition.[57,58] The films may have either compressive or tensile stress, depending on deposition temperature, RF frequency, and pressure. At low frequency and substrate temperature (below 4 MHz and 350°C), the deposited silicon nitride films have compressive stress. This is due to more ion bombardment that ruptures the Si-N, Si-H, and N-H bonds in the nitride layer. If the temperature is too low to anneal out the damage, short-range order in the silicon nitride films will be disrupted and its volume will be expanded, leading to compressive stress.[75] At high frequencies and substrate temperatures, the hydrogen desorption rate is high and ion bombardment is less. Any damage caused by ion bombardment may be annealed out. At high temperatures, hydrogen desorption and crosslinking continues for a time after the actual deposition processing is ended, leading to shrinking of the silicon nitride layer after deposition. This results in a tensile stress in the nitride layer.[58]

Other known and probable causes or effects of process parameters on film properties are:

(1) Films deposited at high power and low pressure are denser and have higher scratch resistance and compressive stress.[51,58]

(2) Higher hydrogen concentration in nitride films lowers the film stress and thermal stability, increases the film etch rate,[58,60,61,76-78] and either improves or degrades the electrical properties of the films, depending on the type of hydrogen bonding and concentration.[60,78-80] The presence of significant Si-H bonding in plasma deposited silicon nitride films appears to reduce the number of silicon dangling bonds.[60] As a result, the metal-nitride-semiconductor (MNS) electrical properties of nitride films with sufficiently more Si-H bonds are found to be significantly improved.[80] However, the overall improvement in the electrical properties and reliability of the film are still below the stringent requirements for utilization in electrically active regions of integrated circuits such as thin gate dielectrics.

(3) In general, the total amount of hydrogen in PECVD films ranges from 9 to 30 atomic percent, depending on deposition conditions.[60] Films deposited with less hydrogen-bearing reactant gas (for example, N_2 vs. NH_3, SiH_2Cl_2 vs. SiH_4) contain less hydrogen under similar process conditions.

(4) Hydrogen in PECVD silicon nitride films generally exist as N-H or Si-H bonds. The Si-H is more sensitive to thermal heating as compared to the N-H bonds. As PECVD silicon nitride films are annealed at temperatures slightly above the deposition temperature, the Si-H bonds are generally broken first and hydrogen evolves out from the film. When the temperature is increased, more N-H bonds are then dissociated. The hydrogen evolution from PECVD during thermal annealing cycles generally affects the film bonding, structure, and subsequently, the electrical properties of the film.[80] For microelectronic applications, care must be taken to avoid rapid annealing the PECVD nitride layer during subsequent fabrication steps which may cause undesirable hydrogen diffusion and possibly device degrada-

tion. Our recent studies show that reactive ion etching of PECVD silicon nitride layers on top of aluminum films in $CF_4 + O_2$ glow discharge may cause the breaking of Si-H bonds in the nitride layer and allow the diffusion of this hydrogen into the aluminum films.[81] This diffusion was found to be one of many possible causes for threshold voltage V_T Shift in FET devices with a PECVD silicon nitride passivation layers.

(5) Most plasma deposited silicon nitride films with refractive indices of around 2.00 are actually SiN_x, where $x < 1.3$.[57,58,60,70] The excess silicon as compared to stoichiometric Si_3N_4 has been attributed to the insertion of large amounts of hydrogen as Si-H and N-H in the film structure. With these hydrogen bondings, both film density and refractive index is lower than stoichiometric Si_3N_4. As a result, excess silicon must be presented to increase the film's refractive index to 2.00. Plasma silicon nitride films with compositions near stoichiometric Si_3N_4 tend to have refractive indices lower than 2.00, except in the case of films deposited at relatively higher temperatures (above 500°C).

(6) All plasma deposited silicon nitride films have slightly Si-rich and hydrogen-poor interfaces with varying thicknesses.[60,70,82] This is due to the intrinsic instability of plasma processing during the initial deposition period. Since the Si-H bond dissociation energy of SiH_4 is lower than that of the NH bond of NH_3 (or N-N of N_2) reactant gases, more silane is broken into reactive species and deposited onto the substrate surface during the initial period of the deposition process. This interface compositional variation will alter both the electrical,[82] and possibly the adhesion properties of the deposited films.

(7) Oxygen contamination during deposition will result in the formation of silicon oxynitride films,[70,72] and these film properties are the subject of continuing interest.

Table 4 shows the overall properties of plasma deposited silicon nitride as compared to high temperature CVD nitride films.

Silicon Oxynitride

Recent studies[70,83,84] indicate that converting a plasma silicon nitride to oxynitride by introducing oxygen will improve the film's thermal stability, cracking resistance, and will decrease stress. Table 5 shows the representative variation of film composition with refractive index, and the average film stress and pinhole density of silicon oxynitride films deposited in a 13.56 MHz parallel plate plasma deposition system at 300°C. For this oxynitride deposition, SiH_4 (1.9% in He), NH_3 and N_2O were used as reactant gases at various flow rate ratios.[70] In general, film refractive index decreases with increasing oxygen concentration in the film. Our studies[13,60,70,80] showed that silicon oxynitride films may have different physical and electrical properties depending on their composition. Variations in oxygen and hydrogen concentration generally have a stronger influence on film properties. Silicon oxynitride films (RI = 1.75-1.80) with oxygen concentration of around 16-20 atomic percent appear to have better physi-

Table 4: Comparison Between Properties of Chemical Vapor Deposited Silicon Nitride With and Without Plasma Enhancement

Property	HT-CVD 900 °C	PE-CVD 300 °C
Composition	Si_3N_4	$Si_xN_yH_z$
Si/N Ratio	0.75	0.8-1.0
Density	2.8-3.1 g/cm^3	2.5-2.8 g/cm^3
Refractive Index	2.0-2.1	2.0-2.1
Dielectric Constant	6-7	6-9
Dielectric Strength	1×10^7 V/cm	6×10^6 V/cm
Bulk Resistivity	$10^{15} - 10^{17}$ ohm·cm	10^{15} ohm·cm
Surface Resistivity	$> 10^{13}$ ohms/Square	1×10^{13} ohms/Square
Stress at 23 °C on Si	$1.2\text{-}18 \times 10^{10}$ dyn/cm^2 (tensile)	$1\text{-}8 \times 10^9$ dyn/cm^2 (Compressive)
Thermal Expansion	4×10^6/°C	$> 4 < 7 \times 10^{-6}$/°C
Color, Transmitted	None	Yellow
Step Coverage	Fair	Conformal
H$_2$O Permeability	Zero	Low-None
Thermal Stability	Excellent	Variable > 400 °C
Solution Etch Rate		
HFB 20-25 °C	10-15 Å/min	200-300 Å/min
49% HF 23 °C	80 Å/min	1500-3000 Å/min
85% H$_3$PO$_4$ 155 °C	15 Å/min	100-200 Å/min
85% H$_3$PO$_4$ 180 °C	120 Å/min	600-1000 Å/min
Plasma Etch Rate		
70% CM$_4$/30% O$_2$, 150 W, 100 °C	200 Å/min	500 Å/min
Na$^+$ Penetration	<100 Å	< 100 Å
Na$^+$ Retained in Top 100 Å	>99%	>99%
IR Absorption		
Si-N Max	~870 cm^{-1}	~830 cm^{-1}
Si-H Minor	—	2180 cm^{-1}

cal and electrical properties compared to other nitride and oxynitride counterparts. It is possible that PECVD oxynitride films with a refractive index around 1.75 may have a stable amorphous bonding structure similar to the high temperature, thermally prepared Si_2N_2O.[85] Our recent Fourier transformed infrared, Auger, X-ray photoelectron spectroscopy, electron spin resonance, and nuclear reaction analyses of both plasma enhanced and low-pressure chemical vapor deposited silicon oxynitride films indicate that the most stable silicon oxynitride films (RI ≅ 1.75-1.80) contain mixed tetrahedrals of SiO_xN_y (x + y = 4) where stable N_2SiO_2 tetrahedral bonding structures may be the most abundant.[86]

128 Thin-Film Deposition Processes and Techniques

Table 5: Properties of Oxynitride Films

Percent Composition*			Refractive Index	Average Stress Dyn/cm^2	Pin Holes #/cm^2
Si	N	O			
71	29	0	1.95	6 × 10^8 T	0.8
43	54	3	1.85	2 × 10^9 C	0.5
31	52	17	1.75	2 × 10^8 C	0.1-0.2
38	26	36	1.65	1.2 × 10^9 C	0.4
35	3.0	62	1.45	2.5 × 10^9 T	

T = Tension
C = Compression

*Composition Analyzed by Auger Analyzer.

Plasma etching of PECVD silicon oxynitride in $CF_4 + O_2$ (8% O_2) glow discharge showed that the etch rate decreased with increasing oxygen concentration in the films.[70] The etch profiles are also affected by the film's oxygen concentration.

Overall, the plasma deposited silicon oxynitride films appear to have many advantages as compared to silicon nitride. Film properties can be modified easily by varying the film composition, i.e., deposition conditions. This class of silicon oxynitride films deserves more attention and will probably be utilized more in microelectronic applications.

Silicon Oxide

Silicon oxide can be prepared by silane (or halide silane) with O_2, N_2O, CO_2, or CO in glow discharge.[87,88] It can also be formed by the decomposition of any oxygenated organosilicon $Si(OR)_4$ where R is any alkali group. For films deposited with nitrogen, carbon and halide-bearing reactant gases, the corresponding contaminants such as nitrogen, carbon and halides are normally observed. Similar to PECVD silicon nitride films, plasma oxide film properties are strongly dependent on deposition conditions such as RF power, frequency, reactant gas types, flow rate ratio, and reactor configuration. Since the reactivity between silane (or halide silane) and oxygen is very high, it is generally difficult to obtain oxide films with good uniformity in a large deposition reactor, especially with diluted silane reactant gas. However, near stoichiometric plasma oxide film (n ≅ 1.46) can be obtained from a silane and oxygen mixture at a low deposition temperature, low power density, and high carrier gas flow rate.[68] Normally, a silane and nitrous oxide (N_2O) gas mixture tends to produce plasma oxide films with better uniformity than an oxygen (O_2) reactant gas. The hydrogen concentration is plasma oxide ranging from 5 to 10 atomic percent and existing in the form of Si-H, Si-O-H, or even H-O-H,[86,87] depending on the deposition conditions. Our electrical measurements showed that plasma oxide films with minimal O-H bonds generally exhibit better MOS electrical properties compared to those that contain more O-H bonds. Plasma oxide films can be deposited with excess silicon to form Si-rich SiO_2 films which can be used as electron injector material for electrical erasable programmable read-only memory (E^2PROM) applications.[90] However, the excess silicon in the deposited films tends to vary as much as 10 atomic percent across a large batch system, thus changing the film's electrical properties and making the films unreliable for manufacturing applications. Single wafer tools reduce the variation, but obviously lower total productivity. Phosphorus and/or boron doped plasma silicon oxide (BPSG, PSG) films have been characterized and used as interlevel metal dielectrics and passivation films.[91-93] Such doped oxides can be remelted or "reflowed" after deposition to produce good step coverage and relatively smooth surfaces.

In a recent study,[94] Smith showed that sidewall tapered PECVD oxide films can be deposited conformally over silicon doped aluminum steps in a parallel plate reactive ion etching system at 13.56 MHz using SiH_4, N_2O, and Ar gases. The improvement in conformality with this process compared to other normal oxide deposition processes is a result of the sputtered etch of the oxide layer by ions (mostly Ar^+) during the deposition process. The topography of the PECVD sidewall tapered oxide layer can be varied by controlling the DC bias and ion bombardment with appropriate conditions.

For both oxide, oxynitride, and nitride films, the step coverage characteristics ranged from conformal, nearly conformal, to nonconformal, depending on deposition conditions. The conformality characteristics have been attributed to the rapid migration of reactive and adsorption species on the deposited surface during the deposition process, while the nonformality is attributed to slow (or an absence of) migration. Adams[87] has studied the conformality of oxide films on various topographies and has showed that excellent to satisfactory step coverage can be obtained with plasma deposited films. Table 6 showed the properties of plasma oxide deposited with SiH_4 and N_2O gases under various conditions. Overall, plasma silicon oxide films have only been used as an insulator for electrically passive regions in microelectronic fabrication. The deposited films are normally used as interlevel dielectrics for metal, passivation layers for FET devices, or lithographic masks. The electrical properties of the films are still unreliable for use as thin gate material in FET devices.

Table 6: Deposition Conditions and Properties For Plasma Oxide Vs. Thermal Oxide[89]

Film Type	N_2O/SiH_4	Temp. (°C)	Pressure (Torr)	Rate (A/min)	Frequency	RI	Stress
PECVD Oxide	65	200	1.0	280	13.6 MHz	1.47	0.5×10^9
	65	300	1.0	320	13.6 MHz	1.47	0.5×10^9
	65	300	0.40	600	57 kHz	1.54	2×10^9
	25	380	0.65	350	400 kHz	1.51	1.1×10^9
	10	800	0.10	50	1.0 MHz	1.46	
Thermal Oxide	–	1000	760	–	–	1.46	2.5×10^9

*Stress in Dynes per Square Centimeter (at Room Temperature) Compressive.

Silicon Films

Silicon films deposited by the plasma enhanced chemical vapor deposition technique exist either as amorphous or as polycrystalline silicon, depending on deposition conditions. In general, the physical, optical, electrical, and photoelectronic properties of PECVD films vary with their bonding structure, hydrogen concentration, and compositional heterogeneities. Variation in PECVD process parameters such as substrate temperature, RF frequency and power, and pressure have a significant effect on the properties of the deposited films.[95-98] Since PECVD amorphous silicon films are the most promising thin film semiconductors for visualization and photovoltaic applications, their properties have been relatively well studied as compared to other silicon base materials.[77] For microelectronic applications, the use of PECVD amorphous silicon films is still very limited as compared to those with polycrystalline structures. Polycrystalline silicon films have been deposited from a mixture of dichlorosilane and argon in a barrel-type plasma deposition system operated at 450 KHz RF frequency.[95] The deposited films are actually silicon hydride (SiH_x) with an amorphous structure at low temperatures. At higher deposition temperatures ($>600°C$), the films have a fine grain polycrystalline structure. The grain size increased from 10 to about 500 nm after annealing above 750°C. Chlorine contamination in the deposited films is about 3×10^{20} cm^{-3} (less than 1 atomic percent). The films can also be doped with phosphorus (by the introduction of PH_3 gases) to form n-type polysilicon conductor films. The electrical properties of annealed n-type polysilicon films at temperatures above 750°C are found to be suitable for MOS' integrated circuit applications.[95] So far, no further report on the use of PECVD silicon for the manufacturing of microelectronic devices has been found. Recent efforts to produce epitaxy by plasma processes have shown promising results[99] and have created more interest. FET devices fabricated on PECVD epitaxial silicon film have shown relatively good electrical characteristics.[100] However, further improvement in film electrical properties, reliability, and tool productivity is needed before the process could be widely used in the fabrication of reliable microelectronic devices.

With more improvements on the way, PECVD silicon films appear to have many promising applications such as thin film transistor material for display panels and stacked 3-dimensional FET device fabrication at reduced process temperatures. For other applications such as photovoltaics, where electrical requirements are less stringent, plasma deposited silicon films with faster deposition rate will offer higher throughput, increase the total productivity, and thus lower the production cost. Recent publications[101] provide more in-depth details on plasma deposited silicon film and their applications.

Other Conductor and Semiconductor Films

Conductors such as tungsten, molybdenum, tungsten silicide, titanium silicide, and tantalum silicide were prepared by plasma assisted processing and characterized for possible uses in microelectronic fabrication.[102-105] Semiconductors such as GaAs and GaSb have also been grown by plasma assisted processing.[106]

PECVD of tungsten, molybdenum, and their silicides have been carried out by Hess et al.[102,103] in a parallel plate, radial flow reactor operated at 4.5 MHz RF frequency. Smooth and pinhole free tungsten films with as-deposited resistivity ranging from 45 to 200 $\mu\Omega$ cm can be deposited using WF_6 and H_2 gas mixture in glow discharge. After annealing at high temperatures up to 950°C, the film resistivity decreased to as low as 8 $\mu\Omega$ cm. The annealed tungsten film resistivity is strongly dependent on the H_2/WF_6 ratios. At high ratio ($H_2/WF_6 = 6$), more fluorine scavenger reactions between H_2 and WF_6 reactive species occurred in the glow discharge, thereby lowering the total fluorine concentration in deposited films and reducing their resistivity. In general, high deposition temperature films show the lowest sheet resistivity. This is probably due to larger grain size, lower defect, and impurity incorporation in the films.

Smooth and stable molybdenum films can also be deposited in an RF glow discharge of H_2 and MoF_6 gas mixture (H_2/MoF_6 ratio = 7). However, large amounts of fluorine contamination ($\cong 20$) atomic percent) were observed in the deposited film. This large amount of fluorine may be due to the formation of stable MoF_3 solids. This contamination in the deposited films can be eliminated by using $Mo(CO)_6$ as the reactant gas.[107] In this study, no oxygen was detected in the films, however, large amounts of carbon contamination (20-30 atomic percent) were observed. The lack of oxygen in the film may be due to the volatility and instability of molybdenum oxides, which prevent their inclusion into the deposited Mo films. Molybdenum films with minimal contamination have been deposited using $MoCl_5$ and H_2 as source gases.[108] Due to the low volatility of $MoCl_5$, heated delivery lines and containers are required to avoid condensation of reactant gases.

Both tungsten and molybdenum silicide films (WSi_x and $MoSi_x$) have been deposited by plasma enhanced techniques.[108,109] In general, the film resistivity varied with film composition (W/Si and Mo/Si ratio). Films with higher silicon concentration tend to have higher resistivity. In the case of PECVD tungsten silicide films, the resistivities also decrease with post deposition annealing treatment. This decrease in resistivity is presumably due to diffusion of halide (F and Cl) and hydrogen contamination in the films.

In a more recent study, titanium silicide ($TiSi_x$) films were deposited in a barrel type, high throughput, plasma enhanced chemical vapor deposition system operated at 50 KHz using $TiCl_4$ and SiH_4 as reactant gases.[104] As-deposited at 380°-450°C, the sheet resistance of the film varied from 5 to 18 ohms/square. After annealing at temperatures as low as 600°-650°C, the sheet resistance decreased to 0.8-1.5 ohms/square. The reported RBS measurement showed that the film was extremely pure with no halide contamination. The film step coverage was good and similar to that obtained from plasma nitride and oxide. Thus, PECVD $TiSi_x$ may be a prime candidate as a low temperature conductor for high density microelectronic devices. Table 7 summarizes the deposition conditions and conductivity of various PECVD conductors.

Overall, the above conductors and semiconductors have problems similar to those of silicon materials, such as contamination (halide, oxygen, hydrogen), variation of film properties with small number of defects, uniformity control, composition, and electrical properties. So far, these problems have not been resolved or refined to the point where PECVD conductors can be used reliably in a manufacturing environment.

Table 7: Deposition Conditions and Resistivity of Plasma Deposited Metal and Metal Silicide Films[80]

Film	Reactants	Electrode Temperature (°C)	Pressure (Torr)	Frequency (MHz)	As-Deposited Sheet Resistivity ($\Omega/\square$)*
W	$WF_6 + H_2$	350	0.2	4.5	2
Mo	$MoF_6 + H_2$	350	0.2	4.5	400
Mo	$MoCl_5 + H_2$	430	-	-	(500 $\mu\Omega$ – cm)
WSi_x	$WF_6 + SiH_4$	230	0.6	13.56	(~ 500 $\mu\Omega$ – cm)
$MoSi_x$	$MoCl_5 + SiH_4$	400	-	-	(800 $\mu\Omega$ – cm)
$TiSi_x$	$TiCl_4 + SiH_4$	450	2	0.05	15-20
$TiSi_x$	$TiCl_4 + SiH_4 + H_2$	350	1	0.3	-
$TiSi_x$	$TiCl_4 + SiH_4$	430	1.5	3	11
$TaSi_x$	$TaCl_5 + SiH_2Cl_2 + H_2$	650	1.5	0.6	(70 $\mu\Omega$ – cm)

*Numbers in Parentheses are Resistivities in $\mu\Omega$ – cm.

EQUIPMENT FOR PLASMA DEPOSITION

In spite of the various complex chemical and physical interactions occurring during plasma processing, equipment for plasma deposition must be optimized to produce processes that are reliable, that have high throughput and minimal particulate contamination, that are user-friendly, and more automated.

Prior to 1974, most plasma deposition equipment was installed in laboratories for research purposes with minimal consideration given to manufacturing environments. In a recent review, Thornton[35] showed a few tool configurations used for plasma assisted processes (Figure 7). Rand,[37], Johnson,[110] and Kumagai[111,112] have also described various configurations and design requirements for many advanced plasma deposition systems. In this section, we will focus our attention on different types of commercially available reactors and some possible future improvements to the equipment.

So far, commercial plasma deposition reactors can be classified in three main configurations, as shown in Figure 8. Radial flow parallel plate as well as LPCVD hot wall tube deposition reactors are used for uniform deposition of thin film over large areas. Barrel type tube reactors are normally used for resist ashing or deposition where uniformity is not of great concern. For dielectric film deposition, alternate current (ac) discharge is used to generate reactive species. The first commercially important plasma deposition apparatus is the Reinberg radial flow, parallel plate reactor, first introduced in 1974 (Figure 9).

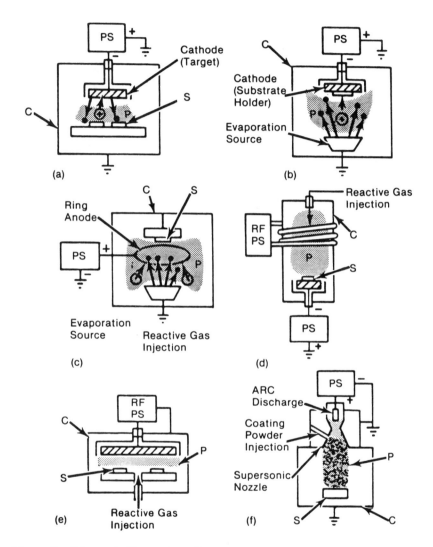

Figure 7: Schematic illustrations (P, plasma; S, substrate; C, vacuum chamber; PS, power supply) of some of the apparatus configurations used in plasma-asisted deposition; (a) sputtering; (b) ion plating; (c) activated reactive evaporation; (d) PACVD, polymerization, nitriding and carburizing; (e) PACVD (the reactive gas may flow radially outward, as shown, or radially inward); (f) low pressure plasma spraying.[35]

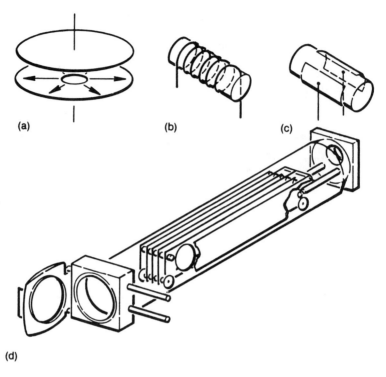

Figure 8: Geometries of production plasma-assisted CVD reactors: (a) planar (radial flow); (b) tube (inductive); (c) tube (capacitive); (d) hot tube.[38]

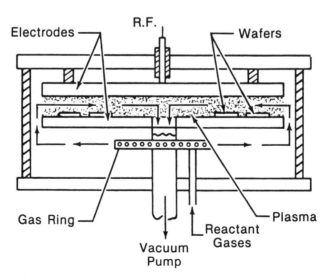

Figure 9: A radial flow parallel plate reactor of the Reinberg design.

Since that time, several commercial plasma deposition systems using two large, horizontal, parallel plate electrodes, such as the Plasma Therm PK-24[113] and the Applied Materials 3300[114] systems have also been developed. These machines are usually operated as air-to-air batch systems with various batch sizes which may accommodate up to 25 100 mm wafers per run. During the deposition, a glow discharge is ignited by the application of RF power and it is generally confined between the two electrodes. Normally, the upper electrode is powered while the lower electrode is grounded. However, other electrode powering schemes will probably be developed to optimize process conditions. For convenience, the wafers are loaded faceup onto the lower electrode and heated to the desired temperature by an electrical resistance heater clamped directly to the lower electrode. Matching networks and controllers are used to tune and maintain constant power and temperature during the deposition process. The gas input system may be installed so reactant gases can enter from the outer edges of the electrode and be pumped out from the center, or vice versa. Different means of gas input may significantly affect the uniformity of the deposited film, especially at low reactant gas flow rates. Recent improvements in reactant gas distribution uniformity, such as the perforated upper electrode and the rotating lower electrode system developed by Applied Materials,[114] have enhanced film uniformity significantly (Figure 10).

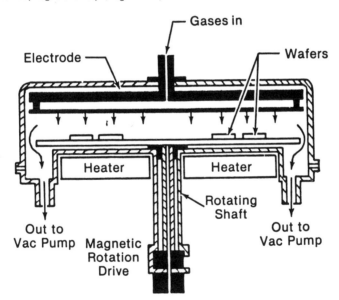

Figure 10: A radial flow parallel plate reactor with perforate electrode *(Applied Material)*.[114]

Load-lock deposition systems, introduced recently, also reduce the amount of particulate contamination and enhance run to run uniformity significantly. In these systems, the deposition chamber remains evacuated; that is, it is not cycled between atmospheric and deposition pressures by the addition of a load/unload chamber which provides an interface to the outside world. This minimizes air contamination and minimizes flaking of deposits from the chamber's internal surfaces. Major manufacturers of these systems include Ulvac,[115] LFE,[116] and ET Equipment.[117] There are variations in the geometry and design between these systems; however, their overall improvement in particulate reduction and thickness uniformity are probably due to the load-lock. Details of the systems have been discussed in recent reports[111] and may also be obtained directly from the vendors.[115-117]

In the late 1970s, the LPCVD batch type hot wall system was introduced by ASM America[118] and Pacific Western.[119] In these systems, the deposition chamber is a resistively heated furnace tube where the wafers are loaded onto both faces of a vertically oriented array of graphite electrodes (Figure 11). Electrodes are alternately connected to either the RF power lead or to ground. In this configuration, uniform temperature profiles inside the tube can be maintained over a large temperature range. The RF excitation technique for these systems is unique because the power is applied in pulses which are a millisecond long at a repetition rate of 50 to 1000 times per second. Some of the main advantages of these systems are:

(1) Minimal particulate contamination from chamber flaking, gas phase particulate contamination from chamber flaking, gas phase particulate nucleation and wafer handling due to the gravitational effect of vertical wafer loading.
(2) Uniform temperature profiles and gas flow rate control provide better batch size uniformity.
(3) High productivity due to large batch size.

More recent improvements in deposition equipment are included in the introduction of the magnetron deposition system from Material Research Corporation[120] and the electron cyclotron resonance system from Anelva.[121] In these systems, external magnetic fields are applied or coupled with the plasma to cause the electrons to travel in a circular motion and to increase their energy. This enhances the dissociation of reactant gas in the glow discharge. Figure 12 shows the typical configuration of an electron cyclotron deposition system. These magnetically enhanced plasma systems may provide high quality films at high deposition rates and lower substrate temperatures as compared to normal plasma deposition processes. However, most of these configurations are single wafer tools or have small wafer loading areas, thus lowering the total productivity. Furthermore, the film thickness uniformity may also be degraded due to the nonuniform nature of the applied magnetic field. A recent review of plasma deposition equipment[123] and design[111,112] provides a more in-depth analysis for those who are interested.

Overall, each system has its own advantages and disadvantages; the specific application will dictate which system should be selected. For example, a researcher may require a system with low throughput and low cost, but with greater versatility in depositing various types of film, such as a greater temperature range and the ability to vary the RF frequency. On the other hand, a manu-

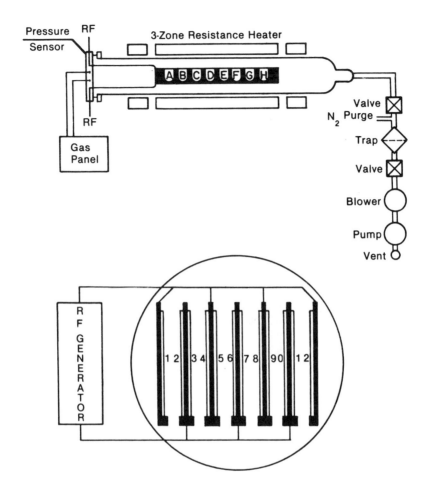

Figure 11: The LPCVD batch type hot-wall plasma deposition system.[38]

facturer may be interested in a system to deposit a specific film with high throughput, good uniformity, and low particulate contamination.

Future plasma equipment should provide high throughput, minimal particulate contamination, lower deposition temperature, ease of cleaning, and finally should provide the ability to tailor film properties by varying process conditions. For example, by being able to adjust the RF frequency in a tool, one may obtain silicon nitride films with either compressive or tensile stresses or provide the ability to change the film's conformal step coverage properties in order to fill trench structures in integrated circuits. Furthermore, systems with more automated process control and minimal human handling will be required for future ultra-large scale integrated circuit (ULSI) fabrication.

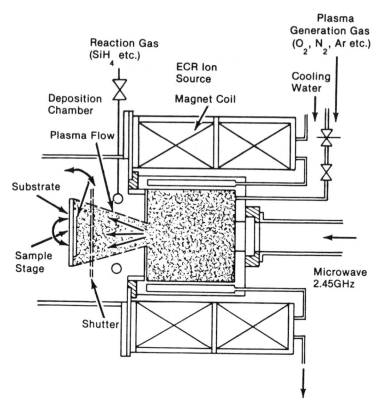

Figure 12: The magnetically enhanced electron cyclotron deposition system.[121]

EFFECTS OF OPERATING PARAMETERS

One of the major problems in plasma processing is parameter control in a large operating volume. The parameters can be categorized to a certain extent, as shown by Kay, et al.[123] in Figure 13. The system is quite complex, making deposition parameters referenced in the literature of little help. The effect of one variable is usually separable from the effects of other variables only over a narrow range. Plasma RF power, flow rate, excitation frequency, substrate surface states, and reactant concentrations, for example, all interact with each other; therefore, generalizations are impossible. In many tools, it is not possible to specify a particular process well enough that it can be routinely scaled up.

In plasma deposition, plasma parameters such as excitation power, frequency, gas flow rate and ratio, pumping speed and even reactor geometry may be varied to obtain desirable film properties. The "optimal" process parameters are normally obtained after many experimental trials. For plasma deposited silicon nitride films, Rand has summarized the results of what is known about the effects of plasma deposition parameters in a chart.[37] However, our experience taught us that the above results may not be true for plasma deposition systems

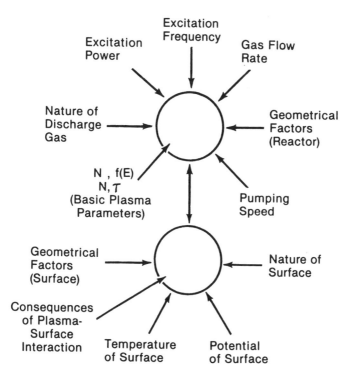

Figure 13: Representation of parameters in plasma deposition process system.[123]

with different configurations. Fortunately, some general guidelines can be used to save time in optimizing process conditions.

(1) Maintaining plasma stability is important for uniform distribution of reactive species in the glow discharge which normally results in better film thickness and compositional uniformity. The plasma is relatively stable when small variations in power within the deposition process do not result in large jump in RF potential and/or dc bias. It also recommended that the glow discharge be confined between electrodes in parallel plate or barrel type reactor systems in order to minimize the interference wall impedance, thus helping to ensure plasma stability.

(2) After attaining plasma stability, plasma density can be varied by changing controllable variables in a deposition system such as RF

power, flow rate, and total pressure. RF power and pressure should not be changed drastically if plasma stability is to be maintained. Plasma stability and density directly affect the uniformity and deposition rates of the film.

(3) Plasma-surface interactions are the least understood and controllable aspects of the deposition process because of the variation of surface states (thickness, temperature, type of surface) during the deposition process, especially during the initial transient deposition period.[82] Therefore, wafer position and temperature should be maintained in a constant state during the deposition process and from run to run. This will minimize the variation in surface states, thus resulting in better film thickness and compositional uniformity.

FUTURE RESEARCH AND DEVELOPMENT

Only a few years ago, the use of plasma deposition for microelectronic products appeared doubtful because the film thickness uniformity was poor and the film properties were not well understood. Since then, much work has been done in this field. Although we still do not have enough knowledge and understanding of the phenomena in the plasma and on the surface immersed in it, the process can now be engineered to be more reliable and scaling up is feasible for manufacturing. In many advanced VLSI production facilities, plasma deposited silicon nitride, oxide, and BPSG films are being used for microelectronic device fabrication. However, maintenance and cleaning are still too frequent. It would be very desirable in the near future to increase the deposition rate without a quality and uniformity penalty. The uniformity and productivity of deposition reactors can be increased by using shaped electrodes, better reactor geometry and gas input systems, and also impedance matching to improve the plasma. Vacuum pumps, screens, and controllers (gas and temperature) can also be improved to minimize undesirable variations during the deposition process. These improvements will enhance both film uniformity and total productivity of the system.

In the near future, new plasma deposition tools, such as the electron cyclotron resonance system[124,125] where excellent film quality can be produced[126] will probably be implemented for production use. Various studies on new experimental deposition systems of remoted plasma enhanced chemical vapor deposition (RPECVD)[127] and gradient-field plasma chemical vapor deposition (GF-PCVD)[128] techniques show promising results in minimizing film contamination, in promoting film stoichiometry, and in reducing radiation damage caused by direct ion-surface interaction. Scaled-up versions of these systems are being developed. Thus, more advanced deposition systems for development and manufacturing may be emerging during the next few years.

For long term development, machines that can handle solid reactants, by vaporizing them into the plasma, should be developed and will permit the low temperature deposition of a variety of novel semiconductor and dielectric films with minimal impurities.

At present, there is an ongoing need to reduce particulate contamination. This can be done by frequent cleaning and minimized gas phase particle nucleations. The load-lock reaction chamber system should reduce the number of particulates significantly. For contamination sensitive semiconducting material, the load-lock system may be a must to minimize contamination. Both gas phase reaction kinetics[34-37,59,129] and mass transfer computations[130,132] of plasma deposition processes have also been developed and correlated with deposition rates and film properties and compositions. It is evident that much more work is necessary in this area. Particularly important are the kinetics of reactive species formation in the plasma, its interaction with surface substrates, and its correlation with film properties such as those shown in recent studies.[27,57,58,60,61,70,81,82]

The role of hydrogen and oxygen in the deposited films must be better understood and their inclusion better controlled before plasma films can be used in more electrically active areas of silicon integrated circuits. The effect of trapped radicals must be minimized to improve electrical thermal stability and moisture resistance. The effect of energetic species bombardment on the film during processing needs to be better understood.

Considering all of the problems discussed, it will probably be a few years before plasma assisted processes are used to full advantage in integrated circuit fabrication.

Acknowledgements

The author wishes to thank F. White for manuscript review. Support from IBM management is also acknowledged.

REFERENCES

1. De Larue, W. and Muller, H.W., *Phil. Trans. Roy. Soc.*, Vol. 169, p. 155 (1978).
2. Roberton, J.K. and Chapp, C.W., *Nature*, Vol. 132, p. 479 (1933).
3. Hays, R.H., *Canadian J. Research*, Vol. A16, p. 191 (1938).
4. Lodge, L.I. and Steward, R.W., *Ibid*, Vol. A26, p. 205 (1948).
5. Wehner, G.K., *Advance in Electron and Electron Physics*, Vol. 7, p. 239 (1955).
6. Letskii, S.N., *Sov. Phys. - Tech. Phys.*, Vol. 27, p. 913 (1957).
7. Anderson, G.S., et al., *J. Appl. Phys.*, Vol. 33, p. 2991 (1962).
8. Atl, L.L., et al., *J. Electrochem. Soc.*, Vol. 110, p. 456 (1963) and Vol. 111, p. 120 (1964).
9. Reinberg, A.R., Abstract No. 6, Extended Abstracts, Vol. 74-1, Electrochem. Soc., Spring 1974 Meeting.
10. Reinberg, A.R., *Ann. Rev. Material Sci.*, Vol. 9, p. 341 (1979).
11. Takasaki, K., Koyamo, K. and Takagi, M., Abstract No. 298, Extended Abstracts, Vol. 80-2, Electrochem. Soc., Fall 1980 Meeting.
12. Suzuki, K., Matsin, J. and Torikai, T., *J. Vac. Sci. and Techn.*, Vol. 20, p. 191 (1982).
13. Underhill, J., Nguyen, V.S., Kerbaugh, M. and Sundling, D., *Proceedings of Soc. Photo Instrument Eng. Sym.*, Vol. 539, pp. 83-89 (1985), and reference therein.

14. Avigal, I., *Solid State Techn.*, pp. 217-224 (October 1983).
15. Bachmann, P., *Proceedings of 7th Int'l. Symposium in Plasma Chemistry*, pp. 7-10, Vol. 1, The Netherlands (1985).
16. Smolinsky, G., Tien, P.K., Riva-Seuseverion, S. and Martin, R.J., *Appl. Phys. Lett.*, Vol. 24, p. 547 (1974).
17. Smolinsky, G., Tien, P.K., Riva-Seuseverion, S. and Martin, R.J., *Appl. Opt.*, Vol. 11, p. 637 (1972).
18. Maisonneuve, M., Segui, Y. and Bui, A., *Thin Solid Films*, Vol. 33, p. 35 (1976).
19. Sachdev, K.G. and Sachdev, H.S., *ibid.*, Vol. 107, p. 245 (1983).
20. Reinberg, A.R., Paper presented at the IUPAC 3rd Int'l. Congress on Plasma Chem., Limoges, France (July 1977).
21. Akik, M., Segui, Y. and Bui, A., *J. Appl. Phys.*, Vol. 51, p. 5055 (1980).
22. Kryszewski, M., Wrobel, A.M. and Tyczkowski, J., *Plasma Polymerization*, M. Shen and A.T. Bell, editors, Amer. Chemical Society Symposium Series, Vol. 108, pp. 219-236, Amer. Chemical Society, Washington, DC (1979).
23. Brosset, D., Bui, A. and Segui, Y., *Appl. Phys. Lett.*, Vol. 33, p. 87 (1978).
24. Segui, Y. and Bui, A., *Thin Solid Films*, Vol. 50, p. 321 (1978).
25. Yasuda, H., *Appl. Polymer Symposia*, No. 22, p. 241, M.A. Golub and J.A. Parker, editors, Wiley, New York (1973).
26. Thompson, L.F. and Smolinsky, G., *J. Appl. Polymer Sci.*, Vol. 16, p. 1179 (1972).
27. Nguyen, S.V., et al., *J. Electrochem. Soc.*, Vol. 132 (8), pp. 1925-1932 (1985).
28. Matsushita, K., et al., *IEEE Trans. on Electron Device*, Vol. 31 (8), p. 1092 (1984).
29. Montasser, K., Hatton, S. and Morita, S., *J. Appl. Phys.*, Vol. 58 (8), pp. 3185-3189 (1985).
30. Sano, M. and Aoki, M., *Thin Solid Films*, Vol. 83, p. 247 (1981).
31. *Proceedings of 7th Int'l. Conf. in Plasma Chem.*, Vol. 1, pp. 123-129, 142-147, 159-164, Int'l. Union of Pure and Applied Chemistry, The Netherlands (1985).
32. *Proceedings of Int'l. Conf. on Ion and Plasma Assisted Techniques*, Amsterdam, The Netherlands, *Thin Solid Films*, Vol. 80, and references therein (June 19, 1981).
33. Yuzuriha, T.Y., Mlynko, W.E. and Hess, D., *J. Vac. Scien, and Technol.*, Vol. A3 (6), p. 2135 (1985).
34. Turban, G., *Pure and Appl. Chem.*, Vol. 56 (2), pp. 215-230 (1984).
35. Thornton, J.A., *Thin Solid Films*, Vol. 107, pp. 3-19 (1983).
36. Winter, H.F., *Topics in Current Chemistry: Plasma Chemistry III*, pp. 65-152, S. Verpek and M. Venugopulan, editors, Springer-Verlag, Berlin (1980).
37. Rand, M.J., *J. Vac. Sci.*, Vol. 16, p. 420 (1979).
38. Sherman, A., *Thin Solid Films*, Vol. 113, pp. 135-149 (1984).
39. Veprek, S., *Thin Solid Films*, Vol. 130, pp. 135-154, and references therein (1985).
40. Bell, A.T., *Topics in Current Chemistry: Plasma Chemistry III*, pp. 44-67, S. Veprek and M. Venugopalan, editors, Springer Verlag, Berlin (1980).

41. *Techniques and Application of Plasma Chemistry*, J.R. Hollahan and A.T. Bell, editors, John Wiley & Son, New York (1974).
42. *Chemical Kinetics of Gas Reactions*, V.N. Kondratiev, Addisson Welsey, Reading, Massachusetts, USA (1964).
43. Von Engel, A., *Ionized Gases*, Claredon, Oxford, 2nd Edition (1965).
44. Von Engel, A., *Electric Plasma, Their Nature and Uses*, Taylor and Francis, London (1983).
45. Hasted, J.B., *Physics of Atomic Collisions*, Butterworths, London, 2nd Edition (1972).
46. de Rosny, G., et al., *J. Appl. Phys.*, Vol. 54, p. 2272 (1983).
47. Nguyen, V.S., *Proceedings of Silicon Nitride Thin Insulating Films*, Vol. 83-8, pp. 453-460, Electrochem. Society Spring 1983 Meeting, San Francisco, California, USA.
48. Toyoshima, Y., et al., *Appl. Phys. Lett.*, Vol. 116, pp. 584-586 (1985).
49. Carter, G. and Calligan, J.S., *Ion Bombardment of Solid*, American Elsevier, New York (1969).
50. Knight, J.C., *Symposia Proceedings of Material Research Society Fall 1984 Meeting*, Vol. 38, pp. 371-381 (1985).
51. Dun, H., Pan, P., White, F.R. and Douse, R.W., *J. Electrochem. Soc.*, Vol. 128, p. 1556 (1981).
52. Wagner, J.J. and Veprek, S., *Plasma Chem. Plasma Processing*, Vol. 2, p. 95 (1982) and Vol. 3, p. 219 (1983).
53. Longeway, P.A., et al., *J. Phys. Chem.*, Vol. 88, p. 73 (1984).
54. Perrin, J., Doctoral Thesis LPNHE/XII/83 at University of Paris VII, France (1983).
55. Matsuda, A., et al., *J. Non-crystalline Solids*, Vol. 59/60, p. 687 (1983).
56. Capitelli, M. and Moninari, E., *Topics in Current Chemistry: Plasma Chemistry II*, p. 59, S. Veprek and M. Venugopalan, editors, Springer-Verlaz, Berlin (1980).
57. Claassen, W.A.P., et al., *J. Electrochem. Soc.*, Vol. 132, p. 893 (1985) and Vol. 130, p. 1249 (1983).
58. Claassen, W.A.P., et al., *Thin Solid Films*, Vol. 129, p. 239 (1983).
59. Nguyen, V.S., *Proceedings of the 9th Int'l Conf. on CVD*, Vol. 84-6, pp. 213-235, Electrochemical Soc. Spring 1984 Meeting, Cincinnati, Ohio, USA.
60. Nguyen, V.S., Lanford, W. and Rieger, P., *Proceedings of 7th Int'l Symposium in Plasma Chemistry,* Vol. 1, pp. 56-61, C.J. Timmermans, editor, The Netherlands (July 1-5 1985) and to be published in *J. Electrochem. Soc.*
61. Lanford, W.A. and Rand, M.J., *Appl. Phys.*, Vol. 49, p. 2473 (1978).
62. Van Haller, I., *Appl. Phys. Letter*, Vol. 37, p. 282 (1980).
63. Konuma, M. and Veprek, S., *Proceedings of 7th Int'l. Symposium in Plasma Chemistry*, Vol. 1, pp. 95-99, C.T. Timmerman, editor, The Netherlands (July 1-5, 1985).
64. Tanaka, K., Hata, N. and Matsuda, A., *Amorphous Semicon. Technol. and Dev.*, Vol. 16, p. 52 (1982).
65. Harshbarger, W.R., VLSI Electronic: *Microstructure Science,* Vol. 8, pp. 411-446, N.G. Einspruch and D.M. Brown, editors, Academic Press, New York (1984).

66. Sterling, H.F., et al., *Solid State Tech.*, Vol. 8, p. 653 (1965).
67. Swan, R.G.G., et al., *J. Electrochem.* Vol. 114, pp. 713-716 (1967).
68. Kotomi, H., Harada, H. and Tsukamoto, K., Report No. 1A-R-5, 29th Meeting of the Japan Soc. of Applied Phys., (April 1982).
69. *Proceedings of Silicon Nitride Thin Insulating Films*, Vol. 83-8, pp. 113-159, V.J. Kapoon and H.J. Stein, editors, The Electrochem. Soc., Pennington, New Jersey, USA (1983).
70. Nguyen, V.S., Pan, P. and Burton, S., *J. Electrochem. Soc.*, Vol. 131, p. 2348 (1984).
71. Rosler, R.S. and Engel, G., *Solid State Technology*, pp. 172-177, (April 1981).
72. Shinha, A.K., et al., *J. Electrochem. Soc.*, Vol. 126, p. 601 (1978).
73. Sharma, G. and Deguzman, J., Extended Abstract No. 247, The Electrochemical Society Spring 1985 Meeting, Toronto, Canada.
74. Allaert, K., et al., *J. Electrochem. Soc.*, Vol. 132, p. 1763 (1985).
75. Eernisse, E.P., *J. Appl. Phys.*, Vol. 48, p. 3337 (1977).
76. Chow, R., et al., *J. Appl. Phys.*, Vol. 53, p. 5360 (1982).
77. Paduscheck, P. and Hopel, C., *Thin Solid Films*, Vol. 110, p. 291 (1983).
78. Fujita, S., et al., *Japan J. Appl. Phys.*, Vol. 33, p. 1144 (1984), and *J. Appl. Phys.*, Vol. 57, p. 426 (1985).
79. Zhow, N.S., et al., *J. Electronic Mat.*, Vol. 14, p. 55 (1985).
80. Nguyen, V.S., Extended Abstract No. B2, pp. 24-25, Electronic Material Conference, Burlington, Vermont, USA (1983) and Extended Abstract, pp. 51-52, Electronic Material Conference, Boulder, Colorado, USA (1985).
81. Nguyen, V.S. and Kim, S.V., Extended Abstract No. 228, Electrochem. Society Fall 1985 Meeting, Las Vegas, Nevada, USA.
82. Nguyen, V.S. and Pan, P., *Appl. Phys. Lett.*, Vol. 54, p. 134 (1984).
83. Takasaki, K., et al., Extended Abstract No. 298, p. 767, Electrochem. Society Fall 1980 Meeting, Hollywood, Florida, USA.
84. Sachdev, S., et al., Extended Abstract, p. 114, Electrochem. Society Fall 1982 Meeting, and Extended Abstract, p. 510, Electrochem. Society Fall 1983 Meeting.
85. Bergen, K., *Special Ceramic*, Vol. 6, pp. 223-224 (1975).
86. Nguyen, V.S., et al., *Proceedings of 4th Int'l. Semiconductor Processing*, ASTM, San Jose, California (January 27-31, 1986).
87. Adams, A.C., et al., *J. Electrochem. Soc.*, Vol. 128, p. 1545 (1981) and Solid State Technol., Vol. 24, p. 135 (1983).
88. Hollahan, J.R., *J. Electrochem. Soc.*, Vol. 126, p. 931 (1974).
89. Gorczyca, T.B. and Gorowitz, B., VLSI Electronic: *Microstructure Science*, Vol. 8, pp. 69-88, Einspruch, N.G., editor, Academic Press, New York (1984).
90. Yokoyama, S., et al., *J. Appl. Phys.*, Vol. 54, p. 7058 (1983).
91. Avigal, I., *Solid State Technol.*, Vol. 24, p. 217 (1983).
92. Tong, J.E., et al., *Solid State Technol.*, Vol. 25, p. 161 (1984).
93. Kern, W. and Schnable, G.L., *RCA Review*, Vol. 43, p. 423 (1982).
94. Smith, G.C. and Purdes, A.J., *J. Electrochem. Soc.*, Vol. 132, p. 2721 (1985).
95. Kamins, T.T., et al., *J. Electrochem. Soc.*, Vol. 129, pp. 2326-2335 and references therein (1982).

96. Boulitrop, F., et al., *J. Appl. Phys.*, Vol. 58, p. 3494 and references therein (1985).
97. Shirafuji, J., et al., *J. Appl. Phys.*, Vol. 58, p. 3661 and references therein (1985).
98. Vanier, P.E., et al., *J. Appl. Phys.*, Vol. 56, p. 1812 and references therein (1984).
99. Donahue, T.J., Burger, W.R. and Reif, R., *Appl. Phys. Lett.*, Vol. 44, p. 346 and references therein (1983).
100. Burger, W.R. and Reif, R., Extended Abstract No. 267, Electrochem. Society Fall Meeting, Las Vegas, Nevada, USA (October 13-18, 1985).
101. *Journal of Non-crystalline Solids*, Vol. 77 and 78, pp. 1-1540, various papers in this issue and references therein (December 1985).
102. Hess, D.W., *VLSI Electronic Microstructure Science*, Vol. 8, pp. 55-68, Einspruch, N.G., editor, Academic Press, New York and references therein (1984).
103. Tang, C.C. and Hess, D.W., *Appl. Lett.*, Vol. 45, pp. 630-635 (1984), and *Solid State Techn.*, pp. 126-128 (March 1983).
104. Rosler, R.S., et al., *J. Vac. Sci. Techn.*, Vol. B2, pp. 733-737 (1984).
105. Hieber, K., et al., *Proceedings of 9th Int'l. Conf. on CVD,* Vol. 84-6, pp. 205-212, Electrochem. Soc. Spring 1984 Meeting, Cincinnati, Ohio.
106. Matsushita, K., et al., *IEEE Trans. on Elect. Dev.*, Vol. 31 (8), p. 1092 (1984).
107. Okuyama, F., *Appl. Phys. A.*, Vol. 28, p. 125 (1982).
108. Tabuchi, A., Tnone, S., Maeda, M. and Tagaki, M., *Japanese Semiconductor Technol. News*, p. 43 (February 1983).
109. Akimoto, K. and Watanabe, K., *Appl. Phys. Lett.*, Vol. 39, p. 445 (1981).
110. Johnson, W.L., *Solid State Technology*, pp. 191-195 (April 1983).
111. Kumagai, H.Y., *Proceedings of Ninth International Symposium on CVD*, Vol. 84-6, pp. 189-204, The Electrochem. Soc., Pennington, New Jersey, USA 08536.
112. Kumagai, H.Y., Paper No. J.S.-FrA1, Presented at 32nd AVS National Meeting (November 1985), to be published in *J. Vac. Sci. and Technol.*
113. Plasma Therm Inc., Kersson, New Jersey 08053, USA.
114. Applied Material Inc., Santa Clara, California 95051, USA.
115. Ulvac North America, Kennebunk, Maine 04043, USA.
116. LFE Corp., Clinton, Massachusetts 01510, USA.
117. E.T. Equipment, Hauppauge, New York 11788, USA.
118. A.S.M. America Inc., Phoenix, Arizona 85040, USA.
119. Pacific Western System Inc., Mountain View, California 94041, USA.
120. Material Research Corp., Orangeburgh, New York 10962, USA.
121. Technical Report on ECR System, Anelva Corp., San Jose, California 95112, USA, and Matsumo, S., et al., *Japanese J. of Appl. Phys.*, Vol. 22, p. L21 (1983), and Vol. 23, p. L534 (1984).
122. Weiss, A.D., *Semiconductor International*, Vol. 6, p. 88 (July 1983).
123. Kay, E., Coburn, J. and Dilks, A., *Topics in Current Chemistry: Plasma Chemistry III*, Vol. 94, pp. 1-40, S. Veprek and M. Venugopalan, editors, Springer-Verlaz, New York (1980).
124. Matsuo, S., et al., *Jap. J. Appl. Phys.*, Vol. 22, pp. 210-212 (1983).
125. Hamasaki, T., et al., *Appl. Phys. Lett.*, Vol. 44, pp. 1049 (1984).

126. Kikkawa, T., et al., Extended Abstract No. 261, The Electrochemical Society Fall 1985 Meeting, Las Vegas, Nevada, USA.
127. Richard, P.D., Markunas, R.J., Lucosky, G., Fountain, G.G., Mansour, A.N. and Tsu, D.V., *J. Vac. Sci. Technol.*, Vol. A3, p. 867 and references therein (1985).
128. Yamazaki, K., Numasawa, Y., Hamano, K., Mizumo, T. and Coleman, J.H., Extended Abstract No. 260, The Electrochemical Society Fall 1985 Meeting, Las Vegas, Nevada, USA.
129. Haller, I., *Appl. Phys. Lett.*, Vol. 37, pp. 282-284 and references therein (1980).
130. Turban, G., et al., *Thin Solid Films*, Vol. 60, pp. 147-155 (1979).
131. Chen, I., *Thin Solid Films*, Vol. 101, pp. 41-53 (1983).
132. Ross, R.C. and Vossen, J.L., *Appl. Phys. Lett.*, Vol. 45, p. 239 (1984).

5
Microwave Electron Cyclotron Resonance Plasma Chemical Vapor Deposition

Seitaro Matsuo

INTRODUCTION

Plasma technology has been successfully applied to semiconductor device fabrication, especially to plasma etching,[1] reactive ion (sputter) etching,[2-5] and plasma deposition processes. As a plasma generation method, RF (radio frequency) discharge is most generally utilized, because a stable and uniform plasma is easily obtained with a relatively large area, and further the plasma is suitable for low temperature processes because of its relatively small heating effect. The gas pressures used range from 10^{-2} to 1 torr, and the plasma density is of the order of 10^{10} cm^{-3}. Therefore, the ionization ratio is from 10^{-6} to 10^{-4}, and the value is very small. In these plasmas, neutral molecules and radicals, most part of the plasma, affect the etching or deposition reactions significantly, although the role of ions are often more essential.

On the other hand, electron cyclotron resonance (ECR) plasma can easily be generated at much lower gas pressures of 10^{-5} to 10^{-3} torr. Using an apparatus modified for semiconductor processes, as will be described, the plasma density is still left at the same order to that of the RF plasma. Therefore, the ionization ratio becomes 10^{-3} to 10^{-1}, larger by about three orders of magnitude than that of the RF plasma. Furthermore, stable plasma generation is possible with reactive gases as in the case of the RF discharge, since no electrode configuration is needed.

The ECR plasma generation method has originally been studied in the field of nuclear-fusion plasma research,[6-9] for the purpose of heating and producing a high temperature plasma. The plasma parameters required for this purpose quite differ from those of the plasma suitable for semiconductor processes. The plasma technology for fusion aims at the control of the plasma itself confined by magnetic fields such as mirror-like configuration, to approach a limit of high temperature and high state density. On the other hand, for semiconductor processes, a moderate plasma generation and supply of reactants to the specimen with little heating are required. The difference of the purposes enables the apparatus construction for semiconductor processes to be much simpler compared to that for fusion.

In semiconductor device fabrication processes, plasma CVD (chemical vapor deposition) technique, which employs plasma reactions by the RF discharge at a low temperature, has become an important research subject in recent years.[10-12] A deposition technique using a microwave discharge and a plasma transport at a low gas pressure with a parallel magnetic field has also been reported.[13] However, in both these techniques, the specimen substrate must still be heated to a temperature of from 250° to 350°C. Furthermore, the quality of the deposited film is inadequate, possibly because raw material gases, such as SiH_4, do not sufficiently decompose, and the deposition reaction on the specimen surface is not complete. These might allow hydrogen and poor molecular bonds to remain in the film.

The ECR plasma deposition apparatus newly developed allows deposition of high quality thin films at room temperatures without the need for thermal reaction. These characteristics are brought about by enhancing the plasma excitation efficiency and by the acceleration effect of ions, using the ECR plasma, and a plasma extraction by a divergent magnetic field.

ECR Plasma Deposition Apparatus

Figure 1 illustrates the ECR plasma deposition apparatus.[14] Microwave power is introduced into the plasma chamber through a rectangular waveguide and a window made of a fused quartz plate. Microwave frequency is 2.45 GHz, and output power is delivered at 50 Hz duty cycle, with the aid of a suitable power supply. The plasma chamber is 20 cm in diameter and 20 cm in height and operates as a microwave cavity resonator (TE_{113}). Magnet coils are arranged around the periphery of the chamber for ECR plasma excitation. The circular motion frequency, electron cyclotron frequency, is controlled by the magnet coils so as to coincide with the microwave frequency (magnetic flux density, 875 G) in a proper region inside of the chamber. These designs are similar to those of the broad-beam ECR ion source developed previously for reactive ion beam etching.[15] The ECR condition enables the plasma to effectively absorb the microwave energy. Thus, highly activated plasma is easily obtained at low gas pressures of 10^{-5} to 10^{-3} torr.

In this apparatus, ions are extracted in the form of plasma stream from the plasma chamber to the specimen chamber, along with a divergent magnetic field, to deposit a film on the specimen substrate. Reactive deposition gases are introduced through two inlet systems, one into the plasma chamber and the other into the specimen chamber. Figure 2 shows a photograph of the apparatus. The plasma chamber and the magnet coils are water-cooled. The vacuum system consists of an oil diffusion pump (2400 ℓ/sec) and a mechanical rotary pump (500 ℓ/min).

Microwave Electron Cyclotron Resonance Plasma CVD   149

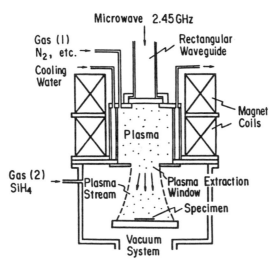

Figure 1: ECR plasma deposition apparatus. Deposition area, 20 cm in diameter. Gas pressure, 10^{-4} to 10^{-3} torr (about 0.01 to 0.1 Pa). Substrate without heating.

Figure 2: Photograph of ECR plasma deposition apparatus. Plasma chamber is arranged inside the magnet coils, and connected to microwave power supply through rectangular waveguide.

DIVERGENT MAGNETIC FIELD PLASMA EXTRACTION

A divergent magnetic field method has been developed for ion extraction in the form of plasma stream from the plasma chamber to the specimen chamber. The intensity of the magnetic field in the specimen chamber is gradually weakened from the plasma chamber to the specimen table, as shown in Figure 3.

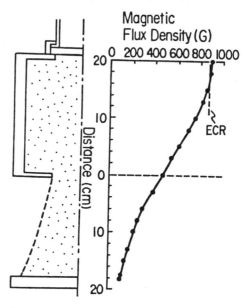

Figure 3: Distribution of magnetic field intensity (magnetic flux density) from the top of plasma chamber to the specimen table.

High energy electrons in circular motion peculiar to ECR plasma are accelerated by the interaction between their magnetic moments and the magnetic field gradient. The accelerated electrons bring about a negative potential toward the specimen table which is electrically isolated from the plasma chamber. Therefore, a static electric field, which acclerates ions and decelerates electrons, is generated along the plasma stream between the plasma chamber and the specimen table so as to satisfy the neutralization condition. The effective ion extraction, transport and bombardment of the specimen surface with a moderate energy ions are thus enhanced during deposition.

Under these conditions, electrons and ions have the same acceleration along the divergent magnetic field as follows:

(1) $$F_i/M = F_e/m$$

where M and m are the respective masses of the ion and the electron. F_i and F_e are the forces on the ion and the electron, respectively. These forces are expressed by the equations:

(2) $$F_i = eE$$

(3) $$F_e = -\mu dB/dz - eE$$

where μ is the magnetic moment of the electron in circular motion, and E is the electric field generated in the plasma stream. The magnetic moment μ is an adiabatic invariant, and given by the kinetic energy of the electron in circular motion W, as follows:

(4) $$\mu = W/B$$

From these relations, the electric field E can be obtained as

(5) $$E = (W_0/eB_0)(-dB/dz)/(1 + m/M)$$

The potential ϕ is obtained, by integration with an approximation, $(1 + m/M) \approx 1$,

(6) $$\phi = -(W_0/e)(1 - B/B_0)$$

where W_0 and B_0 are the electron energy and the magnetic flux density in the plasma chamber, respectively. This equation states that the ion energy is approximately given by the product of the electron energy in the plasma chamber and the ratio of the decreased to the initial magnetic field intensity. Ions are thus accelerated and transported toward the specimen table, and electrons lose the energy of circular motion by the same amount. As a result, deposition reactions induced by ions are enhanced, and heating effects caused by electrons are reduced. The divergent magnetic field plays a role to convert the electron energy of circular motion into the ion energy along the magnetic field. Therefore, the divergent magnetic field method is particularly effective, when it is combined with the ECR plasma generation. An electric potential generation in an ECR plasma, related to a magnetic field distribution, has been previously observed and investigated, though in much higher energy range, in the field of plasma research.[16]

Figure 4 shows a photograph of the plasma stream, extracted from the plasma chamber by the divergent magnetic field method. The plasma extraction window is 10 cm in diameter, and the plasma stream at the specimen table, that is, the deposition area is 20 cm in diameter.

The negative potential generated by the divergent magnetic field was measured using a plane probe, which had a larger area than the plasma stream cross section, from the floating potential. The obtained result is shown in Figure 5 as a function of the distance from the plasma extraction window. The negative potential increases, corresponding to the degree of decrease in the magnetic field intensity. The energy of the accelerated ion through the plasma stream, from the plasma extraction window to the specimen table, is of the order of 10 to 15 eV. Figure 6 shows the dependence of the negative potential at the specimen table on the gas pressure. The gas pressure was controlled by changing the gas flow rate. The negative potential increases rapidly as the gas pressure decreases, due to the increase in the electron mean free path and the electron energy in circular motion. Thus, the potential can be easily controlled in the range from 5 to 30 V by changing the gas pressure, besides the microwave power input.

152 Thin-Film Deposition Processes and Techniques

Figure 4: Photograph of plasma stream, extracted from the plasma chamber.

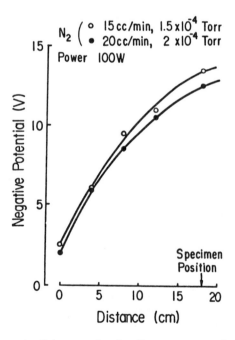

Figure 5: Negative potential generation by divergent magnetic field as a function of distance from the plasma extraction window.

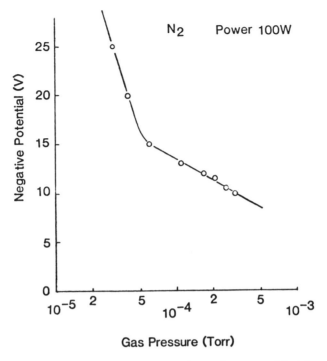

Figure 6: Relation between gas pressure and negative potential at the specimen table.

The plasma potential through the plasma stream was measured directly using an electron emissive probe method,[17] in order to distinguish from each other the respective effects of the electric field in the plasma stream, and the electric field due to the ion sheath in the vicinity of the specimen surface generated by the thermal motion of electrons. The electron emissive probe method utilizes the fact that the usual potential difference between the plasma and the probe surface (ion sheath region) does not occur, when electron exchange between the probe and the plasma is made free, as in the case of a thermionic filament probe. The results are shown in Figure 7, where the substrate potential is chosen as zero. The potential difference due to the ion sheath, the order of 0.3 mm in thickness, is about 10 V. The ion energy incident to the specimen surface is given by the sum of the respective energy gained in the divergent magnetic field and that due to the ion sheath potential difference, and is about 20 to 30 eV. The ions in such an energy range are expected to enhance deposition reactions and to improve the film quality, but not to cause surface damages.

The existence of the electric field in the plasma stream plays an important role to very effectively transport the ions generated in the plasma chamber toward the specimen, in contrast to the conventional methods in which ions are utilized only by the transport through their thermal diffusion. In fact, high ion current density of 3 to 5 mA/cm² is easily obtained at the specimen table position, measured with a negatively biased plane probe.

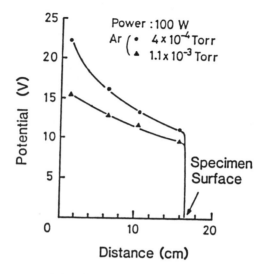

Figure 7: Plasma potential variation along plasma stream, measured with emissive probe method. Substrate potential is chosen as zero. There exists a potential drop due to ion sheath at the surface.

DEPOSITION CHARACTERISTICS

All the experiments on film deposition were carried out without substrate heating. The wafer temperature rise during deposition is shown in Figure 8. The specimen temperature was in the range from 50° to 150°C, due to some heating effect by the plasma. The wafer temperature can easily be kept lower than 50°C by employing a simple wafer cooling scheme. In the above temperature range, the deposition characteristics hardly depend on the temperature for silicon nitride and silicon dioxide. Deposition uniformity is within 5% in the 10 cm diameter middle area.

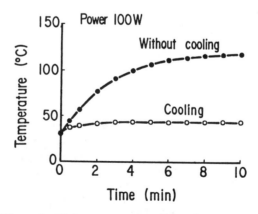

Figure 8: Wafer temperature rise during deposition.

Silicon Nitride Deposition

For silicon nitride (Si_3N_4) film deposition, nitrogen (N_2) and silane (SiH_4) gases are introduced into the plasma chamber and the specimen chamber, respectively. Figure 9 shows the Si_3N_4 deposition characteristics as a function of microwave power, when the introduced gas flow rates are N_2, 10 cc/min and SiH_4, 10 cc/min. The deposition rate increases from about 200 to 300 Å/min, and the refractive index (wavelength, 6328 Å) gradually decreases from about 2.1 to 2.0, with increasing microwave power from 100 to 300 W. The deposition rate is high in comparison with the amount of introduced gas flow rates. This means that the introduced gases are effectively transported and react to form a film, even at a low microwave power. Figure 10 also shows the deposition characteristics,[18] when the deposition rate is increased by increasing the introduced gas flow rates. The deposition rate increases up to about 700 Å/min, and the refractive index markedly decreases with microwave power to 150 W, and then becomes almost constant. This means that the microwave power larger than 150 W is required for sufficiently complete reactions to deposit Si_3N_4 film in this condition. Figure 11 shows the internal stress of the films shown in Figure 10, also as a function of microwave power. The internal stress was measured from the bowing of the silicon substrate caused by film deposition. The stress is compressive for the main, but becomes tensile to some extent at a power of about 150 W. This tendency seems to be related to the variation of refractive index in Figure 10. Figure 12 shows the dependence of the stress on the gas pressure under various conditions. The stress generation seems to be related to the excess energy of ions in reaction, as seen from Figure 6. The internal stress of the Si_3N_4 film can thus be controlled to about zero. The film stress controllability is very advantageous for various applications.

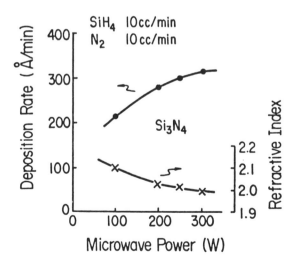

Figure 9: Si_3N_4 deposition characteristics. Deposition rates and refractive indices are shown as functions of microwave power. Gas pressure, 2×10^{-4} torr.

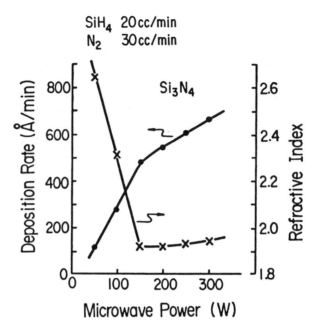

Figure 10: Si_3N_4 deposition characteristics. Gas pressure, 5×10^{-4} torr.

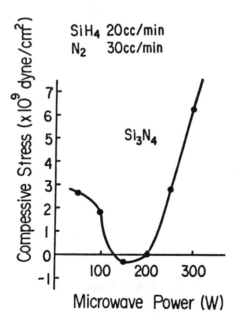

Figure 11: Internal stress of deposited Si_3N_4 films.

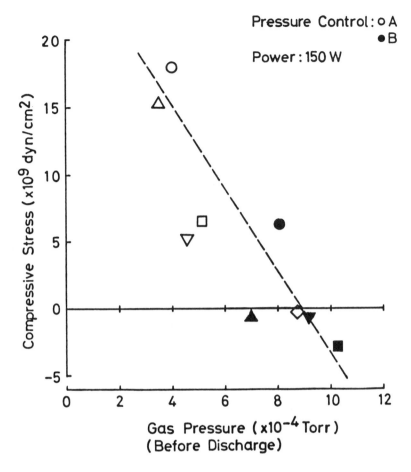

Figure 12: Gas pressure dependence of Si_3N_4 film stresses.

Figure 13 shows the infrared absorption spectrum for the Si_3N_4 film deposited at the microwave power of 150 W and the gas flow rates of N_2, 10 cc/min and SiH_4, 10 cc/min. The Si-N bond peak is clearly observed at the wave number of 845 cm^{-1}, while the Si-H bond peak at about 2100 cm^{-1} is hardly observed. The amount of hydrogen in the film seems very small.

The etch rates of the Si_3N_4 films with a buffered HF solution (BHF, 50% HF:40% NH_4F = 15:85, 20°C) were further examined for film quality evaluation. These rates are shown in Figure 14 as a function of the film refractive index, which was changed by controlling the ratio of the introduced gas flow rates of N_2 and SiH_4. The etch rate reaches the minimum value at the refractive index of about 2.0. The value there is lower than 10 Å/min, which is comparable to those of the high temperature (800°C) CVD films, in spite of the deposition at a low temperature without substrate heating.

158 Thin-Film Deposition Processes and Techniques

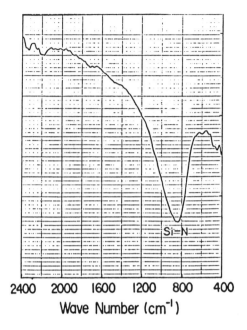

Figure 13: Infrared absorption spectrum for deposited Si_3N_4 film.

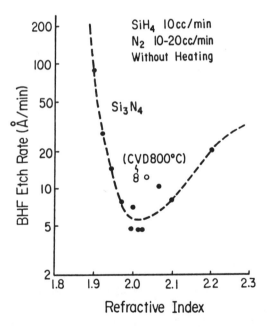

Figure 14: Si_3N_4 film etch rates with BHF solution. BHF, 50% HF:40% NH_4F = 15:85, 20°C.

Silicon Dioxide

Silicon dioxide (SiO_2) can also be deposited by introducing oxygen (O_2) and silane (SiH_4) gases into the plasma and specimen chamber, respectively. Figure 15 shows the SiO_2 deposition characteristics as a function of microwave power, when the introduced gas flow rates are O_2, 10 cc/min and SiH_4, 10 cc/min. The refractive index is almost constant at values from 1.46 to 1.48 in a wide range of microwave power. The deposition rate increases from about 200 to 400 Å/min with increasing microwave power from 50 to 300 W. Figure 16 also shows the SiO_2 deposition characteristics, when the deposition rate was increased by increasing the introduced gas flow rates. High deposition rate over 1000 Å/min is obtained, and the refractive index is almost constant also in this condition, different from the tendency of the Si_3N_4 deposition. The internal stress of the SiO_2 film hardly depends on the deposition condition, giving low values of $2-3 \times 10^9$ dynes/cm^2 (compressive).

Figure 17 shows the infrared absorption spectrum for the SiO_2 film deposited at the microwave power of 100 W and the gas flow rates of O_2, 10 cc/min and SiH_4, 10 cc/min. The Si-O bond peak is clearly observed at the wave number of 1065 cm^{-1}, and Si-H bond peak is hardly observed.

The etch rates of the SiO_2 films with the buffered HF solution were examined and compared to the SiO_2 film prepared by a thermal oxidation method (wet, 1000°C). The result is shown in Figure 18. When the refractive index of the deposited film (n = 1.48) is somewhat larger than that of the thermally oxidized film (n = 1.46), the etch rates of the films almost coincide with each other at various solution temperatures. Figure 19 shows the BHF etch rates as a function of the film refractive index, which was changed by controlling the ratio of the introduced gas flow rates. When the refractive index of the deposited film becomes somewhat larger than the value of 1.46, the etch rate rapidly decreases lower than that of the thermally oxidized film. The degree of the etch rate decrease is seen to be correlated to the intensity of the plasma effect on deposition reactions, according to the order of (1), (2), and (3) in Figure 19.

By introducing an inert gas such as argon (Ar) into the plasma chamber instead of oxygen or nitrogen, silicon films can be deposited. The deposition rate is similar to that of the silicon nitride. Therefore, by introducing a mixture of O_2, N_2 and Ar into the plasma chamber, films of various composition can be easily obtained.

As mentioned above, high quality film deposition of materials, such as silicon nitride and silicon dioxide, at room temperature is possible. The deposition reaction does not require the assistance of a thermal reaction, because both the highly activated plasma and ion bombardment effect with moderate energy sufficiently enhance the film deposition reactions.

Ion Incidence Effects

In order to clarify the contribution of energetic ions to the deposition process, the influence of the existence or absence of ion incidence was examined. Figure 20 shows the distributions of the deposition rate and the BHF etch rates when the substrate was partially masked with a 10 mm gap during a Si_3N_4 deposition, parameterized by the substrate heating temperature. The deposition rates at the shadowed area are much lower than those at the plasma irradiated

160 Thin-Film Deposition Processes and Techniques

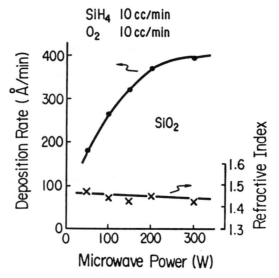

Figure 15: SiO$_2$ deposition characteristics. Deposition rates and refractive indices are shown as functions of microwave power. Gas pressure, 2×10^{-4} torr.

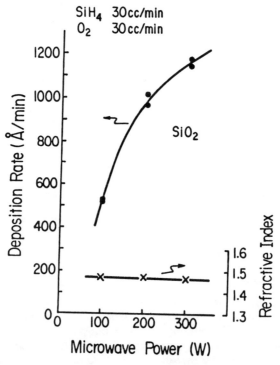

Figure 16: SiO$_2$ deposition characteristics. Gas pressure, 6×10^{-4} torr.

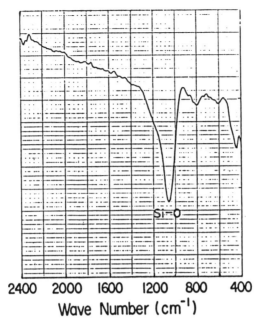

Figure 17: Infrared absorption spectrum for deposited SiO_2 film.

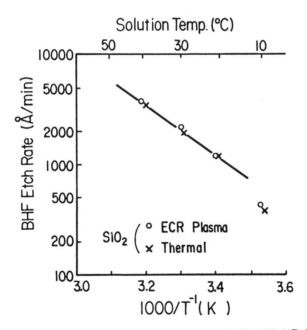

Figure 18: SiO_2 film etch rates with BHF solution. BHF, 50% HF:40% NH_4F = 15:85, 20°C.

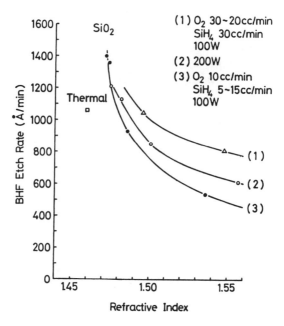

Figure 19: SiO$_2$ film etch rates with BHF solution.

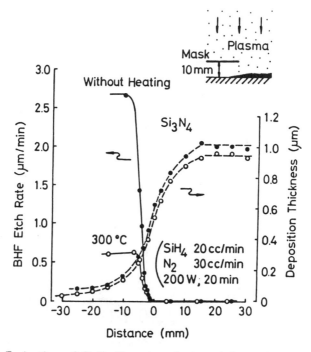

Figure 20: Evaluation of Si$_3$N$_4$ films deposited partially masked with a gap of 10 mm with heating at 300°C and without heating.

area. Further, the BHF etch rates of the film deposited at the shadowed area are much larger in spite of low deposition rates. The effect of the substrate heating at 300°C contributes to improve, to a certain degree, the film quality deposited at the shadowed area, but the film quality is still much inferior to that of plasma irradiated area. The deposition characteristics at the shadowed area seem similar to those of conventional RF plasma CVD, including the temperature dependence.

On the other hand, for the films deposited at the plasma irradiated area, the change in the BHF etch rates by substrate heating is hardly observed. These results indicate that the ion bombardment with a low energy at a low gas pressure combined with a highly activated plasma, even at a low temperature, considerably enhances the deposition reactions, such as hydrogen atom release and molecular bonding reaction.

Material Supply By Sputtering

To extend the advantages of the ECR plasma method to a metallic compound deposition, an ECR plasma deposition apparatus has been developed to add a material supply by sputtering.[19] It realizes low temperature deposition for various compound films by combining plasma extraction with a divergent magnetic field, and the raw material supplies by sputtering.

Figure 21 shows the ECR plasma deposition apparatus. A sputtering target with a shield electrode is placed at the plasma extraction window around the extracted plasma stream. The target is indirectly (radiatively) cooled by watercooling the shield electrode. Sputtering gas (Ar), and material gas (O_2 etc.) are introduced into the plasma chamber, and the specimen chamber. Figure 22 illustrates the deposition mechanism. A DC voltage is supplied to the target plasma stream, so that sputtering will occur with ions from the plasma stream. Sputtered particles are transported with the plasma stream to the specimen substrate. The introduced material gas is ionized and transported in the same way. In this target configuration, a stable operation is realized without abnormal discharges such as spark-over phenomenon, because the target surface is arranged approximately parallel to the magnetic field.

Figure 23 shows the target current characteristics as a function of the target voltage. High target currents over 600 mA are obtained in spite of low gas pressures because of the use of highly ionized ECR plasma. In the range of voltages from 300 to 1000 V which is practically used, the target current is almost constant, determined by the microwave power. Therefore, target currents and target voltages suitable for deposition can be chosen and controlled at the pressures of 10^{-5} to 10^{-3} torr.

The deposition experiments were carried out using a tantalum (Ta) target and an aluminum (Al) target, without substrate heating, to deposit tantalum and aluminum oxide films (Ta_2O_5, Al_2O_3). For deposition, Ar gas was introduced for sputtering into the plasma chamber, and O_2 gas was introduced into the specimen chamber.

Figure 24 shows the Ta_2O_5 deposition characteristics. Oxygen partial pressures were varied under a total gas pressure of 6×10^{-4} torr. The refractive index of deposited film decreases with increasing oxygen partial pressure, and saturates at an index of about 2.1. The deposition rate decreases at higher oxygen partial pressures. In conventional sputtering, fully oxidized films are obtained only in the oxygen partial pressure range where the deposition rates become ex-

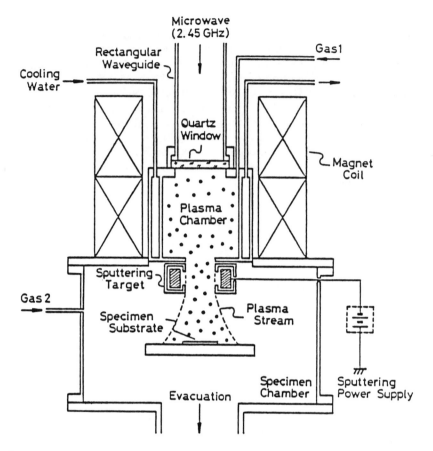

Figure 21: ECR plasma deposition apparatus using material supply by sputtering.

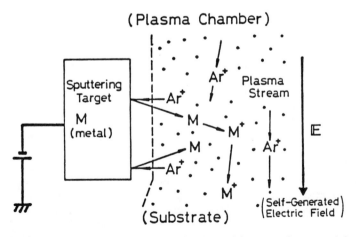

Figure 22: Ilustration of deposition mechanism with sputtering material supply.

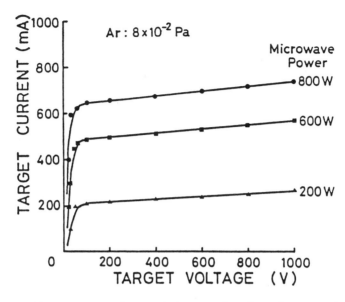

Figure 23: Target current characteristics as a function of target voltage, with microwave power varied. Ar gas pressure, 8×10^{-2} Pa (6×10^{-4} torr).

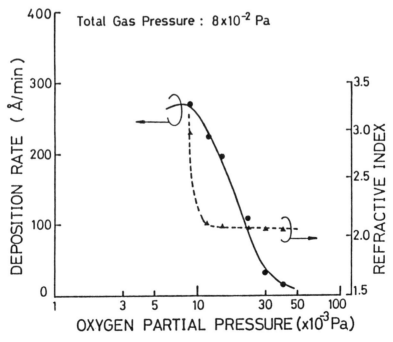

Figure 24: Ta_2O_5 deposition characteristics. Deposition rates and refractive indices are shown as functions of O_2 partial pressure. Total gas pressure, 8×10^{-2} Pa (6×10^{-4} torr).

166 Thin-Film Deposition Processes and Techniques

tremely low. That is, high quality oxide film deposition and high rate deposition are difficult to be performed simultaneously. In this ECR method, high deposition rates over 200 Å/min are obtained even under the condition that fully oxidized film is deposited.

Figure 25 shows the Al_2O_3 deposition characteristics. Deposition characteristics are similar to those of Ta_2O_5 in Figure 24. Aluminum oxide films with a refractive index of 1.65 are obtained under the deposition rate of 350 Å/min. High rate deposition is possible for aluminum oxide in this method with an aluminum target, although conventionally aluminum sputtering is severely affected by the oxygen atmosphere.

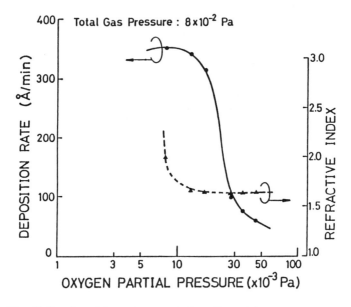

Figure 25: Al_2O_3 deposition characteristics. Deposition rates and refractive indices are shown as functions of O_2 partial pressure. Total gas pressure, 8×10^{-2} Pa (6×10^{-4} torr).

The ECR plasma deposition method employing a sputtering material supply can be applied, combined with introducing various gases such as O_2, N_2, CH_4 and SiH_4, to deposition of various metals, their oxides, nitrides, carbides and silicides. Deposition rate can further be increased by applying a magnetron-mode operation to sputtering the target utilizing a modified magnetic field distribution.[20]

ECR PLASMA CVD SYSTEM

Figure 26 shows a photograph of the ECR plasma CVD system developed for semiconductor device processes (manufactured by ANELVA Corporation).

Figure 26: ECR plasma CVD system with automated cassette-to-cassette operation.

The system is constructed with a lateral ECR plasma source configuration. Silicon wafers, 4 inches in diameter, are loaded and unloaded with a mechanical wafer chucking system. They are processed sequentially through an air-lock chamber with automated cassette-to-cassette operation. A silicon wafer to be processed is placed on a table, surrounded with a fused quartz plate. A wafer-cooling unit can be arranged under the wafer, since no substrate heating is needed in deposition reaction.

It is often important to maintain wafer temperatures sufficiently low during deposition, in the application to semiconductor device fabrication process such as film deposition on a resist-layer or lift-off process.[21] Generally, a wafer placed on a cooled table in vacuum can not be effectively cooled, because the thermal conduction is very poor due to the existence of a microscopic gap between them. To improve the thermal contact between the wafer and a water-cooled table, a wafer-cooling technique utilizing an electrostatic force is employed in the system. An electrode plate covered with a dielectric film is positioned under the wafer, and the wafer itself is used as the second electrode, utilizing that the wafer during processing is automatically grounded (about –20 V) by the plasma stream irradiation. The voltage applied to the first electrode (0-2000 V) does not affect the wafer potential. This wafer-cooling method is simple and reliable.

CONCLUSIONS

A low temperature deposition process utilizing an electron cyclotron resonance (ECR) plasma has been developed. High quality film deposition of various materials is possible, by introducing chemical vapors and/or by using sputtering material supply, at a low temperature without substrate heating. The ECR plasma deposition techniques can be applied to various substrates, such as with poor heat-resistance, utilizing the advantage of the low temperature process. Both effects of the highly activated plasma and of the ion bombardment with energies of about 20 eV at low gas pressures markedly contribute to enhance the deposition reactions and improve the film quality, but appears not to induce surface damage, even for heat-sensitive compound semiconductors.

REFERENCES

1. Abe, H., Sonobe, Y. and Enomoto, T., *Jpn. J. Appl. Phys.*, Vol. 12, p. 154 (1973).
2. Hosokawa, N., Matsuzaki, R. and Asamaki, T., *Jpn. J. Appl. Phys.*, Suppl. 2, Pt. 1, p. 435 (1974).
3. Matsuo, S. and Takehara, Y., *Jpn. J. Appl. Phys.*, Vol. 16, p. 175 (1977).
4. Matsuo, S., *J. Vac. Sci. Technol.*, Vol. 17, p. 587 (1980).
5. Matsuo, S., *Appl. Phys. Letters*, Vol. 36, p. 768 (1980).
6. Consoli, T., Hall, R.B., *Fusion Nucleaire*, Vol. 3, p. 237 (1963).
7. Bardet, R., Consoli, T. and Geller, R., *Fusion Nucleaire*, Vol. 5, p. 7 (1965).
8. Ikegami, H., Ikeji, H., Hosokawa, M., Tanaka, S. and Takayama, K., *Phys. Rev. Letters*, Vol. 19, p. 778 (1967).
9. Sakamoto, Y., *Jpn. J. Appl. Phys.*, Vol. 16, p. 1993 (1977).
10. Sinha, A.K., Levinstein, H.J., Smith, T.E., Quintana, G. and Haszko, S.E., *J. Electrochem. Soc.*, Vol. 125, p. 601 (1978).
11. Lanford, W.A. and Rand, M.J., *J. Appl. Phys.*, Vol. 49, p. 2473 (1978).
12. Adams, A.C., Alexander, F.B., Capio, C.D. and Smith, R.E., *J. Electrochem. Soc.*, Vol. 128, p. 1545 (1981).
13. Tsuchimoto, T., *J. Vac. Sci. Technol.*, Vol. 15, p. 70 (1978).
14. Matsuo, S. and Kiuchi, M., *Jpn. J. Appl. Phys.*, Vol. 22, p. L 210 (1983).
15. Matsuo, S. and Adachi, Y., *Jpn. J. Appl. Phys.*, Vol. 21, p. L 4 (1982).
16. Geller, R., Hopfgarten, N., Jacquot, B. and Jacquot, C., *J. Plasma Phys.*, Vol. 12, p. 467 (1974).
17. Kemp, R.F. and Sellen, J.M., Jr., *Rev. Sci. Instrum.*, Vol. 37, p. 455 (1966).
18. Matsuo, S., Ext. Abs. 16th (1984 Int.) Conf. Solid State Devices and Materials, Kobe, p. 459 (1984).

19. Ono, T., Takahashi, C. and Matsuo, S., *Jpn. J. Appl. Phys.*, Vol. 23, p. L534 (1984).
20. Ono, T., Takahashi, C., Oda, M. and Matsuo, S., *1985 Symp. VLSI Technology*, Kobe, p. 84 (1985).
21. Ehara, K., Morimoto, T., Muramoto, S. and Matsuo, S., *J. Electrochem. Soc.*, Vol. 131, p. 419 (1984).

6

Molecular Beam Epitaxy: Equipment and Practice

Walter S. Knodle and Robert Chow

Molecular beam epitaxy has experienced extremely rapid growth over the last ten years. During this time, the publication rate of technical articles has increased by an estimated order of magnitude reflecting the growing application of the MBE process. Similarly, the number of researchers has grown concomitantly with the technology and a large MBE community has established itself with various specializations according to materials and devices. This large and growing body of research knowledge has advanced and redefined the hardware which in turn has assisted to further advance process development. The entry level scientist is often faced with the formidable task of collecting the hardware familiarity and process foundation necessary to begin MBE research. It is the authors' intent in this chapter to present a broad coverage of the present status of the MBE process and equipment, not dwelling in depth on any one aspect, but instead presenting key references to classic and current articles on important topics. By supplying timely leads to the work of leading researchers, we hope to advance the learning curve of those using this source as a starting point. We sincerely hope that the reader finds this goal achieved.

Two texts worth mentioning as a general reference are: *Molecular Beam Epitaxy*, B. Pamplin, ed., Pergamon Press 1980; and *The Technology and Physics of Molecular Beam Epitaxy*, E. Parker, ed., Plenum Press 1985. There are also regularly published proceedings of value from: the MBE Workshop and the International Conference on MBE (both in the *Journal of Vacuum Science and Technology*), the International Symposium on Gallium Arsenide and Related Compounds, and the IEEE GaAs Integrated Circuit Symposium.

1.0 THE BASIC MBE PROCESS

Although the basic process of MBE, ultra-high vacuum evaporation, had been in practice long before, it was not until the works of Arthur[1] and Cho[2] that a fundamental understanding of the process as applied to compound semiconductor growth evolved. Using mass spectrometric and surface analytical techniques, they studied the film growth process of gallium arsenide at the atomic layer level. Their work set the stage for equipment design specific to UHV epitaxial film growth. Subsequently, Chang, et al[3] and others further advanced the process toward what we recognize today as MBE. The foundation of MBE is rooted in surface analysis and in understanding on an atomic level the nature of epitaxial film growth. Several reviews are of historical and technical significance; for instance, see Cho and Arthur,[4] Chang and Ludeke,[5] Foxon and Joyce,[6] Wood[7] and Kunzel.[8] MBE equipment development still reflects the strong influence of its surface analytical beginnings. Subsequent advances in equipment design were made by Luscher and Collins[9] and others. The development of new and higher performance devices using MBE has advanced its commercial value. Present trends, reflecting the evolution of MBE from basic research to device production, are redefining and refining the equipment for specific material and device requirements.

A functional schematic of a MBE system is shown in Figure 1. It consists of a growth chamber and auxiliary chamber (not present with first generation systems) and a load-lock. Each chamber has an associated pumping system. The load-lock facilitates the introduction and removal of samples or wafers without significantly influencing the growth chamber vacuum. The auxiliary chamber may contain supplementary surface analytical tools not contained in the growth chamber, additonal deposition equipment or other processing equipment. Separating equipment in this manner allows for more efficacious use of the growth chamber and enhances the quality of operations in both the auxiliary and growth chambers.

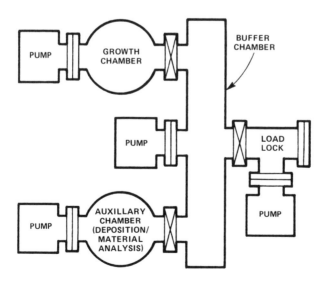

Figure 1: Functional schematic of a basic MBE system.

The growth chamber is shown in greater detail in Figure 2. Its main elements are: sources of molecular beams; a manipulator for heating, translating and rotating the sample; a cryoshroud surrounding the growth region; shutters to occlude the molecular beams; a nude Bayard-Alpert gauge to measure chamber base pressure and molecular beam fluxes; a RHEED (reflection electron diffraction) gun and screen to monitor film surface structure; and a quadrupole mass analyzer to monitor specific background gas species or molecular beam flux compositions.

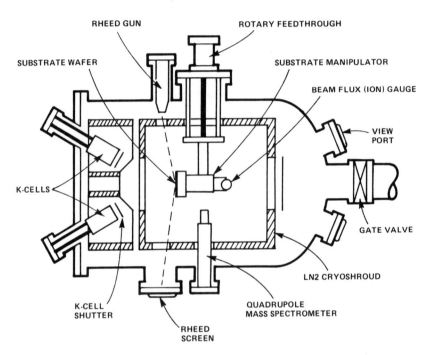

Figure 2: Schematic cross-section of a typical MBE growth chamber.

The auxiliary chamber may be host to a wide variety of process and analytical equipment. Typical surface analytical equipment would be: an Auger electron spectrometer, or equipment for secondary ion mass spectrometry (SIMS), ESCA (electron spectroscopy for chemical analysis) or XPS (x-ray photoelectron spectroscopy). There may be a heated sample station and an ion bombardment gun for surface cleaning associated with this equipment. Process equipment may include sources for deposition or ion beam etching.

The MBE growth process involves controlling, via shutters and source temperature, molecular and/or atomic beams directed at a single crystal sample (suitably heated) so as to achieve epitaxial growth. The gas background necessary to minimize unintentional contamination is predicated by the relatively slow film growth rate of approximately one micron per hour and is commonly in the 10^{-11} Torr range. The mean free path of gases at this pressure and in the beams themselves is several orders of magnitude greater than the normal source-to-sample distance of about 10-20 cm. Hence, the beams impinge unreacted on the sample (indirect paths to the sample are minimized by surrounding the sample with a cryoshroud cooled by liquid nitrogen). Reactions take place predominantly at the sample surface where the source beams are incorporated into the growing film. Proper initial preparation of the sample will present a clean, single crystal surface upon which the growing film can deposit epitaxially. Timely actuation of the source shutters allows film growth to be controlled to the monolayer level. It is this ability to precisely control epitaxial film growth and composition that has attracted the attention of material and device scientists. Some further reviews of the basics of MBE are: Cho,[10] Luscher,[11] and Panish.[12]

Most MBE research has been performed with elements from groups III and V of the periodic table (i.e., Al, Ga, In, As, P and Sb), but much significant work has also been achieved with silicon and germanium, II-VI and IV-VI compounds and various metals. New milestones in material structures have been attained because of MBE, leading to the development of semiconductor devices heretofore difficult or impossible to fabricate and to the invention of entirely new materials[13] and devices previously not imagined. Most of the attention in this review will focus on III-V materials because of their pervasive use in microwave and optoelectronic communication; high speed digital and analog devices; and integrated circuits. Referral to other material systems is given, where appropriate, with references to assist the reader in initiating his (her) own literature search.

2.0 COMPETING DEPOSITION TECHNOLOGIES

While MBE has afforded the fabrication of material and device structures not previously possible, many of the early milestones in compound semiconductor growth were attained using other epitaxial film growth techniques. Bench marks for purity and device performance remain valid today. These alternative deposition technologies can be loosely grouped into liquid phase epitaxial (LPE) and vapor phase epitaxial (VPE) processes. A special category of VPE known as OMVPE (organometallic VPE) or MOCVD (metal-organic chemical vapor deposition) has promise as a production worthy process. Each of these epitaxial processes has distinct advantages and disadvantages discussed below. A comparison of GaAs epitaxial processes is shown in Table 1.

2.1 Liquid Phase Epitaxy

Historically the early compound semiconductor growth was performed using liquid phase epitaxial techniques and many optoelectronic devices are still fabricated using this process. The threshold current values of the best double heterostructure (DH) MBE lasers are typically compared to LPE results. There are several variants of the LPE process, but in the approach most common to multilayer film growth a graphite holder slides a sample between melts of differing composition. A schematic of a graphite slider LPE system is shown in Figure 3a. The temperature and melt composition determine the stoichiometry and deposition rate. Film growth results from the controlled cooling of the supersaturated melt. Different bins are required for each layer of differing alloy com-

Table 1: Comparison of Epitaxial Technologies

	Liquid-phase epitaxy	Vapor-phase epitaxy		Molecular-beam epitaxy
		Chemical vapor deposition	Metal-organic CVD	
Growth rate (μm/min)	~1	~0.1	~0.1	~0.01
Growth temperature (°C)	850	750	750	550
Thickness control (Å)	500	250	25	5
Interface width (Å)	≥50	~65	<10	<5
Dopant range (cm^{-3})	$10^{13}-10^{19}$	$10^{13}-10^{19}$	$10^{14}-10^{19}$	$10^{14}-10^{19}$
Mobility, 77K (cm^2/Vs) (n-type GaAs)	150,000–200,000	150,000–200,000	140,000[a]	160,000[b]

[a] T. Nakanisi, T. Udagawa, A. Tanaka and K. Kamei, *J. Cry. Growth*, Vol. 55, p. 255 (1981).

[b] E.C. Larkins, E.S. Hellman, D.G. Schlom and J.S. Harris, Jr., *App. Phys. Lett.* (to be published).

position. A thorough knowledge of the alloy phase diagram is necessary to accurately control the film composition. Weighed amounts of impurities are added to the melts to control doping. Dopants with relatively high distribution coefficients are not easily controlled.

LPE has the advantages of low capital cost, high deposition rates, high material purity, no toxic gases and a relatively wide selection of dopants. Some disadvantages are: an inability to produce abrupt (monolayer) interfaces and poor large area uniformity; and difficulty in varying stoichiometry and controlling the reproducibility of ternary III-V compounds. Advances in LPE equipment have allowed superlattice structures of 200–300 Å thick layers to be produced (Benchimol, et al.[14]). Despite such progress LPE is not considered amenable to high volume automated manufacturing.

2.2 Vapor Phase Epitaxy and MOCVD

VPE of III-V compounds can be accomplished with different chemistries; hydride, halide or organometallic. The halide and hydride systems are common to the silicon semiconductor industry where epitaxial films are routinely deposited by the hydrogen reduction of chlorosilanes or the pyrolytic decomposition of silane. Gallium arsenide is deposited by passing $AsCl_3$ (with a hydrogen carrier gas) over molten gallium held at 800°C. The resulting gallium chloride and arsenic react at the cooler (700°C) substrate surface to deposit GaAs. Figure 3b

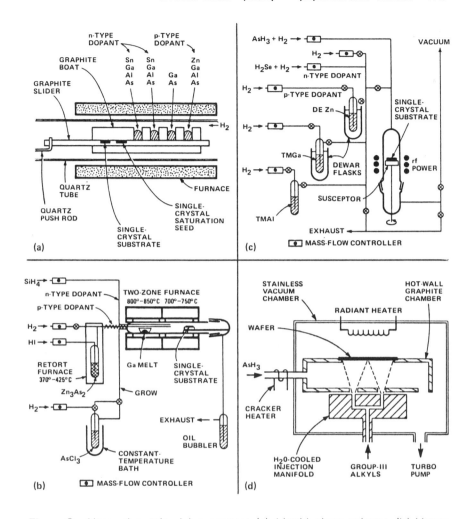

Figure 3: Alternative epitaxial processes: (a) Liquid phase epitaxy, (b) Vapor phase epitaxy (halide), (c) Metal-organic chemical vapor deposition, and (d) Vacuum chemical epitaxy (see Section 7.3.2).

shows a III-V chloride-based VPE system. The reaction chamber is a hot-wall quartz tube. In this diagram, zinc is the p-type dopant. The most limiting disadvantage of III-V halide VPE is the absence of a suitable chemistry for the deposition of aluminum. The deposition of aluminum bearing compounds such as AlAs, AlGaAs and AlGaInP can, however, be accomplished using MOCVD.

Hydride VPE is similar to the halide process and shares virtually identical hardware. Gallium chloride is created by the reaction of HCl gas with molten gallium. Arsenic, generated by cracking arsine, is mixed with a hydrogen carrier gas. The same disproportionation and reduction reactions occur at the heated substrate as in the halide process.

MOCVD offers the most competitive alternative to MBE of the epitaxial film processes. Figure 3c shows a vertical reactor employing trimethylgallium, trimethylaluminum and arsine. All gases are mixed with a hydrogen carrier gas. The organometallics are contained in temperature controlled bubblers. The sample is heated inductively inside a quartz reaction tube (cold-wall system) by external RF coils. In either example, the deposition takes place at or near the heated (550°-750°C) sample surface by pyrolysis. Gas phase parasitic reactions and adduct formation can seriously complicate some growth processes. Growth rates between 1 and 10 microns per hour are common and excellent surface morphology has been demonstrated. A full metallorganic process would use a group V organometallic compound in place of a hydride. For example, trimethylarsine would replace arsine. See Table 1 for a comparison of epitaxial technologies for the deposition of GaAs. Recent developments in MOCVD have allowed researchers to deposit thin (<100 Å) alternating layers with abrupt interfaces. Computer control of gas flows is necessary for high quality superlattice growth. Although film purity and thickness uniformity have been difficult to control, recent progress has been made improving the quality of starting gases and achieving good thickness and doping uniformity (~± 5% and ±10%, respectively) across a three-inch wafer. Triethyl compounds have produced material with lower carbon contamination than the trimethyls.[15] One serious disadvantage of MOCVD is the toxicity of the gases required. In addition, some of the less toxic gases are pyrophoric. Expensive safety precautions are required for operating personnel and to reduce environmental liabilities.

While still inferior in terms of absolute thickness and doping control, MOCVD is considered to be a viable alternative to MBE for many devices. See, for instance, Andre, et al.[16] There has been much recent progress in the development of this still relatively young process. Many structures that at one time were only considered possible using MBE have recently been fabricated using MOCVD. See Takakuwa, et al[17] and Razeghi, et al.[18] Interestingly, the advent of low pressure MOCVD and metal-organic MBE (MOMBE) is bringing both process regimes closer together.

Atomic layer epitaxy (ALE) by hydride[19] and metalorganics[20] and molecular layer epitaxy (MLE)[21] are recent variations of the VPE process that allow sufficient growth control for superlattice structures. While MBE is still superior in terms of growth control, MOCVD has good potential to provide high throughput.[22] However, there are serious safety considerations associated with adopting any process requiring arsine and/or phosphine.

MOCVD is not confined to III-V compounds. Chemistries exist for the deposition of II-VI and IV-VI compounds as well. A review of MOCVD is included in Chapter 7 of this handbook. Further references are presented in *Gallium Arsenide Technology*, Chapter 3, D.K. Ferry, ed., Howard W. Sams & Co. (1985) and in *Semiconductors and Semimetals*, Vol. 22, Part A, Chapter 3, W.T. Tsang, ed., Academic Press (1985).

3.0 MBE-GROWN DEVICES

Although much of the initial MBE research was directed toward the study of surface film growth kinetics, the driving force today is the fabrication of advanced electronic and optoelectronic devices. As the film growth capabilities of MBE became clear, researchers began producing a wide variety of heretofore impossible and unimagined devices. Some of these devices have subsequently been fabricated using other epitaxial techniques, but many were initially conceived in MBE systems. The advantages of generally lower growth temperature and growth rate allow MBE to produce atomically abrupt heterojunctions and doping profiles. The ability to produce these composition variations with material systems of inherently high electron mobility (i.e., GaAs, InP, InGaAs) has permitted the fabrication of very fast devices. Microwave devices with operating frequencies near 100 GHz[23] and high speed digital switching near 5 picoseconds[24] have been achieved. Also, the control of contact layers has reduced parasitic resistances, significantly improving FET performance (e.g., MAG and NF). The ability of MBE to deposit epitaxial metal layers in situ promises to permit the construction of a metal base transistor.[25] The fact that several material systems (e.g., GaAs/AlGaAs) are also optically active allows for the possibility of integrating high speed digital and optical circuits. Recent progress in the growth of GaAs on silicon further increases the options by promising to combine the virtues of these two semiconductor technologies in a monolithic device. MBE silicon devices are discussed in an article by K.L. Wang.[26]

Since it is not the intent of the authors to discuss devices per se, we limit ourselves to three examples relevant to MBE growth: the high electron mobility transistor (HEMT), the heterojunction bipolar transistor (HJBT or HBT) and the multiquantum well (MQW) laser. Tables 2 through 5 catalogue many of the MBE grown devices and circuits.

3.1 Transistors

The transistor, both as a digital switch and as an amplifier, has been key to the development of MBE. Table 2 summarizes the variety of MBE grown transistors. Of particular significance is the high electron mobility transistor (HEMT). The HEMT transistor is often referred to by a number of synonyms such as: MODFET (modulation doped FET), SDHT (selectively-doped heterojunction transistor), and TEGFET (two-dimensional electron gas FET). It is a major force behind making MBE a production worthy process because of its extremely fast switching speed and potential use in supercomputers.[27] Switching speeds near 5 picoseconds and speed-power products of 10 femtojoules have been demonstrated.[24] The HEMT is responsible in large part for the dwindling interest in Josephson junction devices. Its foundations are an outgrowth of superlattice and modulation doping structures made by Dingle, et al[28] and later evolved into the first practical device by Mimura[29] and Hiyamizu.[30] Its performance advantage is due to the separation of ionized donors and charge carriers. Electrons (holes) diffuse from a setback doped layer and are captured in the narrow potential well created at the heterojunction (see Figure 4a). Room temperature mobilities in this two-dimensional layer can approach 10^4 cm^2/Vsec for electrons. HEMT performance is also assisted by in situ MBE grown contact layers which contribute to reduced source and drain parasitic resistances. Reproducible structures require clean, atomically smooth heterojunctions, accurate setbacks, and repeatable doping levels. Several variations of the HEMT have been explored; the inverted HEMT, the double-heterojunction HEMT and the superlattice HEMT.

Table 2: MBE-Grown Transistors

HEMT, MODFET, SDHT, TEGFET, HFET[31]
   Inverted HEMT[32]
   Low-noise HEMT[33]
   GeSi MODFET[34]
   MIS-HFET[35]

HBT[36,49]
   Strained-Layer[37]
   GaAs on Si[38]

HET, MGT[39,40]

MESFET; LOW NOISE[41-43]
   POWER[41,44,45]

BIFET (Buried-Interface FET)[46]

TEBT (Tunneling Emitter BT)[47]

BICFET (Bipolar Inversion Channel FET)[48]

Another important transistor is the heterojunction bipolar transistor (HBT). A cross section is shown in Figure 4b. Since the charge carriers flow vertically in the HBT, device speed is controlled by layer thickness, unlike the FET where the flow is lateral and critical distances are lithographically determined. This fundamental difference implies a potentially higher frequency operating limit for HBTs. HBTs also possess the very desirable characteristic of greater process margins, because turn-on voltages are determined almost entirely by the base material bandgap. Thickness uniformity is thus less critical than compositional reproducibility. In the case of the graded-base HBT, MBE allows the bandgap energy (composition) to vary continuously. MBE's potential to accurately control composition, layer thickness and doping with superior across-the-wafer uniformity makes it ideally suitable for heterojunction bipolar transistor and IC production. MBE HBTs with an f_t of 40 GHz and an F_{max} above 25 GHz have been grown by Asbeck, et al.[49] Other MBE-grown transistors of potential significance, such as the hot-electron transistor, are referenced in Table 2, but will not be discussed here.

A discussion of MBE grown heterojunction bipolar transistors and selectively doped heterostructure transistors is given in Chapters 5 and 6, respectively, of *VLSI Electronics Microstructure Science*, Vol. 11, "GaAs Microelectronics", N.G. Einspruch and W.R. Wisseman, eds., Academic Press (1985). Also important is the article by Kroemer.[50]

3.2 Microwave and Millimeter Wave Devices

The first MBE grown device was a varactor diode[51] followed by the IMPATT diode.[52] These devices demonstrated the relative ease with which MBE could control doping profiles and thickness. High performance mixer and Gunn diodes have also been produced by MBE (see Table 3). In addition to two terminal devices, both low-noise and high power FETs have been successfully fabricated

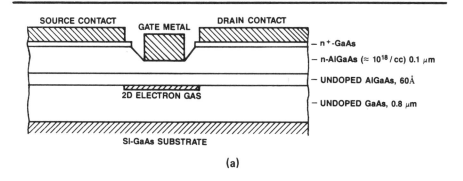

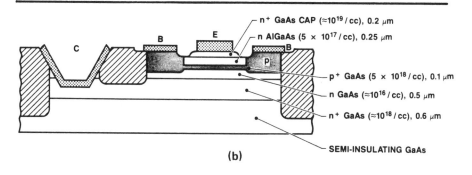

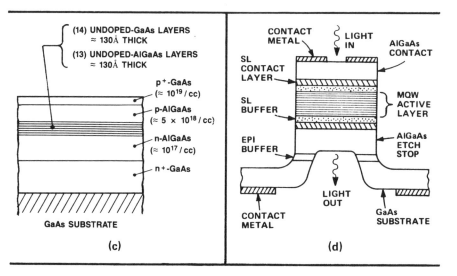

Figure 4: Schematic cross-sections of some MBE-grown devices: (a) High electron mobility transistor, (b) Heterojunction bipolar transistor, (c) Multi-quantum well laser, and (d) QCSE modulator.

using MBE. Several manufactures now produce MBE-grown IMPATTs and low-noise FETs for commercial use.

Table 3: MBE-Grown Microwave Devices

IMPATT[52-54]
MIXER[55-57]
GUNN[58]
VARACTOR[51,56]

3.3 Optoelectronic Devices

Optoelectronic devices require many of the same growth control capabilities (i.e., uniformity, abrupt heterojunctions, doping level and reproducibility) needed for transistors, but in addition there are strict limitations on the levels of deep traps. These traps, acting as recombination centers, can fatally impact the optical properties of LEDs, lasers, photodiodes, etc. Control of grown-in defects, dislocations and unintentional contaminants, all of which can serve as recombination centers, is critical. Considerable progress has been made in the past few years in the quality of MBE grown optoelectronic material. Room temperature cw AlGaAs DH lasers have been demonstrated with threshold currents as low as the best LPE material.[59,60] Tsang[61] had also shown that MBE can produce high quality DH lasers at growth rates near 10 microns per hour with excellent yield and reproducibility. DH lasers with projected mean cw lifetimes of greater than 10^6 hours have been grown. The advantages of MBE are not limited to DH lasers. Table 4 lists some of the many different optoelectronic devices grown by MBE.

One optical device which exemplifies the growth control of the MBE process is the multiquantum well (MQW) heterostructure laser. A cross-sectional sketch is shown in Figure 4c. The MQW structure is composed of approximately 30 alternating layers of two materials (e.g., GaAs and AlGaAs) with layer thicknesses of about 100 nm each. The challenge is to grow this structure with uniform, repeatable layer thickness, repeatable compositions and atomically smooth interfaces free of alloy clusters. Laser Spectra has indicated that MBE can achieve this with thickness control near the monolayer level. Mean lifetimes near 5,000 hours at 70°C have been obtained making MBE grown MQW lasers the longest-lived. Saku, et al[69] has reported the lowest J_{th} for AlGaAs MQW lasers operating below 700 nm and the shortest wavelength room temperature operation of any AlGaAs MQW laser using MBE grown material. Also, Ohmori, et al[70] has achieved the first room temperature cw operation of a GaSb/AlGaSb MQW laser. These results attest to the material quality and growth control capabilities of MBE.

One final class of optoelectronic devices requires mention because of its potential contribution to the long-sought optical computer. This class of devices, based on quantum well structures, uses the Quantum Confined Stark Effect[64] and operates at room temperature. Optical modulators and voltage-tunable detectors can be made and combined into a single device called a SEED (self-electrooptic effect device). SEEDs have been made into linearized optical modulators, optically bistable devices, optical level shifters and OE oscillators. A QCSE modulator is shown in Figure 4d. The device was grown with MBE and has a transparent superlattice buffer layer on either side of a MQW structure.

Table 4: MBE-Grown Optoelectronic Devices*

LASER**
  DFB: Distributed Feedback
  DH: Double Heterostructure
  DBDH: Double Barrier DH
  GRINSCH: Graded-Index Separate Confinement Heterojunction
    BH: Buried Heterostructure
  MQW: Multi-quantum Well
  TJS: Transverse Junction Stripe
  SCH: Separate-Confinement Heterostructure

LED:

PHOTODETECTOR:
  APD: Avalanche:
    Superlattice
      Ge(x)Si(1-x)/Si Strained-Layer[62]
    Graded-gap
    Channeling
  GeSi Waveguide:[63]

SEED: Self Electro-optic Effect Devices[64]

SOLAR CELL:[65]

*A comprehensive review of MBE and III-V optoelectronic devices is given by Tsang, in:*Semiconductors and Semimetals*, Vol. 22, part A, Chapter 2, W.T. Tsang, ed., Academic Press (1985).

**See Norton, et al.[66] and Partin, et al.[67,68] for IV-VI lasers.

3.4 Integrated Circuits

MBE-grown devices have progressed into digital, monolithic microwave, electro-optical and optical curcuits at various levels of integration. Table 5 lists some integrated circuits fabricated from MBE-grown material. OEICs (optoelectronic ICs) have reached the SSI level,[85] optical ICs are at a somewhat lower level of integration, and both HEMT and HBT based LSI ICs have been successfully demonstrated. Texas Instruments[90] has made a MSI HBT 1K gate array and HEMT LSI 1K[79] and 4K[80] SRAMs have been fabricated by Fujitsu (see Figure 5). The 2 ns access time at 77K is the highest ever reported for a 4K RAM. Miller[91] and Andre, et al[16] have addressed the specific MBE equipment challenges presented by HEMT IC requirements. Threshold voltage control to within 30 mV is required across-the-wafer and wafer-to-wafer for LSI RAM HEMT circuits. Abrokwah, et al[92] achieved a MODFET threshold voltage standard deviation of 15% ($<V_t>$ = -2.61 V) across a three-inch wafer while values of 16 mV for E-HEMTs and 24 mV for D-HEMTs have been reported over 2-inch wafers by Mimura, et al.[93] Kuroda, et al[80] obtained threshold voltage standard deviations of 12 mV and 20 mV, respectively, for DCFL E- and D-HEMTs with 1.5 micron gates. These values are about 2.5% of the DCFL logic swing. Current generation MBE systems are capable of meeting the thickness, composition, and doping specifications required for IC material growth on three-inch wafers. Further progress must be made in reducing defect levels, however.

Table 5: MBE-Based Integrated Devices/Circuits

Analog ICs
    115 GHz Oscillator[71]
    HBT Voltage Comparator[72]

Digital ICs
    HEMT Ring Oscillator[24,73,74]
    I-HEMT Ring Oscillator[75]
    HBT ICs[76]
    2-stage Complementary Inverter[77]
    Quantum-well CCD[78]
    1kb SRAM[79]
    4 kb SRAM[80]

OEICs
    LED/Amplifier[81]
    Photoreceiver[82-84]
    Transmitter/receiver[85]
    Laser/FET[86]

Optical IC*
    Optical Switch[87]
    Laser Taper Coupler[88]
    Laser/waveguide[89]

*See R.D. Burnham and D.R. Scifres, "Integrated Optical Devices Fabricated by MBE," in *Prog. Crystal Growth Charact.*, Vol. 2, p. 95 (1979) and E.M. Conwell and R.D. Burnham, "Materials for Integrated Optics: GaAs," *Ann. Rev. Mater. Sci.*, Vol. 8, p. 135 (1978).

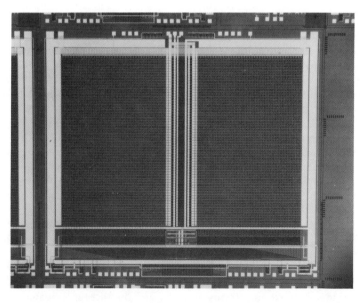

Figure 5: A 4K SRAM made from MBE-grown material.[80]

4.0 MBE DEPOSITION EQUIPMENT

Our attention in this section will focus on III-V MBE equipment. Discussions relevant to silicon and II-VI MBE will be made where appropriate. A system for II-VI MBE is shown in Figure 6.

Figure 6: Photograph of a II-VI MBE system (courtesy of J.P. Faurie, University of Illinois, Chicago).

4.1 Vacuum System Construction

Vacuum system construction practices closely follow those of surface analysis equipment. Standard ultra-high vacuum (UHV) practices are employed (see Dushman,[94] Guthrie[95] or O'Hanlon,[96] for example). All internal materials are carefully selected for minimal background gas contribution and 200°C bakeout temperature compatibility. Materials like stainless steel, quartz and alumina are used where operating temperatures do not exceed 200°C. High purity refractory metals and pyrolytic boron nitride (PBN) are commonly used where temperatures exceed 200°C. Moving parts are typically dry-lubricated with very low vapor pressure molybdenum or tungsten disulfide.

4.1.1 Construction Practices. System chambers are normally fabricated from either 304 or 316 stainless steel using TIG welding techniques. These are the metals of choice for strength, vacuum compatibility and relative ease of fabrication. Following fabrication, chambers are often electropolished and passivated to improve the surface outgassing. Several techniques are practiced and properly prepared stainless surfaces will have an outgassing rate below 5×10^{-11}

184 Thin-Film Deposition Processes and Techniques

Torr l/cm²sec. Post-bake levels in the 4×10^{-12} Torr l/cm²sec range are nominal.[97] There has been some recent success constructing MBE systems from aluminum,[98] but aluminum may not be compatible with all source materials. Nevertheless, base pressures one to two orders below stainless following a 25 hour bake at 156°C are claimed. System chambers are usually isolated by gate valves of various types and sizes. These may be either all-metal valves or Viton sealed. Viton is compatible with repeated 200°C bakeouts if it is not baked under compression. O-rings of other fluorocarbon elastomers with higher temperature limits have been considered. Viton sealed valves have a cost advantage and about a 10X lifetime when compared to all-metal valves. All chamber ports are sealed with metal gaskets.

4.1.2 Multi-Chamber Systems. A typical current generation MBE system will be composed of four separate vacuum chambers: a load-lock for substrate entry and exit, a growth chamber, an auxiliary chamber for analysis or metallization and an intermediate or "buffer" chamber for pregrowth processing (see Figures 7(a)-(b). The load-lock typically achieves base pressures of around 10^{-8} Torr while the auxiliary and buffer chambers reach 5×10^{-10} Torr or better and the growth chamber reaches 5×10^{-11} Torr or better after baking. Multichamber systems allow increased flexibility in configuration and operation. Present load-lock designs allow for the entry of 10 to 16 wafers (or samples) at a time to increase system throughput. Newly introduced wafers can be heated under vacuum to desorb water vapor before proceeding into the intermediate chamber. The separation of growth and analytical instrumentation, which appeared with second generation equipment, was prompted by a need to keep the analytical instrumentation clean and to allow film growth independent of analytical studies. The modular design of current generation MBE systems allows two or more systems to be connected, further increasing the in situ processing capabilities and throughput.

Figure 7a: Varian modular MBE system.

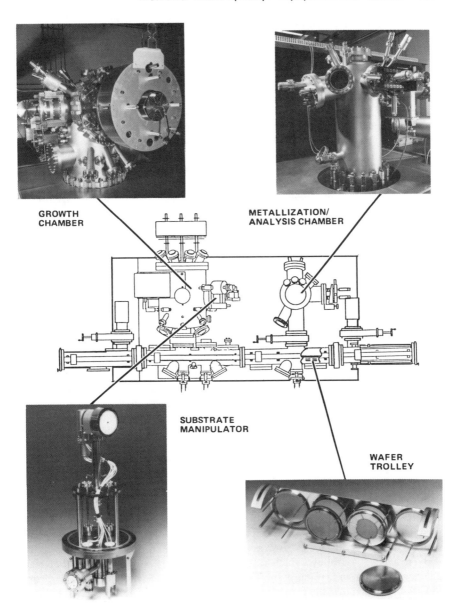

Figure 7b: Key components of Varian modular MBE system.

4.1.3 Pumping Considerations.

MBE systems rely on a family of pumps depending on the film growth materials and on the purpose of the chamber(s) pumped. A typical III-V MBE system will use sorption pumps for roughing because they are clean, simple (reliable) and easily regenerated. After roughing, the chamber(s) are usually transitioned to a sputter ion pump. These pumps are also relatively simple and reliable and capable of maintaining system base pressures in the 1×10^{-11} Torr range for a well-baked system. Titanium sublimation pumping is commonly used to assist the ion pump during initial pumpdown. Closed loop helium cryopumps frequently augment ion pumps on the growth chamber, where they pump gases such as hydrogen and helium somewhat more effectively. Cryopumps are mechanically more complex than ion pumps and therefore have reliability and cost considerations. The growth region is typically surrounded with liquid nitrogen-cooled cryopanels to minimize background gas contaminants. Turbomolecular pumps and trapped diffusion pumps are occasionally found on MBE systems intended for special applications. Such applications arise, for instance, when source elements with high vapor pressures (e.g., phosphorus and mercury) or gas sources (e.g., organometallics and hydrides) are used. These pumps have their drawbacks: turbopumps are less efficient with light element gases and diffusion pumps may backstream oil vapors if operated improperly. Cryopumps have also been used to pump these gases with varying degrees of success. Regardless of the pumping method, high vapor pressure materials such as phosphorus and mercury require a method of collection during bakeout.[99] This exhaust is frequently cryotrapped in a special container which can be removed without venting the system to the ambient. In addition, MBE systems using hydrides are frequently faced with large hydrogen loads and are most commonly evacuated with turbopumps or LN_2 trapped diffusion pumps.

4.1.4 Sample Transfer Techniques.

One challenge for any piece of UHV processing equipment is sample handling. Techniques commonly used at atmospheric pressure such as a vacuum pik and air track are obviously impossible. All existing methods of vacuum transfer are variations of "pick and place." The most common means for bridging the vacuum interface are mechanical feedthroughs and magnetic couplings. Mechanisms using extended metal bellows are undesirable for reliability and cost factors. Rotary motion feedthroughs based on the Scotch yoke mechanism are common. The rotary motion can also be transformed into linear motion through a sprocket and chain. Magnetically-coupled linear and rotary motion feedthroughs are equally common. The MBE system shown in Figure 7a uses magnetic coupling to operate "transfer rods" and "trolleys." A trolley, which can transfer up to 16 wafers at a time between chambers, is shown in Figure 7b. The transfer rods pick individual wafers from the trolley and move them to static stations or manipulators for growth or analysis. Wafer transfer mechanisms will evolve into another generation as automated production systems requiring minimal operator expertise and intervention are developed.

4.2 Sources

Sources are key elements of any MBE system. They must be designed to supply the needed uniformity and material purity. They can often be a source of background gas contaminants. The wide variety of source elements and compounds requires an equally broad assortment of sources.

4.2.1 Thermal Evaporation Sources.

4.2.1.1 Knudsen Cells. Knudsen or K-cells are the standard evaporative source for most MBE systems. A typical K-cell is shown in Figure 8. It consists of a heating element, surrounding heat shield and a crucible thermocouple assembly mounted on a port flange. Some cells include an integral water jacket in addition to heat shields. Construction materials are usually refractory metals such as tantalum and molybdenum and insulators like alumina and PBN. Lower temperature applications allow alumina, but due to the outgassing of contaminants, temperatures above a few hundred degrees usually mandate the use of PBN. The heating elements are wound either non-inductively or operated from DC power supplies to minimize stray B-field contributions that might interfere with analytical instrumentation. Source evaporants are contained inside crucibles which fit into the furnace assembly. Higher vapor pressure materials (e.g., arsenic) may use graphite or quartz crucibles, but PBN is more common particularly at higher temperatures.

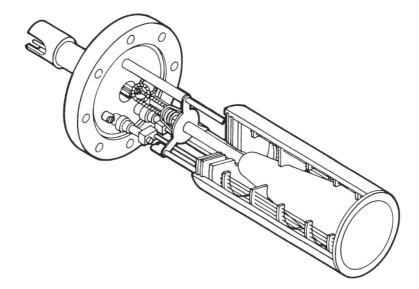

Figure 8: Large capacity K-cell (60 cm^3) for high uniformity across 3-inch wafers.

In an effort to improve the uniformity of deposition, several studies of K-cell effusion have been made.[100-103] The accepted geometry for molten sources (e.g., gallium, indium and aluminum) involves a tapered crucible that will allow the entire melt surface to expose the substrate as the source is depleted. Large diameter sources may not require a taper depending on melt surface-to-substrate separation and sample size. In any event, wafer rotation during growth is also necessary for maximum uniformity. Effusion flux is determined by K-cell temperature which is controlled by a tungsten-rhenium thermocouple feeding a 3-term (PID) temperature controller. The standard thermocouple pair is tungsten-rhenium at either 5% and 26% or 3% and 25% rhenium. The former is somewhat less brittle after repeated high temperature excursions. K-cell flux density is related to temperature as

(1) $$d^2N/dwdt \sim p(T)/(T)^{1/2}$$

In order to achieve gallium flux control within ± 1%, cell temperature must be held to better than ± 0.5°C. This is routinely obtained with existing PID controllers.[9]

It is often desirable to vary film composition in a controlled but rapid manner. Large magnitude, abrupt composition changes (e.g., heterojunctions and certain dopant profiles) can be produced by shuttering two or more K-cells of the appropriate flux. A graded layer of thickness greater than approximately 100 Å can be obtained by using the pulsed MBE technique (see Section 5.2.2) or by varying K-cell temperature. Thin layers or programmed doping profiles (e.g., exponential) require a source with fast thermal response. This can be achieved with a solid, Joule heated source (see 4.2.1.3) or with a modified K-cell. Construction of a fast response cell is similar to a standard cell except that the thermal mass is minimized. Since dopant fluxes are normally quite small, the major thermal mass is the source itself. The small dopant material required allows the cell size to be scaled down relative to the available power thereby decreasing the source response time. Figure 9 shows a typical fast-response dopant source.

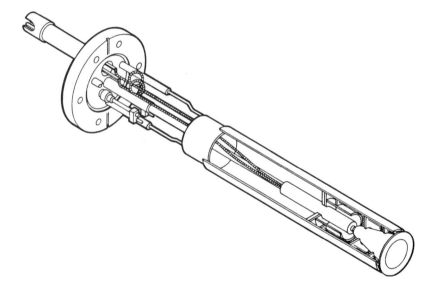

Figure 9: Dopant source designed for fast thermal response.

4.2.1.2 Solid Source Cracking Cells. Some materials, arsenic and phosphorus, for example, will evaporate in more than one molecular form. Often the larger molecule will have a higher vapor pressure and a lower sticking coefficient for a given substrate temperature than the smaller molecules. $As_4(P_4)$, in this example, has a maximum sticking coefficient of ½ while that for $As_2(P_2)$ is near 1. Consequently, the efficiency of source material utilization can be improved by cracking the tetramer into two dimers at the source. There is also evidence that the dimer incorporation results in fewer site defects and better material quality.[104,105] Source construction follows standard K-cell design with the addition of a secondary heat zone at the source exit.[106-108] See Figure 10. Sufficient thermal isolation between the source furnace and the cracking furnace is required to minimize any interaction. The source zone is temperature controlled via a thermocouple and PID controller just as in the K-cell. The cracking zone may not be feedback controlled. Typically, the caracking efficiency will reach a maxium and level off beyond a critical temperature. It is enough to supply constant power somewhat beyond the critical value. Cracking sources often include a large source capacity (about 200 cm^3) to extend the period between refills. Crucible material for the high vapor pressure sources are frequently pyrolytic graphite. Graphite is relatively inexpensive and easily machined and the outgassing of the graphite is acceptably low after vacuum firing. The cracking end of the crucible is generally baffled to improve the cracking efficiency. An efficiency of greater than 90% is typical for arsenic or phosphorus.

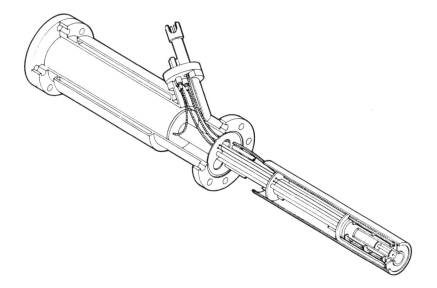

Figure 10: Solid source cracking cell.

4.2.1.3 Solid Sources Using Direct Heating. Joule heating of materials for evaporation is not a new technique, but it has been appropriately reconsidered by T.N. Jackson, et al[109] to achieve silicon doping. A segmented wafer serves as the source until each leg is open. The wafer is easily replaced when expended. The advantages accrue from the small thermal mass which allows for a more rapid variation of flux and greater heating efficiency. See Figure 11.

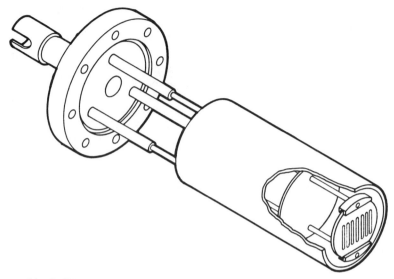

Figure 11: Solid silicon source using Joule heating (following the method of T.N. Jackson, et al.[109]).

4.2.1.4 Ionizing Sources. We discriminate between ionizing and implantation sources which are mentioned in Section 4.2.3. Naganuma and Takahashi[110] used electron impact ionization at the exit mouth of a K-cell to accelerate zinc atoms to energies near 1 keV. In this way, they increased the sticking coefficient sufficiently to incorporate this p-type dopant into GaAs. Gases have been similarly ionized. Takahashi[111] has introduced ionized hydrogen during the MOMBE growth of GaAs to control doping. Ionized hydrogen yielded p-type material, unionized n-type and no hydrogen yielded n+ GaAs. It was proposed that this technique be used to grow nipi superlattice structures.

4.2.1.5 Load-Locked Sources. One attempt to solve the problem created by finite source lifetime is the load-locked source. In its simplest form the source must be retracted behind a valve which is then sealed allowing the source to be removed and refilled. The difficulty is finding a valve which can maintain adequate growth chamber vacuum integrity during source bakeout and is suitably long-lived. Two valves with differential pumping is better, but then cost and reliability factors increase. Since most systems require some periodic maintenance, an accepted alternative is to employ more than one, large volume source of the high expenditure materials (e.g., arsenic). These sources are often capable of running for four to six months without refill allowing other issues to trigger system downtime.

4.2.1.6 Continuous Sources (Mercury). Sources which can be maintained in a liquid state (e.g., mercury) allow for continuous operation simply by adjusting the reservoir quantity and height.[112,113] See Figure 12. This can be a fortunate solution to the relatively short source life of high vapor pressure, low sticking coefficient materials. Unfortunately, the number of materials amenable to this technique is quite limited.

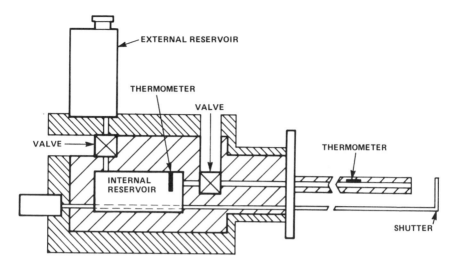

Figure 12: A continuous feed mercury source (K.A. Harris, et al.[113]).

4.2.2 Electron Beam Heated Sources. Materials with very high melting temperatures are often evaporated using electron beam heating. Such sources are common in silicon MBE. Commercially available sources typically use magnetic deflection to direct an intense electron beam into a water-cooled hearth. The evaporant charge is shaped to fit the hearth. Scanning allows the beam to sweep over the charge surface. This improves material usage and extends source life. Materials commonly evaporated in this manner are tungsten, cobalt, nickel, silicon, and germanium. Metal silicides deposited by coevaporation require two e-gun sources. The high temperatures involved often result in significant radiant heating of the substrate which may aggravate substrate temperature control. Flux control may be accomplished by a quartz crystal monitor or electron excited emission spectroscopy (i.e., Inficon Sentinel). Using optimized geometries, excellent uniformity over a rotating substrate can be obtained.[114,115]

4.2.3 Implantation Sources. Implantation sources are differentiated from simple ionizing sources (in this work) by the existence of mass separation. Implantation sources are considerably larger (by virtue of the bending magnets) and more complex than the ionizing sources of Section 4.2.1.4. Some of the earliest applications were made by Bean and Dingle[116] who increased the sticking coefficient of Zn in GaAs; and Ota[117] who implanted arsenic in silicon during MBE. Shimizu, et al[118,119] has used mass selection with low-energy (100-200 eV) group-V implantation to grow GaAs, InP and InGaAsP. De Jong, et al[120] have used 10 keV As implantation to make silicon modulation doping structures.

4.2.4 Gas Sources. Gas sources are not new to MBE. Arsine was employed by Calawa[121] to improve the material quality of gallium arsenide. Later, Chow and Chai[122] using phosphine also experimented with gas sources. Serious consideration of gas sources has accompanied the successes of Panish, et al[123] and Tsang.[124] These processes are discussed in Section 6.4. There are two broad approaches. In one case, the gas or gases are cracked at the source orifice. This is common for the hydrides: arsine and phosphine. In the other case, the gases are simply directed toward the heated substrate where pyrolysis occurs. Any heating of this latter source is only sufficient to prevent condensation at the source. Refractory materials or quartz are used in the source construction depending on the gas and temperature range of operation. Gas sources are amenable to flux distribution manifolds and are likely candidates for multiple wafer processing. In addition, there is the possibility that large area uniformity can be achieved without wafer rotation.

4.2.5 Source Shutters and the Source Flange. One of the main virtues of MBE originates from the fact that fast acting shutters coupled with a slow growth rate allow for monolayer film growth control. Shutter motion may be either the flipping or rotating of a refractory metal blade between the source exit orifice and the substrate. Shutter actuation times below 0.1 second are nominal. Refractory metal shutters, once outgassed, maintain relatively clean outer surfaces in the closed position since they are heated by the source radiation. Source temperature transients can occur when the radiation losses are changed upon shutter opening and closing. These transients have been minimized by source modification[125] and by positioning the shutter blade obliquely to the source axis. It is common practice to mount 6 to 10 sources and their shutter mechanisms on a large vacuum flange. The sources are usually surrounded by a cooled baffle filled with either a water-alcohol mixture or liquid nitrogen to thermally isolate the sources from one another and minimize the consequent outgassing. A photograph of a typical source flange assembly is shown in Figure 13.

4.3 Sample Manipulation

Proper sample (substrate) manipulation during growth is critical to producing high quality, uniform epitaxial layers. The substrate manipulator is responsible for holding, orienting, heating and rotating the sample (or wafer). A typical substrate manipulator is shown in Figure 7b.

4.3.1 Sample Mounting. Traditionally, MBE samples have been mounted to molybdenum holders using indium solder. At film growth temperatures, the indium is liquid and provides adequate attractive force and good thermal conduction. Bonding free of voids maintains a minimal temperature gradient between the block and sample. Initial sample mounting is often done on a hot plate inside a glove box. This technique is difficult with large size samples, especially when they are brittle III-V materials. Post-deposition processing difficulties caused by the roughened back surface are common. Precoating the sample back surface with silicon nitride has been one solution.[126] Samples not amenable to this technique have been mechanically held (e.g., refractory metal retaining clips). The success of mechanical holders has largely eliminated these concerns.[127-130] Direct, free substrate heating is preferred and capable of ± 5°C uniformity across an undoped 3-inch GaAs wafer at 750°C. Proper sample holding will produce minimal mechanical damage at temperatures over 850°C.

4.3.2 Sample Temperature Control. Sample temperature can be set and maintained by either thermocouple or pyrometer feedback.[131] Thermocouples are most commonly used because they are inexpensive and fairly reproducible. Although they do not provide accurate absolute sample temperature they allow adequate maintenance of sample temperature. Calibration is achieved using fixed points such as alloy eutectic temperatures and sometimes the oxide desorption temperature of gallium arsenide. Pyrometer emissivities are also set using these

Figure 13: A source flange showing shutters and chilled, surrounding baffle.

points. Two-color pyrometers can, in principle, avoid this calibration and accommodate changes in viewport transmittance. Sample emissivity may vary with material and doping level and large variations must be adjusted for individually. Heater power input is regulated by a PID controller as with source furnaces.

4.3.3 Sample Rotation Contol. Standard practice with large substrates (>2 inches diameter) is to rotate the wafer during film deposition. Rotational speeds of 5 to 20 RPM are typical for layers greater than about 100 Å and growth rates around one micron per hour. Very thin or single plane uniform layers require higher speeds possibly over 100 RPM. Most MBE systems are equipped with a sample manipulator that allows the substrate to be continuously oriented about two axes. Remote, small-angle positioning is provided to assist RHEED pattern adjustment.

4.4 System Automation

The complexities of MBE grown devices has made some amount of automation mandatory.[3,132,133] Extended superlattices, for example, would be virtually impossible without the accurate sequencing of source shutters. The reproducibility of structures run-to-run would likewise be fortuitous. MBE system automation revolves around film growth control and the monitoring of certain functions (i.e., chamber pressures). Material structures are typically controlled from a "recipe" program that sets the sample temperature and rotation speed, the source temperatures and shutter sequencing from a given material structure.[134] Pre-growth calibration provides the proper constants to relate thickness and doping level to shutter timing and source temperature. The control or monitoring of other system facilities can vary greatly depending on the user's needs. The system controller can range from a small microprocessor to a mini-mainframe. More sophisticated systems allow several databases to be established: wafer recipe and growth history; system pressure history and RGA signature; source flux versus temperature history; and RHEED oscillation profiles, for example. A typical system configuration is shown in Figure 14. Basic system software is usually supplied by the MBE equipment manufacturer, but specific applications may be written by the user.

4.5 Performance Parameters

Typical performance parameters for current generation MBE equipment are shown in Table 6. In addition to absolute performance specifications, there is a need for repeatability: the ability to maintain these specifications wafer-to-wafer and run-to-run over the system operating cycle. This is an especially important criteria for production systems. It is not uncommon for specifications to vary over this period, but they must remain within an allowed window. Normally, specifications with respect to thickness uniformity, growth originated defect levels, doping and composition uniformity are valid throughout the maximum specified uptime cycle.

5.0 PRINCIPLES OF OPERATION

A large number of films have been grown by MBE as listed in Table 7. This section covers the process portion of MBE growth, concentrating on GaAs, and Si and II-VI compounds to a lesser degree. The common analytical techniques to achieve and maintain high quality epitaxy are mentioned. A major concern of any MBE user is safety, of which the preliminaries will be included also in this section. For more specific process details, the readers may reference MBE reviews on III-V compounds,[4,8,167] Si,[168-171] II-VI compounds,[172-174] Zn-chalcogenides[175] and IV-VI compounds,[176] and specific papers on GaAs[177] and AlGaAs.[178]

Molecular Beam Epitaxy: Equipment and Practice   195

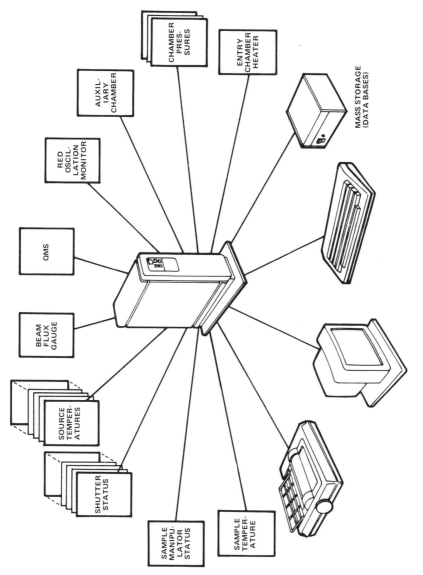

Figure 14: Functional diagram of computer controlled MBE system.

Table 6: Typical Performance Parameters

Thickness Uniformity[a] (3-inch dia. wafer)	<±2%
Composition Control[b] ($Al_xGa_{1-x}As$)	±1%
Doping Control[a,c] (3-inch dia. wafer)	<±2%
Defect Density[d] (Oval Defects)	<500/cm^2
Throughput[e]	75/week

[a] Saito and Shibatomi[135] at Fujitsu have obtained thickness and carrier concentration (doping uniformities of less than ±1% over a 3-inch wafer with optimized geometry.

[b] The composition of III-V binary compound semiconductors is largely controlled by surface reaction kinetics. Stoichiometry will be achieved given the minimum III/V flux ratio for a given growth temperature. Higher order alloys are more difficult to control in general because of incorporation competition between the group-V elements.

[c] Control of the absolute doping level depends largely on the accuracy of the monitoring method. Once a correlation between source temperature and dopant flux is established from Hall measurements, for example, repeatability is excellent due to the slow expenditure of source material.

[d] Dislocation density is excluded. Oval defects can be reduced by careful sample, system and source material preparation. Levels below 500/cm² are routinely achievable with K-cell gallium sources if particulates are controlled.

[e] Throughput has become more important as more MBE systems fill device production roles. Current single wafer MBE equipment is capable of producing five 3-inch wafers per eight hour shift for one micron of material. Assuming full operation five days per week will result in 75 wafers per week maximum. Few systems produce more than half this number per week. Actual throughput depends on many factors such as device structure, asset utilization and operator skill. Future production equipment will reduce the number of uncontrolled variables (see Section 7.1).

Table 7: MBE-Grown Materials (Film: Substrate)

III-V
    AlAs: GaAs
    AlGaAs: GaAs, Si
    AlGaSb: GaAs
    AlSb: GaSb
    GaAs: GaAs, Ge,[136] Si[137]
    GaAsSb: GaAs,[138] InP, InAs,[139] GaSb[139]
    GaP: Si, GaP
    GaSb: GaAs, GaSb[140]
    InAlAs: InP
    InAlP: InGaP
    InAs: GaAs, GaSb
    InAsSb: GaSb, InSb, GaAs
    InGaAs: InP, InAs, GaAs,[139] GaSb[139]
    InGaAlAs: InP[141]
    InGaAlP: GaAs[142]
    InGaAsP: InP
    InGaAsSb: GaSb, GaAs[143]
    InGaP: InAlP
    InGaSb: GaAs[144]

(continued)

Table 7: (continued)

    InP: InP
    InSb: GaAs

IV
  Si: Si
  Ge: Si
  GeSi: Si[145]

II-VI
  CdMnTe: CdTe, GaAs[148]
  CdS: InP
  CdTe: GaAs,[146] InP, InSb[147]
  CdZnS: GaAs
  CdZnTe: GaAs[146]
  HgCdTe: CdTe, GaAs,[146] InSb, ZnCdTe
  HgMnTe: GaAs[146]
  HgTe: CdTe
  HgZnTe: GaAs[146]
  ZnMnSe: ZnSe, GaAs[149]
  ZnS: GaP
  ZnSe: GaAs,[133,151,152] InP, Si[153]
  ZnSeTe: GaAs[150]
  ZnTe: InP

IV-VI
  PbEuSeTe: PbTe[68]
  PbSnSe: BaF_2, PbSe, CaF_2
  PbS: BaF_2, PbSe
  PbSe: BaF_2, PbSe
  PbTe: BaF_2
  PbYbSnTe: BaF_2,[154] PbTe[154]

Insulators
  BN: Si
  BaF_2: InP,[155] CdTe[155]
  CaF_2: Si,[156] GaAs,[157] InP[158]
  $CaSrF_2$: GaAs[159]
  LaF_3: Si[160]
  $SrBaF_2$: InAs,[161] InP[158]
  SrF_2: GaAs[157]

Metals
  Al: GaAs, InP[162]
  Ag: InP[166]
  Au: GaAs
  $CoSi_2$: Si[165]
  Fe: GaAs[164]
  Mo: GaAs[163]
  $NiSi_2$: Si[25]
  Sn: GaAs
  W: GaAs[163]

5.1 Substrate Preparation

The intent of substrate cleaning preparations is to create a suitable surface for epitaxial growth. The initial growth surface must be relatively free of contaminants and atomic imperfections. There are two reasons why such a surface is difficult to create for MBE samples. First, there is no easy manner to remove the top 3 or 4 atomic layers of the surface immediately prior to epitaxial growth, as is done in VPE . MBE cleaning preparations are combinations of chemical steps done in a low particulate, noncontaminating and atmospheric environment, with the final cleaning steps done in vacuum. Second, a clean MBE sample must be transferred from an atmospheric situation to the UHV growth chamber. During this transfer period, the clean reactive surface may pick up contaminants. Nevertheless, many cleaning preparations have been developed for substrates of a variety of materials which repeatedly yield device quality epitaxial layers. Naturally, the simplest surface cleaning preparations are tried before progressing to more involved and complex preparations (i.e., chemical cleaning to sputter cleaning to high temperature desorption).

5.1.1 III-V Substrate Cleaning. Many detailed investigations for optimizing cleaning preparations for MBE III-V substrates were done on GaAs[4,179] although variations of these preparations are applied successfully towards other III-V substrates such as: InP,[180] InAs,[181] InSb,[181,182] GaSb[140] and etc. A generic cleaning preparation consists of the following steps:

1. A degrease to remove residual waxes from the polishing step;

2. A chemical etch in concentrated acid, such as HCl or H_2SO_4, to remove other surface contaminants;

3. Immersion in a stagnant solution $H_2SO_4:H_2O_2:H_2O$ and Br: CH_3OCH_3 are just 2 examples) to etch back the surface of the substrate;

4. A rinse in deionized (18 MOhm) water to form a thin and protective oxide cap on the etched substrate; and

5. Thermal desorption of the oxide layer in UHV.

The fourth reoxidation step is critical. Munoz-Yague, et al[183] demonstrated that the proportions of $H_2SO_4:H_2O_2:H_2O$ can be optimized for the specific substrate orientation and dopant concentration. Massies, et al[184] proposes that for step 4, immersing the substrate in static deionized water or drying the substrate in an oxygen environment produces a more consistent oxide rather than just rinsing.

Carbon is the residual contaminant most difficult to remove and keep off a clean surface. Another cleaning preparation exposes the substrates to UV radiation and ozone as an alternative for the third and fourth cleaning steps.[180,185,186] Two types of lamps are used, one which emits frequencies for exciting and dissociating the residual C-contamination and the other which creates ozone and atomic oxygen. The atomic oxygen reacts with the excited molecules, producing more volatile molecular species such as CO_2. Additional exposure time to these lamps regrows an oxide layer.

5.1.2 Silicon Substrate Cleaning. A generic cleaning preparation of Si substrates for molecular beam homo- and hetero- epitaxy is less definitive. The literature contains a number of cleaning preparations which etch off and regrow the SiO_2 in various wet chemical solutions. The purpose of these investigations is to regrow an SiO_2 layer which thermally desorbs in situ at a low temperature. Lower temperatures are desirable because slip lines form during the high temperature cleaning of Si surfaces.[187] There has been significant progress since the original work[188-190] which report a thermal desorption temperature-time of 1200°C-1 minute. Recent work[191] reports thermal desorption temperature-times of about 750°C-45 minutes.

Alternative cleaning preparations of Si substrates include reactive beam desorption, Si-radiation and sputtering. Wright, et al[192] originally demonstrated reactive beam desorption by aiming a molecular beam of Ga at the SiO_2 layer. Two reactions occur at a 800°C substrate temperature which convert the SiO_2 layer into more volatile oxides of the metal and SiO. Wang[193] has since used Al and Yang, et al[194] has proposed the use of In as the reactive beam. In Si-radiation, a Si beam is used to etch the SiO_2 at 800°C.[195] Sputter-cleaning of Si substrates is not recommended because the 1200°C anneal typically introduces slip lines.[187]

5.1.3 II-VI Substrate Cleaning. In general, the sputter-cleaning and annealing the substrate has been successfully applied by those who grow the II-VI class of materials by MBE. The substrate preparation of II-VI homoepitaxial growths are described by Faurie, et al[196] for CdTe and Park for ZnSe.[197] In both cases, the substrates were first degreased, the CdTe substrate then went through a $BrCH_2OCH_3$ etch whereas the ZnSe substrate bypassed this etch, both were then sputter-cleaned and annealed in situ.

Hetero-epitaxy of Cd-Te on III-V substrates has also been demonstrated. To grow CdTe on InSb, the reported cleaning preparation is a hot isopropyl alcohol rinse followed by sputter-cleaning and annealing in situ.[198] In a TEM study of CdTe grown on InSb, thermal cleaning of the InSb substrate was compared to sputter-cleaning. The results showed thermally induced In precipitation at the interface.[182] The generic cleaning preparation was used on GaAs for hetero-epitaxy of CdTe on GaAs. However, the thermal desorption of the oxide does not take place in an As_4 beam.[199]

5.2 Growth Procedure

Prior to the initiation of the III-V growth, it is common practice to thermally outgas all the sources (except for Group V materials) at temperature greater than their eventual operating temperatures. After a period of time, the sources are reset to their operating temperature. Beam flux measurements are taken with either a beam flux monitor (ion gauge), quartz crystal monitor, or a residual gas analyzer when the group III sources equilibrate at the operating temperature. The beam flux measurement procedure using on ion gauge is as follows. Position an ion gauge at or near the substrate growth position. With all other shutters closed, open and close the shutter on the source in question, noting the pressure readings. The beam flux measurement is the difference between the pressure readings with the shutter opened and closed. The beam flux readings (beam equivalent pressure) indicate the arrival rate of the group III species, and thus are a measure of the growth rate at a given substrate temperature.

After the group V flux is set about 10-20 times greater than the group III flux, the thermal desorption of the passivating oxide may commence. Growth is initiated by exposing the clean, heated sample to the group III and V beams via opening of the shutters.

Dopants are incorporated similarly by shutter actuation of the dopant source. Calibration of the dopant concentration as a function of source temperature may be done empirically.

5.2.1 Thermal Transient. MBE K-cell assemblies consist of a crucible of charge and an externally controlled shutter closing over the mouth of the crucible. Initially, thermal equilibrium is obtained prior to growth with the shutter closed. However, the initiation of growth by opening the shutter introduces a temperature perturbation onto the charge because now, the effusion furnace radiates energy out an open instead of a closed end. Typically, the exposed surface of the charge regains thermal equilibrium within a minute. This shutter-induced thermal transient affects directly the amount of material flux from the charge. In superlattice structures, where some layers require less than one minute of deposition time, the thermal transients are obstacles to growing a constant composition of the group III metals within a single superlattice period.

Thermal transients have been reduced by a variety of methods. One simple method is to increase the separation distance between the mouth of crucible and the shutter, or to vary the angle of the shutter with respect to the plane of the crucible opening. Another method is to increase the separation distance between the top surface of the charge and the shutter by not charging the crucible fully.[200] A second concentric crucible is then used to shape the beam for uniformity and to retain some of the energy. The third method employs a computer which compensates for the thermal transients with prespecified amounts of power to the effusion cell.

5.2.2 Doping Control. A number of dopant sources have been used for MBE of GaAs. C.E.C. Wood[201] has reviewed these dopants, and describes the incorporation mechanisms as currently understood. The dopants are thermally evaporated or sublimated from an effusion cell, unless otherwise noted. Briefly, Si, Ge, Sn, Te,[202,203] Se[204] and SnTe[205] are n-type dopants for GaAs, where Si is the most common. P-type dopants for GaAs are Be,[206] Mg,[207] Mn,[208] implanted Mg^+,[209] implanted Mn^+,[210] and implanted Zn^+,[211] where Be is the most common. Gas sources containing Sn^{212} and Si^{213} molecules were also studied.

In MBE of Si, the group V species which are used as n-type dopants are As and Sb. The technique to dope Si with As is by low energy (0.5 to 1.0 keV) ion implantation.[117] Evaporation,[214,215] low energy ion implantation[216] and secondary implantation with Si^{217} have been used to dope Si by Sb. In addition, low energy ion implantation is used to obtain large area dopant uniformity. Increased Sb doping control and concentration level has been demonstrated by negatively biasing the Si substrate during growth,[218,219] a technique named "potential enhanced doping."

The group III species used for acceptors in Si-MBE are In, Ga, Al and B. The former three dopants are thermally evaporated from an effusion cell.[188,220] Boron may be deposited during growth by low energy ion implantation,[221] evaporated from a B_2O_3 source or sublimated from a saturated B-doped Si charge.[222]

The doping type and concentrations of MBE-HgCdTe compounds is critically related to the growth conditions. For example, a 6% temperature change or an increase in the Hg flux by a factor of 2 converts the doping from n-type to p-type.[196] HgCdTe may be conventionally doped n-type with In from an effusion cell.[223]

A study of various doping profiles predicted that a rectangular-step doping profile results in an FET with linear I-V characteristics when the width becomes infinitesimally small.[224] MBE offers the capability of incorporating various doping profiles, and was suited for this particularly stringent requirement. Planar doping was demonstrated by interrupting the MBE growth briefly and depositing Ge on GaAs at the appropriate time in the growth sequence.[225,226]

Malik, et al[227,228] applied planar doping techniques to create new types of rectifying diodes. A p^+ planar-doped layer was grown between two n^+ regions, separated by insulating layers of thickness L_1 and L_2. The barrier height, V_{BO} at zero bias is calculated from:

$$(2) \qquad V_{BO} = [(L_1 + L_2)/(L_1 L_2)] [eNx/E]$$

where eNx is the charge density of the planar doped region and E is the permitivity. The barrier height may be adjusted from zero to the semiconductor band-gap. The asymmetry in the I-V characteristics may be independently varied by the position of the thin P^+ layer between the n^+ layers.

Subharmonic mixer diodes with symmetric and asymmetric I-V characteristics were grown and characterized.[229,230] Other device structures which used planar doping techniques are a 3-terminal switch with a variable threshold voltage independent of bias voltage[231] and an HEMT for reduced persistent photoconductivity.[232]

5.2.3 Compositional Control. Compositional control in MBE is a function of flux ratios, substrate temperature and chemical species, since all these factors affect the surface incorporation kinetics on a substrate. In many cases, the fluxes are controlled via the source temperature to grow the desired composition. Beam flux monitoring prior to growth establishes reference points for future growths. In III-V alloy growths, the group-III species have sticking coefficients of one at substrate temperatures below 600°C. The beam flux of the group-III species, corrected for instrument-induced errors such as ionization efficiencies, are proportional to the desired atomic composition.

Stoichiometric control of x in $III_x III_{1-x} V$ alloys results in control of the band-gap. For instance, choosing an x between zero to one in $Al_x Ga_{1-x} As$, a material with a room temperature band-gap from 2.16 eV to 1.35 eV may be grown. Other examples are given in Figure 15 where specific compositions may be grown lattice-matched to suitable substrates.[233]

A heterojunction transistor was proposed which takes advantage of a band-gap induced electric field.[234] A phototransistor based on this principle was grown by MBE with a graded band-gap structure.[235] Reviews of band-gap engineering and graded band-gap devices have been written.[236,237]

Band-gap grading may also be applied advantageously for contact layers. An $In_x Ga_{1-x} As$ layer was grown on an n-type GaAs from $0 < x < 1$, and then the final InAs surface was metallized, creating ohmic contacts.[238]

When multilayered structures of various III-V alloy compositions are needed, a set of sources are dedicated to a particular layer composition. However, the fixed number of source ports limits the number of compositions which may be grown in a multilayered structure. This is especially true for thin layers, where the

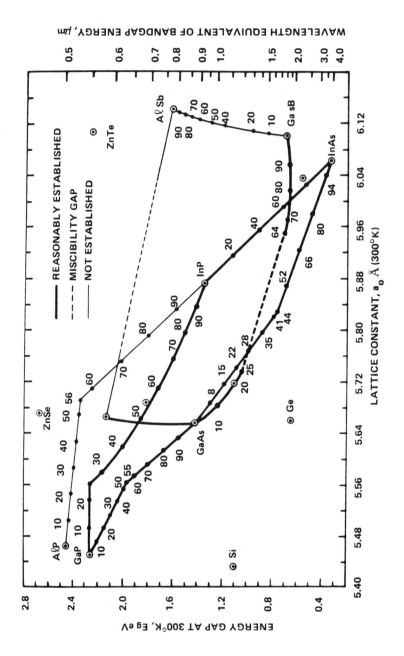

Figure 15: Energy gap versus lattice constant for some III-V materials at 300°K (R. Sahai, et al.[233]).

short growth times do not allow a set of sources to reach thermal equilibrium for the succeeding layer composition.

A novel solution was proposed called "pulsed molecular beam epitaxy."[239] The group III beams are mechanically chopped such that the beam arrives at the substrate in pulses. The number of pulses are coordinated, say between Al and Ga, summing up atomically to the correct composition in one monolayer. The beam chopping time periods are chosen to be less than the time to grow one monolayer. Multilayered structures of AlGaAs[240] and InGaAlAs[241] have been grown successfully.

5.2.4 Interrupted Growth. High quality heterojunctions were attained when the growth is interrupted prior to the succeeding epitaxial layer, allowing time for the growth front to smooth over. In contrast, traditional heterojunction structures were grown in a continuous mode to avoid excessive impurity incorporation. The growth interruption technique utilizes RHEED to monitor the surface smoothness. During continuous growth, the RHEED intensity of a given spot oscillates with the period of a monolayer growth time.[242-244] The RHEED intensity maximum occurs when the surface growth front is smoothest. These RHEED intensity oscillations were also correlated to an MBE growth mechanism on GaAs[245] via computer simulations.[246] The growth conditions for the heterojunctions were optimized through computer-simulated results. To date, GaAs/AlGaAs[247,248] and GaAs/InAs[249] interfaces have been studied.

5.2.5 In Situ Metallization. Although other aspects of MBE metallization are of interest to researchers, we limit ourselves to metallization with respect to device structures in this treatment.

The benefits of in situ metallization accrue from a clean, well ordered interface. This produces reproducible Schottky barriers and low resistance nonalloyed contacts. Reduced contact resistance is critical to low noise microwave performance. Stable contacts are especially important for InP devices where the barrier height is inherently low resulting in higher leakage currents. Di Lorenzo, et al[250] were the first to demonstrate in situ nonalloyed ohmic subsequent in situ AuGe/AgAu top metallization. Reproducible low resistance contacts resulted. Cho, et al[251] have been able to reduce the noise temperature of microwave mixer diodes by growing epitaxial Al Schottky barriers. Ohno, et al[252] were able to produce an Al gate InGaAs MESFET by depositing epitaxial Al over 600 Å of $Al_{.48}$, $In_{.52}$, As lattice-matched to $Ga_{.47}$, $In_{.53}$ As. This created an 0.8 eV gate barrier height reducing the gate-to-drain leakage. Finally, McLean, et al[253] have successfully grown in situ epitaxial Al on GaAsSb. The ability of MBE to grow low resistance ohmic contacts and reproducible Schottky barriers is an important element of IC development.

MBE metallization has also contributed to the successful demonstration of new metal base devices. Cobalt and nickel silicide metallic layers have been grown epitaxially between Si allowing the first practical application of an epitaxial metal-base transistor (Rosencher, et al).[254] Similarly, Rosencher, et al[255] have demonstrated the first permeable-base transistor (PBT) with an epitaxial $CoSi_2$ grating. Derkitus, et al[256] have observed transistor action from W-GaAs metal gate transistors following the work of Harbison, et al.[257] The 80-150 Å thick W was grown nonepitaxially on the single crystal GaAs. These devices are distinct from PBTs because the pore size is smaller than the depletion length.

5.3 In Situ Analysis

Growing epitaxial films in UHV is advantageous because a number of analytical techniques may be used in situ which yield information on crystal structure, chemical bonding and composition. The physical mechanisms of reflection high energy electron diffraction (RHEED), x-ray photoelectron spectroscopy (XPS), Auger electron spectroscopy (AES) and secondary ion mass spectroscopy SIMS will be described briefly. A description of residual gas analysis (RGA) will also be included in this section.

5.3.1 Reflection High Energy Electron Diffraction.

In reflection high energy electron diffraction (RHEED), a high energy beam of electrons aimed at a glancing angle, with respect to the sample, reflects off the surface-most layers of atoms. The reflected beam is scattered and impinges onto a visual monitor (fluorescent screen or camera). If the top surface is periodic with a period greater than the wavelength of the beam (i.e., lattice spacing is greater than the wavelength of the electron beam) diffraction occurs. The well-known Bragg law, $\lambda = 2d \sin \theta$, relates the wavelength, λ, surface periodicity, d, and diffraction angle, θ, for achieving constructive interference (diffraction). The electron energy is related to the wavelength by the de Broglie equation: $\lambda = h/p = 12.25/(V^{1/2})$ Å, where V is in eV.

RHEED patterns displayed on the visual monitor range from diffuse rings to spots to streaks. A diffuse ring usually indicates the presence of an amorphous layer, i.e., an oxide on the surface. A pattern of spots indicates thermal faceting, pitting, and/or a rough surface. When streaks appear, the general interpretation is the surface is smooth and of good crystalline perfection. The intensities and position of the streaks also yield crystal structure. As the growth proceeds on the substrate, the top surface becomes smoother and the RHEED streaks narrow. A more detailed discussion on the application of RHEED to MBE is given by Cho, et al[4] and Ploog, et al.[258]

5.3.2 X-ray Photoelectron Spectroscopy.

X-ray photoelectron spectroscopy (XPS) is also known as electron spectroscopy for chemical analysis, ESCA. A review of this surface analytical technique has been described by Riggs, et al.[259] Briefly, monoenergetic x-ray photons with energy $h\nu$ are aimed towards a sample. The spatial resolution is coarse with respect to AES because of the difficulty in focusing x-rays. The atoms within the first 25 Å absorb these photons and emit excited electrons with a characteristic binding energy, E_b. An electron energy spectrometer is used to measure the kinetic energy, KE, of the emitted electrons. The energies are related by $E_b = h\nu - KE + WF$, where WF is the spectrometer work function. The XPS sensitivities are comparable to that of AES, and XPS identifies all elements except for H. In addition to obtaining elemental information, an atom residing in a molecule has forces acting upon its electron density. The changes in electron density induce a shift in the binding energy, enabling the determination of chemical information. Classically, XPS utilized these chemically induced shifts in the binding energy to observe various oxidation states of metals.[260] In MBE, XPS was similarly used to evaluate cleaning procedures on GaAs and InP substrates.[261]

5.3.3 Auger Electron And Secondary Ion Mass Spectroscopy.

Auger electron spectroscopy (AES) is a quick technique for determining thin film composition. Sensitivities range from 0.05 to 5.0 atomic % across the Periodic Table.[262] The sampling depths of 5-20 Å may be combined with sputtering to obtain composition depth profiles. Spatial resolution of 0.03 micron to 30 microns are obtained. In MBE, the common application is to determine residual amounts of oxygen and carbon after the last in situ surface cleaning step.

The three-step sequence of energy transitions leading to the emission of an Auger electron initiate with an incoming electron beam knocking out a core electron, say in the K-shell for example. The resultant vacancy is filled by an electron from a higher orbited shell (L_1 or other higher orbital shells). The electron transition from the L_1 orbit releases an amount of energy ($E_K - E_{L_1}$) to another electron in the L_2 shell. If the energy is sufficiently high, the L_2 electron is emitted as an Auger electron. The three electron process results in measured energies characteristic of each particular element. The Auger electron energy, E, is related to the transition according to $E = E_K - E_{L_1} - E_{L_2} + \phi$, where ϕ is the work function of the spectrometer. Hydrogen and helium are obviously not detectable by the Auger process. A general review article on AES has been written[263] and one specific to MBE by Ploog, et al.[258]

Secondary ion mass spectroscopy (SIMS) can be more sensitive than AES by 2 to 3 orders of magnitude. SIMS is therefore applicable for determining contamination and dopant levels.[258] Sputtering may also be used with SIMS to determine concentration depth profiles of specific elements. Otherwise, the top 10 Å of the surface is sampled.

The physical mechanism of SIMS involves a primary ion beam (typically O^+ or Cs^+) to sputter a sample, generating a flux of secondary neutral and ionized species. The secondary ions are accelerated through as mass spectrometer and counted. A general review on SIMS is given by McHugh.[264]

5.3.4 Residual Gas Analysis. Residual gas analysis (RGA) is performed by mass spectrometry. Although RGA does not directly analyze the epitaxial layers, it does indicate the type of epitaxial environment and the composition of the beam flux. The RGA spectra can show the presence of air, water, or nitrogen leaks, and the presence of oxide in the source flux. Quantifying the RGA signals into partial pressures may be done with calibrated leak rates, or by accounting for the effects of ionization efficiency, filter transmission efficiency, and electron multiplier gain for various gases.[265] However, the hot filament of the RGA ionizer may alter the concentration of various molecular species and give misleading concentrations.

5.4 Materials Evaluation

The materials characterization techniques for MBE films are those commonly used for high purity semiconductor films. These techniques help in maintaining the film quality and identifying impurities. The control of impurities, desired and unwanted, is of utmost importance for developing new materials with unique properties. The techniques which will be described below are optical microscopy, Hall effect, capacitance-voltage, photoluminescence and deep level transient spectroscopy.

5.4.1 Optical Microscopy. At times, impurities or crystal defects may cause microscopic defects, such as oval defects. A Normarski (differential interference contrast option) optical microscope is an excellent tool for the initial and rapid evaluation of these defects. Also, optical microscopy gives morphological information which may be related to various growth conditions. There are chemical solutions for etching GaAs[266,267] which selectively etch various types of crystalline defects.

5.4.2 Hall Effect.

The average transport properties of semiconductors are determined using the Hall effect. The measured parameters of Hall coefficient, R_H, and Hall resistivity, ρ, are related to the free carrier concentration, n, by:

(3) $$n = r_H/eR_H,$$

where r_H is the Hall coefficient scattering factor and assumed to be unity, and to the mobility by:

(4) $$\mu = R_H/\rho$$

For n-type GaAs, the concentration of acceptors, N_a, and donors, N_d, may be derived from solving:

(5) $$n_{77} = N_d - N_a$$

and

(6) $$N_d + N_a = f(\mu_{77})[\ln_e(6.94 \times 10^{17}/n_{77}) - 1]$$

where μ_{77} and n_{77} are the mobility and free carrier concentrations at 77°K, and $f(\mu_{77})$ is from Figure 9 of Stillman, et al.[268]

The Hall effect measurements were simplified by van der Pauw. By keeping the contacts small and at the circumference of the sample, and the sample thickness uniform and without geometric holes, a sample of arbitrary shape may be used for the Hall effect measurement.[269] With only four sample contacts, ρ may be determined from a single resistance measurement and R_H from a magnetically-induced change in resistivity.[270] A review which discusses the Hall effect has been written by Blood, et al.[271] In addition to measuring transport properties in single epitaxial layers, high mobility structures such as HEMTs are also characterized by the Hall effect.

5.4.3 Capacitance-Voltage.

From the capacitance-voltage (C-V) method, a dopant profile may be obtained. The method requires a Schottky barrier to be made by a small area, A, contact on the sample. A reverse bias voltage, V, is applied to the barrier and the capacitance is measured. The depletion depth X_d is derived from:

(7) $$X_d = \epsilon A/C$$

where ϵ is the dielectric constant. The dopant concentration, n, at X_d is derived from:

(8) $$n(X_d) = \frac{C_3}{\epsilon e A^2}(dC/dy)^{-1}$$

where e is the electronic charge. However, the maximum depth is limited by electrical breakdown at the surface of the sample. For n-type GaAs, X_d equals 20 microns and 0.02 micron for dopant concentrations of 1×10^{15} cm^{-3} and 1×10^{18} cm^{-3}, respectively. Also, there are resolution trade-offs between X_d and n.[272]

Because of the problems mentioned above, a C-V technique was developed to contact and etch the sample in an electrochemical cell.[273] The etch depth, X_e, is determined by the amount of material removed according to Faraday's law of electrolysis:

$$(9) \qquad X_e = \frac{M}{ZFDA} \int Idt$$

where M is the molecular weight of the sample, D is the density of the sample, Z is the atomic charge transferred and equals 6 for GaAs, F is Faraday's constant, A is the dissolution area, I is the current and t is the time. The actual depth is the sum of X_d and X_e. In addition to depth profiling the sample, the electrochemical C-V technique may be expanded with a laser source to study the photovoltage characteristics of the sample. Concentration profiles of AlGaAs/-GaAs layers and the Al alloy content may be determined by this technique.[272]

5.4.4 Photoluminescence Spectroscopy. Photoluminescence spectroscopy (PL) is a nondestructive optical technique, ideally suited for evaluating high purity semiconductors because electrical contacts are not required for these high resistivity materials. In PL, a laser beam with a photon energy greater than the band-gap energy of the sample excites electron-hole pairs (excitons) within the sample. The optical emission from the resultant recombination of these excitons with shallow defect centers is scanned with a monochromator and detected by a photomultiplier. The samples may be cooled to 4.2°K for high resolution (0.1 meV) studies.

PL is used to determine molar concentrations in ternary III-V compounds because of the near band-gap energy information. Whereas Hall and C-V techniques give impurity transport and distribution information, PL has been used to empirically identify certain shallow acceptors in GaAs[274] and InP.[275] Concentrations of donors in the 1×10^{15} cm^{-3} and acceptors in the 1×10^{14} cm^{-3} range are resolvable by PL.[274] A section on PL, in the study of GaAs and InP, is included in a review article on high purity III-V semiconductor materials.[276] A review on high resolution PL was written by Reynolds, et al[277] on MBE grown multi-quantum well structures.

5.4.5 Deep Level Transient Spectroscopy. Deep level transient spectroscopy (DLTS) is a fast and sensitive transient capacitance technique.[278] A Schottky barrier is prepared on the sample. At a given temperature, a pulse generator applies a bias voltage onto the barrier, injecting carriers across the barrier and filling deep level traps. The capacitance, C, of the barrier is monitored as the barrier returns to thermal equilibrium and the traps empty. The deep trap concentration, N_T, is determined by:

$$(10) \qquad \Delta C/C = N_T/2N_B$$

where N_B is the background doping concentration. A typical value of $\Delta C/C$ is about 1×10^{-6} and of N_B is about 1×10^{15} cm^{-3}, giving a resolvable N_T of 1×10^9 cm^{-3}.[262] Judicious selections of the time window over which ΔC is measured facilitates the emission rate data interpretation of a given trap. The emission rate, e, varies exponentially with the inverse temperature:

$$(11) \qquad e = K \exp(-\Delta E/T)$$

208 Thin-Film Deposition Processes and Techniques

where the preexponential factor K is assumed to be independent of temperature and ΔE is the enthalpic activation energy. By varying the pulse height, the deep trap concentration distribution may be profiled. By varying the pulse duration, the capture rates of the traps may be derived. Presently, there are commercial DLTS units available.[279-282]

In the initial DLTS study of MBE GaAs, nine traps were identified and labeled MO through M8.[283] Since then, the four traps, M1 through M4, have been identified as related to MBE of GaAs epitaxy and are possibly impurity-defect complexes.[284] In the same study, Fe and Cu induced deep traps were identified in MBE of GaAs. Another trap, labelled M00, was recently found.[285]

The deep level traps found in MBE grown AlGaAs bear little correlation to those of GaAs.[286] The major trap is the one with $\Delta E = 0.78$ eV, which had strong concentration dependence on the substrate growth temperature[287] and is also identified in VPE material.[286] In the latter study, four of the six traps found in AlGaAs had a concentration dependence on the Al alloy composition, indicating that the deep traps are some sort of Al-contributing impurity. Another study was done on deep levels in MBE grown AlGaAs as a function of growth parameters, Si doping concentration and Al alloy composition from $0 < x < 1$. Two major traps were identified as ME6, prominent at $x < 0.2$, and ME3, prominent at $x < 0.2$. The authors proposed that the ME6 trap is created by a donor-vacancy complex (DX) center responsible for persistent photoconductivity.[288] The ME3 trap is thought to originate from a group-III vacancy or an antisite defect.[289]

5.5 Safety

In the MBE process, toxic materials are used as the source evaporants and also created during growth. Specifically in GaAs growths, inorganic As and the Be dopant material are known to be carcinogenic for humans. Toxicity levels of these and other materials may be found in Sax.[290] A study of GaAs toxicity on rats was published in 1984.[291] The data suggest that GaAs dust should be considered as a source of As exposure. MBE operators and service personnel should consult industrial hygienists for the appropriate respirators, protective clothing and handling techniques for As and GaAs. The U.S. Department of HEW[292] has a recommended standard for handling As in an industrial environment.

Air, surface wipes and biological monitoring should be done to insure that the As levels are below OSHA recommendations.[290] All the monitoring should at least be performed in conjunction with potentially hazardous operations, such as recharging of the As source at the source flange. Air and wipe samples may be done more frequently, for example, during sample loading at the load-lock area. Biological monitoring is typically performed with an urine analysis for levels of As and its metabolized compounds. The urine analysis becomes more complex if the personnel had eaten shellfish within 48 hours of the test.[293]

6.0 RECENT ADVANCES

MBE is an incredibly fertile process. Its evolution into materials engineering and advanced device production is exceptional. Some current areas of research, relevant to the future viability of MBE, deserve mentioning.

6.1 RHEED Oscillation Control

Although intensity oscillations in RHEED patterns were recorded over six years ago by C.E.C. Wood their nature was not well understood at that time and consequently their potential use was not perceived. The first published explanation is credited to Harris, et al[294] who observed pattern oscillations when GaAs growth was resumed on a tin rich surface and to Wood.[295] Since then RHEED oscillations have been observed with AlGaAs and Ge[296] and InGaAs[297] on GaAs substrates and recently RHEED oscillations for silicon on silicon[298,299] have been reported. Two approaches to measuring intensity variations are shown in Figure 16a. Examination of the specular beam in an off-Bragg condition reveals an increased sensitivity to the step density (and terrace width). A minimum in intensity corresponds to a maximum density of surface steps. The oscillations damp with time at a rate depending on the initial surface conditions and the final step density. One period of oscillation corresponds to the growth of one monolayer (Ga + As). An illustration of specular beam intensity versus time when the gallium beam is exposed is shown in Figure 16b. Since the gallium flux determines the growth rate under As-stabilized growth, the period of oscillation allows the gallium flux to be determined. Growth under Ga-stabilized conditions allows the arsenic incorporation rate to be measured and thus the As/Ga ratio.[300] RHEED oscillations can also be used to determine the Al mol fraction in the growth of AlGaAs. Thus, the technique can be used to calibrate source fluxes without resorting to the beam flux gauge thereby avoiding difficulties associated with determining the gauge sensitivity.

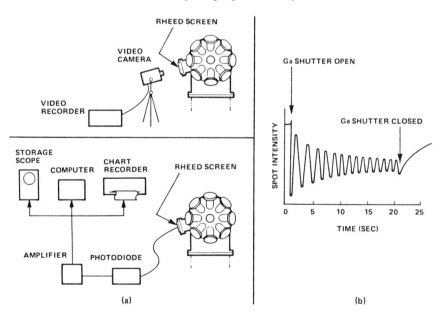

Figure 16: (a) Two possible arrangements for analyzing RHEED oscillations; (b) Typical variation in spot intensity with gallium shutter opened.

6.2 GaAs on Silicon

There are several potential advantages to growing GaAs on silicon. For example, the speed and optical properties of GaAs can be combined with the superior thermal conductivity of silicon. In addition, one can imagine silicon logic circuits with GaAs optoisolators for interconnect simplification and noise immunity. Recent results have fueled enthusiasm for the monolithic integration of GaAs and silicon devices. Initial attempts to grow GaAs devices on Si involved the use of a thin nucleating layer of germanium.[301,302] Wang[303] was the first to show that the antiphase disorder could be suppressed without using a Ge buffer. Despite the relatively large lattice mismatch (~4%) and the probability of antiphase domains considerable success has been achieved growing directly on (100) silicon substrates cut between 3 and 4 degrees off-axis toward (110). GaAs MESFETs,[304] MODFETs,[305] lasers[306,307] and HBTs[38] have been successfully grown on silicon. Although there is considerable debate over the mechanism which allows the strain accommodation, many researchers are reporting positive results. Superlattice buffer layers are often employed to accommodate the mismatch and retard the outdiffusion of impurities and dislocation threading[308] from the substrate. A judicious scheduling of processing steps has allowed the production of silicon MOSFETs and GaAs MESFETs on the same wafer.[309] VLSI circuits with central HEMT logic elements and peripheral silicon I/O drivers are conceivable. If viable and economically competitive, GaAs on silicon will present immediate problems to MBE equipment vendors who will be required to supply very high uniformity over 4 inch and possibly 6 inch wafers. Besides the challenge this presents for conventional K-cell sources, there are also difficulties associated with heating and handling such wafer sizes in vacuum.

6.3 Oval Defect Reduction

Morphological defects have been a serious concern since device-quality, low defect density films were sought. Surface defects can degrade the material electronic properties and present lithographic problems. III-V surface defects have many different origins; the substrate itself, pregrowth surface particulates, the film growth process, and postgrowth handling.[310] Careful substrate selection, preparation and handling have reduced defects to the point where those due to the growth itself are dominant.[311-315] The so-called "oval defect" has emerged as the principal defect contributor. A typical oval defect is shown in Figure 17. Oval defects vary from 1 to 20 microns in length and are characterized by a major axis aligned along a <110> direction. Each defect appears to be bounded by (111) stacking faults that originate on a gallium-rich inclusion. Oval defect densities between 10^3 and 10^4 are not uncommon. Recent studies of oval defects have revealed two major sources: substrate surface particulates[316] and the film growth process. Careful substrate cleaning[317] and improved equipment design can greatly reduce the particulate contribution. It is the latter cause that has now gained the most attention. Oval defects originating during growth increase in density with film thickness and have been attributed to the presence of gallium oxides[318-320] and the spitting of gallium droplets from the source K-cell.[311,321] Neither mechanism can be differentiated on the basis of the observed increase in defect density with gallium source temperature. Various approaches that reduce the gallium oxides in the system have been shown to reduce the density of oval defects.[322,323] Defect densities down to 100 ±50 cm^{-2} have been claimed using conventional K-cell sources. Tsang[324] has reported no oval defects using gas sources. The successful production of large area integrated circuits will require densities below 10 to 50 cm^{-2} depending on the mean defect size.

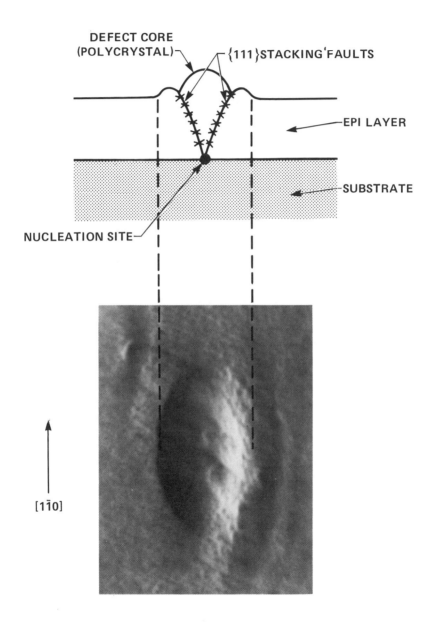

Figure 17: Typical oval defect and cross sectional drawing.

6.4 Chemical Beam Epitaxy/Gas Source MBE

There is growing interest in gas sources within the MBE community, especially among users contemplating production applications. Several approaches are under evaluation: mixed gas and conventional sources and all-gas processes using hydride and metalorganic sources or only organometallic sources. The high toxicity of the group-V hydrides is an issue of concern and the development of reduced toxicity alternatives is mentioned in Section 7.4. Some researchers have combined group-III alkyls with conventional and cracked elemental group-V sources.[325,326,336] Gas sources have also been used for other reasons. An improvement in GaAs material quality has been found by Calawa[327] and Pao, et al[328] following the introduction of hydrogen during growth. Tokumitsu, et al[329] have added ionized hydrogen to trimethylgallium (TMG) and As_4 to reduce the carrier concentration by two orders of magnitude. Also, Briones, et al[330] have tried to use silane as a dopant source with limited success.

6.4.1 Hydride MBE. The introduction of gaseous sources into MBE systems dates back to Morris and Fukui[331] who used arsine and phosphine to grow GaAs, GaP and GaAsP. Calawa[121] demonstrated material improvements with the substitution of cracked arsine for solid arsenic as the group V source. Chow and Chai[332] reported on the cracking efficiency of phosphine under similar conditions. In each case, the group V hydride was passed through a heated source section where it was decomposed according to the reaction:

(12) $$4GH_3 \rightarrow G_4 + 6H_2$$

Further heating results in the cracking:

(13) $$G_4 \rightarrow 2G_2$$

Within a given temperature range G and G_4 are minor contributors to the total gas pressure when compared to G_2 and hydrogen. Significant progress has been made by Panish[333] who used "high" pressure (200-2000 Torr) sources to grow InGaAsP. Several types of laser heterostructures were successfully grown. More recently, Temkin, et al[334] and Vandenberg, et al[335] have grown GaInAs(P)/InP quantum well structures using a single high pressure hydride source. The group-V ratio was set by varying the individual hydride pressures. Panish refers to this technique as Gas Source MBE or GSMBE.

6.4.2 Metalorganic MBE. Another related approach has been taken by Tsang[124] at AT&T Labs. Tsang explored an all-alkyl gas process using TMG or triethylgallium (TEG) and trimethylarsine (TMAs) to grow GaAs, InP and InGaAs. The results of material quality analysis were disappointing probably due to the varying purity of the group-V alkyls. Better results were obtained using arsine in place of TMAs. Tsang refers to this process as Chemical Beam Epitaxy or CBE. Combining TEG and As_4 has yielded some of the best MOMBE grown GaAs yet seen. Kondo, et al[336] obtained a background impurity concentration of $8 \times 10^{14}/cm^3$ at room temperature. AlGaAs has been grown by the addition of triethylaluminum (TEAl). Tsang and Miller[337] report material quality comparable to the best MBE and MOCVD. GaAs/AlGaAs single and multi-quantum well structures were grown with superior quality PL line shape. Further, Kawaguchi, et al[338,339] have produced the highest quality MBE InP to date using TEIn and phosphine. A 77°K mobility of 105,000 cm^2/Vsec at a carrier concentration of 9×10^{13}/cm^3 was achieved. This is comparable to the best LPE and MOCVD material. Lastly, Tsang, et al[340] using TEG, TMIn and arsine have grown InGaAs on InP with the highest quality epilayers of any technique.

Gas source MBE promises to solve two of the biggest roadblocks to production MBE: throughput and defect density. Gas sources offer the obvious advantage of being essentially inexhaustible thus reducing the downtime associated with conventional source replenishment. Also, there should be advantages in terms of large (<3-inch) diameter wafer uniformity that accrue from gas distribution designs. Although this has not yet been demonstrated, excellent compositional uniformity has been achieved without wafer rotation using gas sources.[341] Further, it appears that the growth rate can be increased from 1 micron per hour to somewhere between 3 and 6 microns per hour and possibly higher depending on the material system in use. Finally, the growth contribution to oval defects appears to be eliminated with gas sources. Tsang[324] has reported the absence of oval defects in GaAs and InGaAs grown with TMG or TEG and TMAs whereas oval defects occurred when elemental gallium sources were used with either elemental As$_4$ or cracked TMAs. Assuming that satisfactory gas quality can be consistently supplied, the biggest drawback to gas source MBE would be the safety issues associated with the highly toxic hydrides and the pyrophoric alkyls. The required safety monitoring and protection equipment are nontrivial capital costs for many potential users.

6.5 Superlattice Structures

Superlattice structures were among the first MBE structures conceived.[342] Such structures exemplified the film growth capabilities of the MBE process and were virtually impossible to fabricate by any other technique at that time. Superlattice structures can be divided into two general classifications: compositional (or heterostructure, including modulation doped superlattices) and doping (or homostructure). Epitaxial superlattices of lattice mismatched materials have been grown when the layers are sufficiently thin and are known as strained-layer superlattices (SLSs).

6.5.1 Strained-Layer Superlattices. When materials of sufficiently large lattice mismatch are grown on top of one another, a dislocation network is formed at the interface to accommodate the resulting strain. If the overgrowth is sufficiently thin, however, the accommodation will be absorbed entirely by a variation from the bulk lattice constant. Alternating thin layers of two such materials will result in a superlattice whose electronic properties can be adjusted within certain limits by pair choice and layer thickness.[343] For example, direct gap SLSs have been grown from the indirect-gap couple: GaP and GaAs$_{0.4}$P$_{0.6}$ and low current level optoelectronic devices have been proposed.

Several III-V SLSs have been grown: GaAsP/GaAs, GaAsP/GaP, and InGaAs/GaAs. Recently, InSb/InAs$_{.26}$Sb$_{.74}$ MBE SLSs have been successfully grown by Lee, et al.[344] Ge$_x$Si$_{1-x}$/Si SLSs have been grown by Bean, et al[345] and investigated for their unique optical properties. A Sb modulation-doped Si/Ge$_{.45}$Si$_{.55}$ SLS grown by Jorke and Herzog[346] yielded a 5x enhancement in room temperature mobility. MBE HgTe/ZnTe SLSs were successfully fabricated for the first time by Faurie, et al[347] and CdTe/ZnTe SLSs have been grown by Monfroy, et al.[348]

6.5.2 Superlattice Buffer Layers.
Superlattice layers have received considerable attention for their potential to improve device performance. It is proposed that the superlattice prevents the propagation of threading dislocations and substrate contaminants into the active overlayers. SL buffer layers may eliminate the need for a thick (about 1 micron) undoped GaAs layer. Fujii, et al[349] has reported GaAs/AlGaAs GRINSCH lasers with a 175 A/cm^2 minimum threshold current density, the lowest reported for a similar cavity length. The buffer layer consisted of 5 couples of 150 Å GaAs and 150 Å $Al_{0.7}Ga_{0.3}As$ spaced 1.5 microns below the quantum well active layer. Noda, et al[350] used a GaAs/AlAs superlattice buffer to improve the interface of a self-aligned structure laser, thus reducing the threshold current by 8-12%. Schaff and Eastman[351] have improved GaAs power FET performance using a 100 layer, 270 Å period AlGaAs/GaAs SL buffer layer. Morioka[352] has succeeded in growing GaAs on indium doped substrates using a 2 micron, 20 period superlattice of 100 Å GaAs and 100 Å AlAs. The PL improved substantially over that normally seen from material grown on 1% indium doped substrates and was comparable to chrome-doped substrates at 0.1% indium. Superlattice buffers have also been used to assist the growth of GaAs on silicon (see Section 6.2).

6.5.3 Superlattice Device Structures.
Superlattice structures have played prominently in the development of new devices. MQW lasers, avalanche photodiodes and superlattice HEMTs have been mentioned earlier, for example. New devices exploiting the tunable electronic and optical properties afforded by doping superlattices have begun to emerge. Schubert, et al[353] and Vojak, et al[354] have reported on efforts to extend the useful optical limit of AlGaAs/GaAs above 0.9 micron. Other materials have also been considered. InAsSb SLSs have been proposed for 8-12 micron detectors[344,355] because of the potential advantages of such devices over HgCdTe. Temkin, et al[356] and Pearsall, et al[357] have extended the spectral response of silicon photodetectors to 1.3 microns using a Ge_xSi_{1-x}/Si SLS. The approximately 6000 Å thick SL served both as the absorbing medium and to prevent the propagation of threading dislocations. Pearsall, et al[357] succeeded in showing an avalanche multiplication of 50 at 1.1 microns while Temkin, et al[356] have measured a photoconducting gain of 40 at 1.3 microns and projected a 200 Mb/s data rate over 25 km. Faurie, et al[358] has proposed new IR detector materials from HgTe/CdTe superlattices. These SLs should have a weaker energy gap temperature dependence than comparable HgCdTe detectors operating in the 8 to 12 micron range.

FUTURE DEVELOPMENTS

7.1 Production Equipment

The demand for production capable MBE equipment has developed naturally from device demand. Devices which were spawned in the research lab have become desirable and mass producible. The MBE equipment required to demonstrate such devices is considerably different from that needed to produce it in volume. Whereas, second generation MBE machines were designed to offer maximum versatility, a production machine can be defined more specifically since a particular material system or device is sought. The emphasis shifts from configurational flexibility to reliability, both of the process and the hardware. Production equipment, based on an established process, must maintain that process wafer-to-wafer. The justification for any piece of production equipment is a low ratio of the cost per device to the value added per device. Step yield, maintenance costs and downtime are contributors to the cost per device.

Some devices allow such a value added that MBE is without question the process of choice. In cases where the device structure relaxes process constraints, other approaches such as MOCVD may be competitive or superior to MBE on a cost per device basis. Current efforts among MBE equipment vendors are focused on increasing wafer throughput and reducing system downtime and capital costs. At present, the major markets are for AlGaAs/GaAs analog (microwave) and digital devices. The next larger markets are for other III-V materials for microwave and optoelectronic devices. Many of the devices in these markets require sophisticated processing and have a high added value. Second generation equipment was an attempt to address these markets and such machines are today producing device material for commercial use. Multichamber processing, multiple wafer load lock chambers with preprocessing capabilities, three-inch wafer compatibility and high uniformity have all contributed to demonstrating MBE's production worthiness.[92,126] Other advances are needed for the next generation of production MBE equipment.

The critical issues for production MBE are throughput and material quality. Batch processing, at least on the scale traditionally achieved using CVD, will probably never be attained by MBE, but small three to five wafer loads are certainly feasible. It is possible, in principle, to maintain the superior thickness and doping uniformity specifications over several wafers simultaneously. Compositional control should likewise be excellent across such a small batch. This may be accomplished using conventional sources, but it is likely that gas sources will be required. Recently, growth rates in the 3 to 5 microns per hour range have been reported using gas sources.[124] Excellent morphology is also claimed. The question of source gas purity remains an issue, however. There must be a reliable supply of high purity gases in order to maintain consistent material quality.

In order to reduce die costs, future production MBE equipment will be expected to meet critical uptime levels. Gas sources could further this goal by virtually eliminating the downtime associated with refilling conventional effusion cells. Load-locked, large volume cells are the only alternative where factors such as toxicity preclude the use of gas sources by certain device manufacturers. Combining the throughput advantages of gas sources with cassette-to-cassette automated wafer handling and real time in situ growth monitors should allow throughputs in the 20 to 50 wph range for one micron of material growth. Uptime requirements will also demand reliable hardware and advanced automation. Sophisticated system automation will monitor all aspects of the system, reducing the necessary operator skill level and providing diagnostics to assist service. Eventually, expert systems and robotics will eliminate operators entirely.[359] Third generation III-V production systems are currently under development by all major equipment manufacturers.[115,360] Although the market for silicon MBE is less well defined, production MBE equipment for large diameter silicon wafers has been developed.[361] A high-throughput production silicon MBE machine capable of handling 8-inch wafers is shown in Figure 18.

Figure 18: A silicon MBE production system with 200 mm wafer capability (courtesy of J.C. Bean, AT&T Laboratories).

7.2 In Situ Processing

The UHV environment of MBE is considered by some to be the ultimate condition for processing microelectronic devices. For example, see Harbison.[362] Instead of surrounding all process steps with a clean environment, the ambient is contained within one piece of equipment. Such a scheme is a natural extension of existing efforts to control process conditions more tightly thereby increasing yield and reproducibility. Particulate contamination due to wafer handling and air exposure can be virtually eliminated with proper designs. An all-vacuum process could eliminate the need for class 10 and better cleanrooms. Such an integrated process is still far in the future, but UHV deposition, etching, implantation, annealing and lithography have been demonstrated. Table 8 lists some examples. Current modular MBE systems lend themselves to the addition of additional processing (see Figure 5). In situ cleaning, annealing, metallization and analysis (inspection) have, in fact, been common to second generation equipment. There are current efforts to add laser and focused ion beam (FIB) processing. The joining of focused ion beam and MBE equipment has been put to practice by Miyauchi and Hashimoto.[363] They successfully demonstrated maskless FIB implantation of multilayer structures and showed that device quality was significantly improved by in situ versus interrupted growth of an imbedded stripe DH laser.

Table 8: In Situ Processing Elements

> Metallization
> Schottky contacts[251,253,364]
> Non-alloyed contacts[250,365,366]
> Shadow Masking[367-369]
> Selective Epitaxy[370,371]
> Ion Implantation[117,138,372]
> Focused Ion Beam Processing[363]
> Laser Annealing[383]

7.3 Process Developments

Several epitaxial processes closely related to conventional MBE or MOMBE have been tried or suggested and are mentioned briefly below. These processes are rather less developed than MBE and it is unclear what their future role will be, but it is important to remember that at one time MBE was in a similar state.

7.3.1 Ionized Cluster Beam Epitaxy. An epitaxial growth technique with comparable hardware to MBE is ionized cluster beam epitaxy or ICBE. Film growth takes place on heated substrates in UHV using MBE-type source geometries. The fundamental difference is in the design of the effusion cells. First, the exit nozzle is restricted so that large aggregates of between 100 and 1,000 atoms are formed by adiabatic expansion of the source vapors into the vacuum. Second, some percentage of aggregates are charged by electron bombardment and accelerated toward the substrate (as with ionized sources, see Section 4.2.1.4) at 1-10 keV. The substrate is bombarded by a mixture of ionized and neutral clusters. The virtues of this approach are higher surface mobility (lower growth temperatures) and higher sticking coefficients (higher maximum doping levels). Many metals and compound semiconductors have been epitaxially deposited on a variety of substrates. Much of the experimental work has occurred in Japan under Takagi and Yamada.[373-375] In the United States, work is underway at Rensselaer Polytechnic Institute with support from the Semiconductor Research Institute. Commercial ICBE equipment has been available, but in principle any MBE equipment could perform ICBE with a properly configured source.

7.3.2 Vacuum Chemical Epitaxy. Another epitaxial technique, vacuum chemical epitaxy (VCE), shares aspects of MOMBE and MOCVD. The process has been developed mainly by Fraas and co-workers at Chevron[376] and is shown schematically in Figure 3d. Total pressure during growth is between 1 and 100 mTorr. This is in the region between MOMBE and low pressure MOCVD. System base pressures near 10^{-8} Torr are achieved with a turbo pump. Growth occurs through the pyrolysis of a group-III alkyl and a hydride at the heated wafer. The hydride may be cracked at the gas inlet or a solid source cracker can be used (just as in MBE). Ternary and quaternary alloys can be deposited by premixing the alkyls and, separately, the hydrides. The reduced growth pressure results in an alkyl mean free path greater than the source-to-wafer distance (as in MBE). The hydride mean free path can vary depending on whether it is cracked, but it is generally in molecular flow across the wafer. The hot graphite reactor walls assist hydride decomposition and increase Group-V material incorporation efficiency. A typical growth rate is 5 microns per hour and Fraas has grown several III-III-V and III-V-V alloys. A cold-wall, diffusion-pumped version of VCE has been used by Sugiura, et al[377] to grow InSb. The role of VCE relative to MOMBE and MOCVD has yet to be determined, but it is, in principle, a production capable technique.

7.3.3 Irradiation Assisted MBE. Several researchers are exploring the process advantages of assisting film growth by bombarding the substrate with either electrons, ions[378] or photons during deposition. Kondo and Kawashima[379] enhanced the lateral epitaxial growth of gallium arsenide over tungsten with hydrogen and gallium ion bombardment. Using gallium and hydrogen ions, they obtained single crystal layers at temperatures where polycrystalline growth normally occurs. The photo-assisted work follows similar ideas under development at higher pressures during CVD. If practical, benefits may arise from a lack of substrate heater (less background contamination) and a tunable reaction chemistry using different laser frequencies. Takahashi[380] has reported photo-assisted MOMBE using an ArF excimer laser at 193 nm. A roughly linear relationship between pulse frequency and growth rate was found. Superlattice growth using switched lasers of different wavelength is proposed. Nishizawa, et al[381] have combined photoepitaxy and molecular layer epitaxy to grow GaAs using TEG and arsine. Elsner, et al[382] at IBM have proposed pulsed laser-assisted MBE of GaAs using a 694 nm ruby laser. The laser-assisted MBE growth of silicon has also been reported.[383] In this instance, the deposition was at room temperature and subsequently annealed with a 694 nm ruby laser. Films up to 4000 Å thick yielded a (2 x 1) LEED pattern on Si.[100]

7.4 Toxic Gases and Environmental Concerns

The use of hydrides and organometallics in MBE brings along the attendant hazards and requirements for safety. The hydrides of As and P are extremely toxic, and phosphine is additionally pyrophoric. The American Conference of Government-Industrial Hygienists have set threshold limit values (TLV) of 0.3 ppm and 0.05 ppm, averaged over an 8-hour period, for phosphine and arsine respectively. Although the metal organics are not considered as toxic, they are pyrophoric. Due to the pyrophoric nature of the metal organics and low TLVs of the hydrides, users must exercise engineering as well as administrative safety controls wherever these gases are stored, handled, used and disposed. Each area contains a potential toxic gas emergency. Engineering controls include designs to dilute potential gas leaks from fittings, cylinders, regulators, etc., and to incorporate adequate sensors and burn boxes. Administrative controls include the training of personnel and providing protective gear for personnel, instituting safety and emergency procedures and protecting the surrounding community from accidental gas releases. The latter includes toxic neutralization systems, such as burn boxes and scrubbers. Currently under study are less toxic sources of gaseous phosphorus[384] for OMVPE which may prove suitable for MBE.

REFERENCES

1. Arthur, J.R., Jr., *J. Appl. Phys.*, Vol. 39, p . 4032 (1968).
2. Cho, A.Y., *J. Vac. Sci. and Technol.*, Vol. 8, p. S31 (1970).

3. Chang, L.L., Esaki, L., Howard, W.E., Ludeke, R. and Schul, G., *J. Vac. Sci. Technol.*, Vol. 10, p. 655 (1973).
4. Cho, A.Y. and Arthur, J.R., *Prog. in Solid-State Chemistry*, Vol. 10, p. 157 (1975).
5. Chang, L.L. and Ludeke, R., *Epitaxial Growth, Part A.*, p. 37, J.W. Mathews, ed., Academic Press (1975).
6. Foxon, C.T. and Joyce, B.A., *Current Topics in Material Science*, Vol. 7, E. Kaldis, ed., North-Holland, Amsterdam/New York (1980).
7. Wood, C.E.C., *Physics of Thin Films*, G. Hass, ed., Vol. 11, p. 35, Academic Press (1980).
8. Kunzel, H., *Physica* Vol. 129 B & C, p. 66 (1985).
9. Luscher, P.E. and Collins, D.M., *Prog. Crystal Growth Charact.*, Vol. 2, p. 15 (1979).
10. Cho, A.Y., *J. Vac. Sci. Technol.*, Vol. 16, p. 275 (1979).
11. Luscher, P.E., *Solid State Technol.*, Vol. 20, p. 43 (1977).
12. Panish, M.B., *Science*, Vol. 208, p. 916 (1980).
13. Schuller, I.K., *Phys. Rev. Lett.*, Vol. 44, p. 1597 (1980).
14. Benchimol, J.L., Slempkes, S., N'Guyen, D.C., Le Roux, G., Bresse, J.F. and Primot, J., *J. Appl. Phys.* Vol. 59, p. 4068 (1986).
15. Kobayashi, N. and Makimoto, T., *Jpn. J. Appl. Phys.*, Vol. 24, p. L824 (1985).
16. Andre, J.P., Wolny, M. and Rocchi, M., 12th Int. Symp. GaAs and Related Compounds, Karuizawa, Japan, 1985, *Inst. Phys. Conf. Ser.*, Vol. 79, p. 379 (1986).
17. Takakuwa, H., Tanaka, K., Mori, Y., Arai, M., Kato, Y. and Watanabe, S., *IEEE Trans. Electron Dev.*, Vol. ED-33, p. 595 (1986).
18. Razeghi, M., Maurel, P., Omnes, F., Ben Armor, S., Dmowski, L. and Portal, J.C., *Appl. Phys. Lett.*, Vol. 48, p. 1267 (1986).
19. Usui, A. and Sunakawa, H., *Jpn. J. Appl. Phys.*, Vol. 25, p. L212 (1986).
20. Bedair, S.M., Tischler, M.A., Katsuyama, T. and El-Masry, N.A., *Appl. Phys. Lett.*, Vol. 47, p. 51 (1985).
21. Nishizawa, J., Abe, H. and Kurabayashi, T., *J. Electrochem. Soc.*, Vol. 132, p. 1197 (1985).
22. Tandon, J.L. and Yeh, Y.C.M., *J. Electrochem. Soc.*, Vol. 132, p. 662 (1985).
23. Haydl, W.H., Smith, R.S. and Bosch, R., *IEEE Elec. Dev. Lett.*, Vol. EDL-1, p. 224 (1980).
24. Shah, N.J., Pei, S.S., Tu, C.W. and Tiberio, R.C., *IEEE Trans. Electron Dev.*, Vol. ED-33, p. 543 (1986).
25. Tung, R.T., Gibson, J.M. and Levi, A.F.J., *Appl. Phys. Lett.*, Vol. 48, p. 1264 (1986).
26. Wang, K.L., Novel Devices by Si-Based Molecular Beam Epitaxy, *Solid State Technol.*, Vol. 28, p. 137 (1985).
27. Solomon, P.M. and Morkoc, H., *IEEE Trans. Electron Dev.*, Vol. ED-31, p. 1015 (1984).
28. Dingle, R., Stormer, H.L., Gossard, A.C. and Wiegmann, W., *Appl. Phys. Lett.*, Vol. 33, p. 665 (1978).
29. Mimura, T., Hiyamizu, S., Fujii, T. and Nanbu, K., *Jpn. J. Appl. Phys.*, Vol. 19, p. L225 (1980).

30. Hiyamizu, S., Mimura, T., Fujii, T., Nanbu, K. and Hasimoto, H., *Jpn. J. Appl. Phys.*, Vol. 20, p. L245 (1981).
31. Sheng, N.H., Chang, M.F., Lee, C.P., Miller, D.L. and Chen, R.T., *Elec. Dev. Lett.*, Vol. EDL-7, p. 11 (1986).
32. Cirillo, N.C., Jr., Shur, M.S. and Abrokwah, J.K., *Elec. Dev. Lett.*, Vol. EDL-7, p. 71 (1986).
33. Kamei, K., Kawasaki, H., Hori, S., Shibata, K., Higashiura, M., Watanabe, M.O. and Ashizava, Y., 12th Int. Symp. GaAs and Related Compounds, Karuizawa, Japan, 1985, *Inst. Phys. Conf. Ser.*, Vol. 79, p. 541 (1986).
34. Pearsall, T.P. and Bean, J.C., *IEEE Elec. Dev. Lett.*, Vol. EDL-7, p. 308 (1986).
35. Arai, K., Mizutani, T. and Yanagawa, F., *Jpn. J. Appl. Phys.*, Vol. 24, p. L623 (1985).
36. Chang, M.F., Asbeck, P.M., Miller, D.L. and Wang, K.C., *Elec. Dev. Lett.*, Vol. EDL-7, p. 8 (1986).
37. Sullivan, G.J., Asbeck, P.M., Chang, M.F., Miller, D.L. and Wang, K.C., *Electron. Lett.*, Vol. 22, p. 419 (1986).
38. Fischer, R., Chand, N., Kopp, W., Morkoc, H., Erickson, L.P. and Youngman, R., *Appl. Phys. Lett.*, Vol. 47, p. 397 (1985).
39. Bell, T.E., ed., The Quest for Ballistic Action, *IEEE Spectrum*, Vol. 23 p. 36 (1986).
40. Reddy, U.K., Chen, J., Peng, C.K. and Morkoc, H., *Appl. Phys. Lett.*, Vol. 48, p. 1799 (1986).
41. Cho, A.Y., DiLorenzo, J.V., Hewitt, B.S., Niehaus, W.C., Schlosser, W.O. and Radice, C., *J. Appl. Phys.*, Vol. 48, p. 346 (1977).
42. Bandy, S.G., Collins, D.M. and Nishimoto, C.K., *Electron. Lett.*, Vol. 15, p. 218 (1979).
43. Omori, M., Durmmond, T.J. and Morkoc, H., *Appl. Phys. Lett.*, Vol. 39, p. 566 (1981).
44. Wataze, M., Mitsui, Y., Shimanoe, T., Nakatani, M. and Mitsui, S., *Electron. Lett.*, Vol. 14, p. 759 (1978).
45. Wood, C.E.C., DeSimone, D. and Jadaprawira, S., *J. Appl. Phys.*, Vol. 51, p. 2074 (1979).
46. Drummond, T.J., Koop, W., Arnold, D., Fischer, R., Morkoc, H., Erickson, L.P. and Palmberg, P.W., *Electronic Lett.*, Vol. 19, p. 986 (1983).
47. Xu, J. and Shur, M., *IEEE Electron Device Letters*, Vol. EDL-7, p. 416 ((1986).
48. Taylor, G.W., Simmons, J.G., Mand, R.S. and Cho, A.Y., *J. Vac. Sci. Technol.*, Vol. B4, p. 603 (1986).
49. Asbeck, P.M., Gupta, A.K., Ryan, F.J., Miller, D.L., Anderson, R.J., Liechti, C.A. and Eisen, F.H., *IEDM Tech. Digest*, Vol. 84, p. 864 (1984).
50. Kroemer, H., *Proceedings of the IEEE*, Vol. 70, p. 13 (1982).
51. Cho, A.Y. and Reinhart, F.K., *J. Appl. Phys.*, Vol. 45, p. 1812 (1974).
52. Cho, A.Y., Dunn, C.N., Kuvas, R.L. and Schroeder, W.E., *Appl. Phy. Lett.*, Vol. 25, p. 224 (1974).
53. Hierl, T.L. and Luscher, P.E., *Proc. 2nd Int. Symp. on MBE and Clean Surface Techniques*, Tokyo, Japan, p. 147 (1982).
54. Shih, H.D. and Bayraktaroglu, B., *J. Vac. Sci. Technol.*, Vol. B2, p. 269 (1984).

55. Linke, R.A., Schneider, M.V., and Cho, A.Y., *IEEE Trans. on Microwave Theory and Tech.*, Vol. MTT-26, p. 935 (1978).
56. Harris, J.J. and Woodcock, J.M., *Electron. Lett.*, Vol. 16, p. 317 (1980).
57. Nagle, J.P., Hing, L.A., Kerr, T.M. and Summers, J.G., *J. Vac. Sci. Technol.*, Vol. B4, p. 631 (1986).
58. Haydl, W.H., Smith, R.S. and Bosch, R., *Appl. Phys. Lett.*, Vol. 37, p. 556 (1980).
59. Tsang, W.T., *Appl. Phys. Lett.*, Vol. 34, p. 473 (1979).
60. Tsang, W.T., *Appl. Phys. Lett.*, Vol. 36, p. 11 (1980).
61. Tsang, W.T., *Appl. Phys. Lett.*, Vol. 38, p. 587 (1981).
62. Temkin, H., Bean, J.C., Pearsall, T.P., Olsson, N.A. and Lang, D.V., *Appl. Phys. Lett.*, Vol. 49, p. 155 (1986).
63. Pearsall, T.P., Temkin, H., Bean, J.C. and Luryi, S., *Elec. Dev. Lett.*, Vol. EDL-7, p. 330 (1986).
64. Miller, D.A.B., *The Second Int. Conf. on Modulated Semi. Structures, Collected Papers*, Kyoto, Japan, p. 459 (1985).
65. Katsumoto, S., Yamamoto, A. and Yamaguchi, M., *Jpn. J. Appl. Phys.*, Vol. 24, p. 636 (1985).
66. Norton, P., Knoll, G. and Bachem, K.-H., *J. Vac. Sci. Technol.*, Vol. B3, p. 782 (1985).
67. Partin, D.L., Majkowski, R.F. and Swets, D.E., *J. Vac. Sci. Technol.*, Vol. B3, p. 576 (1985).
68. Partin, D.L., *Appl. Phys. Lett.*, Vol. 43, p. 996 (1983).
69. Saku, T., Iwamura, H., Hirayama, Y., Suzuki, Y. and Okamoto, H., *Jpn. J. Appl. Phys.*, Vol. 24, p. L73 (1985).
70. Ohmori, Y., Suzuki, Y. and Okamoto, H., *Jpn. J. Appl. Phys.*, Vol. 24, p. L657 (1985).
71. Tserng, H.Q. and Kim, B., *IEEE GaAs IC Symposium*, Monterey, CA, p. 11 (1985).
72. Wang, K.C., Asbeck, P.M., Chang, M.F., Miller, D.L. and Eisen, F.H., *IEEE GaAs IC Symposium*, Monterey, CA, p. 99 (1985).
73. Pei, S.S., Shah, N.J., Hendel, R.H., Tu, C.W. and Dingle, R., *IEEE GaAs IC Symposium*, p. 129 (1984).
74. Cirillo, N.C., Jr., Abrokwah, J.K. and Jamison, S.A., *IEEE GaAs IC Symposium*, p. 167 (1984).
75. Kinoshita, H., Nishi, S., Akiyama, M. and Kaminishi, K., *Jpn. J. Appl. Phys.*, Vol. 24, p. 1061 (1985).
76. Asbeck, P.M., Miller, D.L., Anderson, R.J., Deming, R.N., Chen, R.T., Liechti, C.A. and Eisen, F.H., *IEEE GaAs IC Symposium*, p. 133 (1984).
77. Mizutani, T., Fujita, S. and Yanagawa, F., Int. Symp. GaAs and Related Compounds, Karuizawa, Japan, 1985, *Inst. Phys. Conf. Ser.*, Vol. 79, p. 733 (1986).
78. Goodhue, W.D., Burke, B.E., Nichols, K.B., Metze, G.M. and Johnson, G.D., *J. Vac. Sci. Technol.*, Vol. B4, p. 769 (1986).
79. Kobayashi, N., Notomi, S., Suzuki, M., Tsuchiya, T., Nishiuchi, K., Odani, K., Shidatomi, A., Mimura, T. and Abe, M., *IEEE GaAs IC Symposium*, Monterey, CA, p. 207 (1985).

80. Kuroda, S., Mimura, T., Suzuki, M., Kobayashi, N., Nishiuchi, K., Shibatomi, A. and Abe, M., *IEEE GaAs IC Symposium*, Monterey, CA, p. 133 (1984).
81. Wada, O., Sanada, T., Hamaguchi, H., Fujii, T., Hiyamizu, S. and Sakurai, T., *Jpn. J. Appl. Phys.*, Vol. 22, Supp. 22-1, p. 587 (1983).
82. Barnard, J., Ohno, H., Wood, C.E.C. and Eastman, L.F., *IEEE Elect. Dev. Lett.*, Vol. EDL-2, p. 7 (1981).
83. Wake, D., Scott, E.G. and Henning, I.D., *Electron. Lett.*, Vol. 22, p. 719 (1986).
84. Wang, H. and Ankri, D., *Electron. Lett.*, Vol. 22, p. 391 (1986).
85. Matsueda, H., Hirao, M., Tanaka, T., Kodera, H. and Nakamura, M., Int. Symp. GaAs and Related Compounds, Karuizawa, Japan, 1985, *Inst. Phys. Conf. Ser.*, Vol. 79, p. 655 (1986).
86. Fujii, T., Yamakoshi, S., Nanbu, K., Wada, O. and Hiyamizu, S., *J. Vac. Sci. Technol.*, Vol. B2, p. 259 (1984).
87. Sakano, S., Inoue, H., Nakamura, H., Katsuyama, T. and Matsumura, H., *Electronics Lett.*, Vol. 22, p. 594 (1986).
88. Reinhart, F.K. and Cho, A.Y., *Appl. Phys. Lett.*, Vol. 31, p. 457 (1977).
89. Ishikawa, H., Takagi, N., Ohsaka, S., Hanamitsu, K., Fujiwara, T. and Takusagawa, M., *Appl. Phys. Lett.*, Vol. 36, p. 520 (1980).
90. Yuan, H.T., McLevige, W.V. and Shih, H.D., *VLSI Electronics Microstructure Science* Vol. II, Chapter 5, GaAs Microelectronics.
91. Miller, D.L., *IEEE GaAs IC Symposium*, p. 37 (1984).
92. Abrokwah, J.K., Cirillo, N.C., Jr., Heux, M.J. and Longerbone, M., *J. Vac. Sci. and Technol.*, Vol. B2, p. 252 (1984).
93. Mimura, T. and Abe, M., *IEEE GaAs IC Symposium* Monterey, CA, p. 207 (1985).
94. Dushman, S., *Scientific Foundations of Vacuum Technique*, Wiley, New York (1962).
95. Guthrie, A., *Vacuum Technology*, Wiley, New York (1963).
96. O'Hanlon, J.F., *A User's Guide to Vacuum Technology*, Wiley, New York (1980).
97. Strausser, Y., *Review of Outgassing Results*, Varian Reports VR-51, Varian Assoc., Palo Alto, CA (1969).
98. Miyamoto, M., Itoh, T., Komaki, S., Narushima, K. and Ishimaru, H., *Proc. 8th Symp. on ISIAT*, Tokyo, Japan (1984).
99. Tsang, W.T., Miller, R.C., Capasso, F. and Bonner, W.A., *Appl. Phys. Lett.*, Vol. 41, p. 467 (1982).
100. Shen, L.V.L., *J. Vac. Sci. Technol*, Vol. 15, p. 10 (1978).
101. Svensson, S.P. and Andersson, T.G., *J. Vac. Sci. Technol.*, Vol. 20, p. 245 (1982).
102. Curless, J.A., *J. Vac. Sci. Technol.*, Vol. B3, p. 531 (1985).
103. Jackson, S. C., Baron, B.N., Rocheleau, R.E. and Russell, T.W.F., *J. Vac. Sci. Technol.*, Vol. A3, p. 1916 (1985).
104. Kunzel, H. and Ploog, K., *Appl. Phys. Lett.*, Vol. 37, p. 416 (1980).
105. Tsang, W.T., Ditzenberger, J.A. and Olsson, N.A., *IEEE Elect. Dev. Lett.*, Vol. EDL-4, p. 275 (1983).
106. Krosor, B.S. and Bachrach, R.Z., *J. Vac. Sci. Technol.*, Vol. B1, p. 138 (1983).

107. Huet, D., Lambert, M., Bonnevie, D. and Dufresne, D., *J. Vac. Sci. Technol.*, Vol. B3, p. 823 (1985).
108. Lee, R.L., Schaffer, W.J., Chai, Y.G., Liu, D. and Harris, J.S., *J. Vac. Sci. Technol.*, Vol. B4, p. 568 (1986).
109. Jackson, T.N., Kirchner, P.D., Pettit, G.D., Rosenberg, J.J., Woodall, J.M. and Wright, S.L., U.S. Patent 4,550,047 (1985).
110. Naganuma, M. and Takahashi, K., *Appl. Phys. Lett.*, Vol. 27, p. 342 (1975).
111. Takahashi, K., 12th Int. Symp. on GaAs and Related Compounds, Karuizawa, Japan, 1985, *Inst. Phys. Conf. Ser.*, Vol. 79, p. 73 (1986).
112. Summers, C.J., Meeks, E.L. and Cox, N.W., *J. Vac. Sci. Technol.*, Vol. B2, p. 224 (1984).
113. Harris, K.A., Hwang, S., Blanks, D.K., Coor, J.W., Jr., Schetzina, J.F. and Otsuka, N., *J. Vac. Sci. Technol.*, Vol. A4, p. 2061 (1986).
114. Bean, J.C., Conf. on Physics of VLSI, Palo Alto, CA (1984).
115. Tabe, M., *J. Vac. Sci. Technol.*, Vol. B3, p. 975 (1985).
116. Bean, J.C. and Dingle, R., *Appl. Phys. Lett.*, Vol. 35, p. 925 (1979).
117. Ota, Y., *J. Appl. Phys.*, Vol. 51, p. 1102 (1980).
118. Shimizu, S., Tsukakoshi, O., Komiya, S. and Makita, Y., *J. Vac. Sci. Technol.*, Vol. B3, p. 554 (1985).
119. Shimizu, S., Tsukakoshi, T., Komiya, S. and Makita, Y., 12th Int. Symp. on GaAs and Related Compounds, Karuizawa, Japan, 1985, *Inst. Phys. Conf. Ser.*, Vol. 79, p. 91 (1986).
120. de Jong, T., Douma, W.A.S., Doorn, S. and Saris, F.W., *Materials Lett.*, Vol. 1, p. 157 (1983).
121. Calawa, A.R., *Appl. Phys. Lett.*, Vol. 38, p. 701 (1981).
122. Chow, R. and Chai, Y.G., *Appl. Phys. Lett.*, Vol. 42, p. 383 (1983).
123. Panish, M.B., Temkin, H. and Sumski, S., *J. Vac. Sci. Technol.*, Vol. B3, p. 657 (1985).
124. Tsang, W.T., *J. Vac. Sci. Technol.*, Vol. B3, p. 666 (1985).
125. Maki, P.A., Palmateer, S.C., Calawa, A.R. and Lee, B.R., *J. Vac. Sci. Technol.*, Vol. B4, p. 564 (1986).
126. Hwang, J.C.M., Brennan, T.M. and Cho, A.Y., *J. Electrochem. Soc.*, Vol. 130, p. 493 (1983).
127. Oe, K. and Imamura, Y., *Jpn. J. Appl. Phys.*, Vol. 24, p. 779 (1985).
128. Erickson, L.P., Carpenter, G.L., Seibel, D.D., Palmberg, P.W., Pearah, P., Kopp, W. and Morkoc, H., *J. Vac. Sci. Technol.*, Vol. B3, p. 536 (1985).
129. Hellman, E.S., Pitner, P.M., Harwit, A., Liu, D., Yoffe, G.W., Harris, J.S., Jr., Caffee, B. and Hierl, T., *J. Vac. Sci. Technol.*, Vol. B4, p. 574 (1986).
130. Mars, D.E. and Miller, J.N., *J. Vac. Sci. Technol.*, Vol. B4, p. 571 (1986).
131. Wright, S.L., Marks, R.R. and Wang, W.I., *J. Vac. Sci. Technol.*, Vol. B4, p. 505 (1986).
132. Stanchak, C.M., Morkoc, H., Witkowski, L.C. and Drummond, T.J., *Rev. Sci. Instrum.*, Vol. 52, p. 438 (1981).
133. Cooper, A.J., Sahin, A.S., Jones, P.L., King, D. and Moore, D., *Vacuum*, Vol. 34, p. 925 (1984).
134. Wunder, R., Stall, R., Malik, R. and Woelfer, S., *J. Vac. Sci. Technol.*, Vol. B3, p. 964 (1985).

135. Saito, J. and Shibatomi, A., *Fujitsu Sci. and Tech. J.*, Vol. 21, p. 190 (1985).
136. Wang, W.I., *J. Vac. Sci. Technol.*, Vol. B3, p. 552 (1985).
137. Kroemer, H., Polasko, K.J. and Wright, S.C., *Appl. Phys. Lett.*, Vol. 36, p. 763 (1980).
138. Kerr, T.M., McLean, T.M., Westwood, D.I., Grange, J.D., *J. Vac. Sci. Technol.*, Vol. B3, p. 535 (1985).
139. Chang, C.-A., Ludeke, R., Chang, L.L. and Esaki, L., *Appl. Phys. Lett.*, Vol. 31, p. 759 (1977).
140. Kodama, M., Hasegawa, J., Kimata, M., *J. Electrochem. Soc.*, Vol. 132, p. 659 (1985).
141. Silberg, E., Chang, T.Y., Ballman, A.A. and Caridi, E.A., *J. Appl. Phys.*, Vol. 54, p. 6974 (1983).
142. Asahi, H., Kawamura, Y. and Nagai, H., *J. Appl. Phys.*, Vol. 53, p. 4928 (1982).
143. Tsang, W.T., Chiu, T.H., Kisker, D.W., Ditzenberger, J.A., *Appl. Phys. Lett.*, Vol. 46, p. 283 (1985).
144. Yano, M., Takase, T. and Kimata, M., *Jpn. J. Appl. Phys.*, Vol. 18, p. 387 (1979).
145. Bean, J.C., Sheng, T.T., Feldman, L.C., Fiory, A.T. and Lynch, R.T., *Appl. Phys. Lett.*, Vol. 44, p. 102 (1984).
146. Faurie, J.P., Reno, J., Sivananthan, S., Sou, I.K., Chu, X., Boukerche, M. and Wijewarnasuriya, P.S., *J. Vac. Sci. Technol.*, Vol. A4, p. 2067 (1986).
147. Farrow, R.F.C., Wood, S., Greggi, J.C., Jr., Takei, W.J., Shirland, F.A., *J. Vac. Sci. Technol.*, Vol. B3, p. 681 (1985).
148. Kolodziejski, L.A., Sakamoto, T., Gunshor, R.L. and Datta, S., *Appl. Phys. Lett.*, Vol. 44, p. 799 (1984).
149. Kolodziejski, L.A., Gunshor, R.L., Bonsett, T.C., Venkatasubramanian, R. and Datta, S., *Appl. Phys. Lett.*, Vol. 47, p. 169 (1985).
150. Yao, T., Makita, Y. and Maekawa, S., *J. Cryst. Growth*, Vol. 45, p. 309 (1978).
151. Park, R.M., Mar, H.A., Salansky, N.M., *J. Vac. Sci. Technol.*, Vol. B3, p. 676 (1985).
152. Yao, T., Makita, Y. and Maekawa, S., *Appl. Phys. Lett.*, Vol. 35, p. 97 (1979).
153. Mino, N., Kobayashi, M., Konagai, M., Takahashi, K., *J. Appl. Phys.*, Vol. 58, p. 793 (1985).
154. Partin, D.L., *J. Vac. Sci. Technol.*, Vol. B1, p. 174 (1983).
155. Farrow, R.F.C., Sullivan, P.W., Williams, G.M., Jones, G.R. and Cameron, D.C., *J. Vac. Sci. Technol.*, Vol. 19, p. 415 (1981).
156. Schowalter, L.J., Fathauer, R.W., Goehner, R.P., Turner, L.G., DeBlois, R.W., *J. Appl. Phys.*, Vol. 58, p. 302 (1985).
157. Sullivan, P.W., Bower, J.E., Metze, G.M., *J. Vac. Sci. Technol.*, Vol. B3, p. 674 (1985).
158. Tu, C.W., Sheng, T.T., Read, M.H., Schlier, A.R., Johnson, J.G., Johnston, W.D., Jr. and Bonner, W.A., *J. Electrochem. Soc.*, Vol. 130, p. 2081 (1983).

159. Siskos, S., Fontaine, C. and Munoz-Yague, A., *J. Appl. Phys.*, Vol. 56, p. 1642 (1984).
160. Sinharoy, S., Hoffman, R.A., Rieger, J.H., Takei, W.J., and Farrow, R.F.C, *J. Vac. Sci. Technol.* Vol. B3, p. 722 (1985).
161. Sugiyamma, K., *J. Appl. Phys.*, Vol. 56, p. 1733 (1984).
162. Houzay, F., Moison, J.M., Bensoussan, M., *J. Vac. Sci. Technol.*, Vol. B3, p. 756 (1985).
163. Bloch, J., Heiblum, M., Komem, Y., *Appl. Phys. Lett.*, Vol. 46, p. 1092 (1985).
164. Prinz, G.A. and Krebs, J.J., *Appl. Phys. Lett.*, Vol. 39, p. 397 (1981).
165. Arnaud D'Avitaya, F., Delage, S., Rosencher, E. and Derrien, J., *J. Vac. Sci., Technol.*, Vol. B3, p. 770 (1985).
166. Cullis, A.G. and Farrow, R.F.C., *Thin Solid Films*, Vol. 58, p. 197 (1979).
167. Cho, A.Y., *Proc. of NATO Advanced Study Institute*, L.L. Chang and K. Ploog, eds., pp. 191-226, Martinus Nijhoff, Dordrecht, Netherlands (1986).
168. Konig, U., Kibbel, H. and Kasper, E., *J. Vac. Sci. Technol.*, Vol. 16, p. 985 (1979).
169. Bean, J.C. and Sadowski, E.A., *J. Vac. Sci. Technol.*, Vol. 20, p. 137 (1982).
170. Shiraki, Y., *J. Vac. Sci. Technol.*, Vol. B3, p. 725 (1985).
171. Saris, F.W. and DeJong, T., *Proc. of NATO Advanced Study Institute*, L.L. Chang and K. Ploog, eds., pp. 263-292, Martinus Nijhoff, Dordrecht, Netherlands (1986).
172. Faurie, J.P., Sivananthan, S. and Reno, J., *J. Vac. Sci. Technol.*, Vol. A4, p. 2096 (1986).
173. Farrow, R.F.C., *Proc. of NATO Advanced Study Institute*, L.L. Chang and K. Ploog, eds., pp. 227-262, Martinus Nijhoff, Dordrecht, Netherlands, (1986).
174. Smith, D.L. and Pickhardt, V.Y., *J. Appl. Phys.*, Vol. 46, p. 2366 (1975).
175. Yao, T., *Res. Electrotech. Lab.*, Vol. n845, p. 1 (1985).
176. Holloway, H., and Walpole, J.N., *Prog. Crystal Growth Charact.*, Vol. 2, p. 49 (1979).
177. Hwang, J.C.M., Temkin, H., Brennan, T.M. and Frahm, R.E., *Appl. Phys. Lett.*, Vol. 42, p. 66 (1983).
178. Heiblum, M., Mendez, E.E. and Osterling, L., *J. Appl. Phys.*, Vol. 54, p. 6982 (1983).
179. Massies, J. and Contour, J.P., *J. Appl. Phys*, Vol. 58, p. 806 (1985).
180. Scott, E.G., Wake, D., Livingstone, A.W., Andrews, D.A. and Davies, G.J., *J. Vac. Sci. Technol.*, Vol. B3, p. 816 (1985).
181. Noreika, A.J., Francombe, M.H. and Wood, C.E.C., Jr., *J. Appl. Phys.*, Vol. 52, p. 7416 (1981).
182. Wood, S., Greggi, J., Jr., Farrow, R.F.C., Takei, W.J., Shirland, F.A. and Noreika, A.J., *J. Appl. Phys.*, Vol. 55, p. 4225 (1984).
183. Munoz-Yague, A., Piqueras, J. and Fabre, N., *J. Electrochem. Soc.*, Vol. 128, p. 149 (1981).
184. Massies, J. and Contour, J.P., *Appl. Phys. Lett.*, Vol. 46, p. 1150 (1985).
185. Vig, J.R., *J. Vac. Sci. Technol.*, Vol. A3, p. 1027 (1985).

186. McClintock, J.A., Wilson, R.A. and Beyer, N.E., *J. Vac. Sci. Technol.*, Vol. 20, p. 241 (1982).
187. Joyce, B.A., *Surf. Sci.*, Vol. 35, p. 1 (1973).
188. Becker, G.E. and Bean, J.C., *J. Appl. Phys.*, Vol. 48, p. 3395 (1977).
189. Henderson, R.C., *J. Electrochem. Soc.*, Vol. 119, p. 772 (1972).
190. Ota, Y., *J. Electrochem. Soc.* Vol. 124, p. 1796 (1977).
191. Ishizaka, A., Nakagawa, K. and Shiraki, Y., *2nd International Symp. on MBE and Related Clean Surface Techniques*, Tokyo, pp. 183–186 (1982).
192. Wright, S. and Kroemer, H., *Appl. Phys. Lett.*, Vol. 36, p. 210 (1980).
193. Wang, W.I., *Appl. Phys. Lett.*, Vol. 44, p. 1149 (1984).
194. Yang, H.T. and Mooney, P.M., *J. Appl. Phys.*, Vol. 58, p. 1854 (1985).
195. Kugimiya, K., Hirofuji, Y. and Matsuo, N., *Jpn. J. Appl. Phys.*, Vol. 24, p. 564 (1985).
196. Faurie, J.P., Million, A. and Piaguet, J., *J. Cryst. Growth*, Vol. 59, p. 10 (1982).
197. Park, R.M., Mar, H.A. and Salansky, N.M., *J. Vac. Sci. Technol.*, Vol. B3, p. 1637 (1985).
198. Farrow, R.F.C., Jones, G.R., Williams, G.M. and Young, I.M., *Appl. Phys. Lett.*, Vol. 39, p. 954 (1981).
199. Nishitani, K., Ohkata, R. and Murotani, T., *J. Electron. Mater.*, Vol. 12, p. 619 (1983).
200. Maki, P.A., Palmateer, S.C., Calawa, A.R. and Lee, B.R., *J. Electrochem. Soc.*, Vol. 132, p. 2813 (1985).
201. Wood, C.E.C., *Proc. of NATO Advanced Study Institute*, L.L. Chang and K. Ploog, eds., pp. 149–189, Martinus Nijhoff, Dordrecht, Netherlands (1986).
202. Cho, A.Y. and Hayashi, I., *Appl. Phys. Lett.*, Vol. 42, p. 4422 (1971).
203. Wood, C.E.C. and Joyce, B.A., *J. Appl. Phys.*, Vol. 49, p. 4854 (1978).
204. Wood, C.E.C., *Appl. Phys. Lett.*, Vol. 33, p. 770 (1978).
205. Collins, D.M., *Appl. Phys. Lett.*, Vol. 35, p. 67 (1979).
206. Ilegems, M., *J. Appl. Phys.*, Vol. 48, p. 1279 (1977).
207. Cho, A.Y. and Panish, M.B., *J. Appl. Phys.*, Vol. 43, p. 5118 (1972).
208. Ilegems, M. and Dingle, R., *Int. Symp. GaAs and Related Compounds*, 1974, Deauville, France (1975).
209. Mannoh, M., Nomura, Y., Shinozaki, K., Mihara, M. and Ishi, M., *J. Appl. Phys.*, Vol. 59, p. 1092 (1986).
210. Ilegems, M., Dingle, R. and Rupp, L.W., *J. Appl. Phys.*, Vol. 46, p. 3059 (1975).
211. Naganuma, M. and Takahashi, K., *Appl. Phys. Lett.*, Vol. 27, p. 342 (1975).
212. Kowalczyk, S.P. and Miller, D.L., *J. Vac. Sci. Technol.*, Vol. B3, p. 1534 (1985).
213. Briones, F., Golmayo, D., González, L. and de Miguel, J.L., *J. Vac. Sci. Technol.*, Vol. B3, p. 568 (1985).
214. Metzger, R.A. and Allen, F.G., *J. Appl. Phys.*, Vol. 55, p. 931 (1984).
215. Ota, Y., *J. Electrochem. Soc.*, Vol. 126, p. 1761 (1979).
216. Sugiura, H., *J. Appl. Phys.*, Vol. 51, p. 2630 (1980).

217. Jorke, H., Herzog, H.-J. and Kibbel, H., *Appl. Phys. Lett.*, Vol. 47, p. 511 (1985).
218. Kubiak, R.A.A., Leong, W.Y. and Parker, E.H.C., *J. Electrochem. Soc.*, Vol. 132, p. 2738 (1985).
219. Kubiak, R.A.A., Leong, W.Y. and Parker, E.H.C., *Appl. Phys. Lett.*, Vol. 46, p. 565 (1985).
220. Knall, J., Sundgren, J.-E., Green, J.E., Rockett, A. and Barnett, S.A., *Appl. Phys. Lett.*, Vol. 45, p. 689 (1984).
221. Swartz, R.G., McFee, J.H., Voshehenkov, A.M., Finegan, S.M. and Ota, Y., *Appl. Phys. Lett.*, Vol. 40, p. 239 (1982).
222. Ostrom, R.M. and Allen, F.G., *Appl. Phys. Lett.*, Vol. 48, p. 221 (1986).
223. Boukerche, M., Reno, J., Sou, I.K., Hsu, C. and Faurie, J.P., *Appl. Phys. Lett.*, Vol. 48, p. 1733 (1986).
224. Williams, R.E. and Shaw, D.W., *IEEE Trans. Electron Dev.*, Vol. ED-25, p. 600 (1978).
225. Wood, C.E.C., Judaprawira and Eastman, L.F., *IEDM Tech. Digest*, p. 388, talk 16.3, Washington, D.C. (1979).
226. Wood, C.E.C., Metze, G., Berry, J. and Eastman, L.F., *J. Appl. Phys.*, Vol. 51, p. 383 (1980).
227. Malik, R.J., AuCoin, T.R., Ross, R.L., Board, K., Wood, C.E.C. and Eastman, L.F., *Electron. Letts.*, Vol. 16, p. 836 (1980).
228. Malik, R.J., Board, K., Eastman, L.F., Wood, C.E.C., AuCoin, T.R., Ross, R.L. and Savage, R.O., *IEDM Tech. Digest*, p. 456, talk 17.7, Washington, D.C. (1980).
229. Malik, R.J. and Dixon, S., Jr., *IEEE Electron. Dev. Lett.*, EDL-3, p. 205 (1982).
230. Dixon, S., Jr. and Malik, R.J., *IEEE Trans. Microwave Theory and Tech.*, Vol. MTT-31, p. 155 (1983).
231. Szubert, J.M. and Singer, K.E., *J. Vac. Sci. Technol.*, Vol. B3, p. 794 (1985).
232. Hiyamizu, S., Sasa, S., Ishikawa, T., Kondo, K. and Ishikawa, H., *Jpn. J. Appl. Phys.*, Vol. 24, p. L431 (1985).
233. Sahai, R., Harris, J.S., Eden, R.C., Bubulac, L.O. and Chu, J.C., *Critical Reviews in Solid State Sciences*, p. 567 (1975).
234. Kroemer, H., *RCA Rev.*, Vol. 18, p. 332 (1957).
235. Capasso, F., Tsang, W.T., Bethea, C.G., Hutchinson, A.L. and Levine, B.F., *Appl. Phys. Lett.*, Vol. 42, p. 93 (1983).
236. Capasso, F., *Surf. Sci.*, Vol. 142, p. 513 (1984).
237. Capasso, F., *Physica*, Vol. 129B, p. 92 (1985).
238. Woodall, J.M., Freeout, J.L., Pettit, G.D., Jackson, T. and Kirchner, P., *J. Vac. Sci. Technol.*, Vol. 19, p. 626 (1981).
239. Kawabe, M., Matsuura, N. and Inuzuka, M., *Jpn. J. Appl. Phys.*, Vol. 21, p. 439 (1981).
240. Kawabe, M., Matsuura, N. and Inuzuka, M., *Jpn. J. Appl. Phys.*, Vol. 21, p. L447 (1982).
241. Fujii, T., Nakata, Y., Sugiyama, Y. and Hiyamizu, S., *Jpn. J. Appl. Phys.*, Vol. 25, p. L254 (1986).
242. Neave, J.H. and Joyce, B.A., *Appl. Phys.*, Vol. A31, p. 1 (1983).

243. Sakamoto, T., Kawai, N.J., Nakagawa, T., Ohta, K. and Kojima, T., *Appl. Phys. Lett.*, Vol. 47, p. 617 (1985).
244. Van Hove, J.M., Lent, C.S., Pukite, P.R. and Cohen, P.I., *J. Vac. Sci. Technol.*, Vol. B1, p. 741 (1983).
245. Madhukar, A. and Ghaisas, S.V., *Appl. Phys. Lett.*, Vol. 47, p. 247 (1985).
246. Ghaisas, S.V. and Madhukar, A., *J. Vac. Sci. Technol.*, Vol. B3, p. 540 (1985).
247. Madhukar, A., Lee, T.C., Yen, M.Y., Chen, P., Kim, J.Y., Ghaisas, S.V. and Newman, P.G., *Appl. Phys. Lett.*, Vol. 46, p. 1148 (1985).
248. Voillot, F., Madhukar, A., Kim, J.Y., Chen, P., Cho, N.M., Tang, W.C. and Newman, P.G., *Appl. Phys. Lett.*, Vol. 48, p. 1009 (1986).
249. Grunthaner, F.J., Yen, M.Y., Fernandez, R., Lee, T.C., Madhukar, A. and Lewis, B.F., *Appl. Phys. Lett.*, Vol. 46, p. 983 (1985).
250. DiLorenzo, J.V., Niehaus, W.C. and Cho, A.Y., *J. Appl. Phys.*, Vol. 50, p. 951 (1979).
251. Cho, A.Y., Kollberg, E., Zirath, H., Snell, W.W. and Schneider, M.V., *Electron. Lett.*, Vol. 18, p. 424 (1982).
252. Ohno, H., Barnard, J., Wood, C.E.C. and Eastman, L.F., *IEEE Electron. Dev. Lett.*, Vol. 1, p. 154 (1980).
253. McLean, T.D., Kerr, T.M., Westwood, D.I., Blight, S.R., Page, D. and Wood, C.E.C., 12th Int. Symp. GaAs and Related Compounds, 1985, Karuizawa, Japan, 1985, *Inst. Phys. Conf. Ser.*, Vol. 79, p. 349 (1986).
254. Rosencher, E., Delage, S., Campidelli, Y., and Arnaud D'Avitaya, F., *Electron. Lett.*, Vol. 20, p. 762 (1984).
255. Rosencher, E., Glastre, G., Vincent, G., Vareille, A. and Arnaud D'Avitaya, F., *Electron. Lett.*, Vol. 22, p. 700 (1986).
256. Derkits, G.E., Jr., Harbison, J.P., Levkoff, J. and Hwang, D.M., *Appl. Phys. Lett.*, Vol. 48, p. 1220 (1986).
257. Harbison, J.P., Hwang, D.M., Levkoff, J. and Derkits, G.E., *Appl. Phys. Lett.*, Vol. 47, p. 1187 (1985).
258. Ploog, K. and Fisher, A., *Appl. Phys.*, Vol. 13, p. 111 (1977).
259. Riggs, W.M. and Parker, M.J., *Methods of Surface Analysis*, chap. 4, A.W. Czanderna, ed., Amsterdam, Elsevier Scientific Publ. (1975).
260. Nordling, C., Sokolowski, E. and Siegbahn, K., *Ark. Fys.*, Vol. 13, p. 483 (1958).
261. Contour, J.P., Massies, J. and Saletes, A., *Jpn. J. Appl. Phys.*, Vol. 24, p. L563 (1985).
262. McGuire, G.E., Church, L.B., Jones, D.L., Smith, K.K. and Tuenge, D.T., *J. Vac. Sci. Technol.*, Vol. A1, p. 732 (1983).
263. Joshi, A., Davis, L.E. and Palmberg, P.W., *Methods of Surface Analysis*, chap. 5, A.W. Czanderna, ed., Amsterdam, Elsevier Scientific Publ. (1975).
264. McHugh, J.A., *Methods of Surface Analysis*, chap. 6, A.W. Czanderna, ed., Amsterdam, Elsevier Scientific Publ. (1975).
265. Uthe Technology International, Sunnyvale, CA, USA.
266. Abraham, M.S. and Buiocchi, C.J., *J. Appl. Phys.*, Vol. 36, p. 2855 (1965).
267. Stirland, D.J., Int. Symp. GaAs and Related Compounds, 1976, Edinburg, Scotland (1977).

268. Stillman, G.E. and Wolfe, C.M., *Thin Solid Films*, Vol. 31, p. 69 (1976).
269. van der Pauw, L.J., *Philips Res. Repts.*, Vol. 13, p. 1 (1958).
270. van der Pauw, L.J., *Philips Res. Repts.*, Vol. 20, p. 220 (1958).
271. Blood, P. and Orton, J.W., *Rep. Prog. Phys.*, Vol. 41, p. 11, (1978).
272. Blood, P., *Semiconductor Sci. and Technol.*, Vol. 1, p. 7 (1986).
273. Ambridge, T. and Faktor, M.M., *J. Appl. Electrochem.*, Vol. 5, p. 319 (1975).
274. Ashen, D.J., Dean, P.J., Hurle, D.T.J., Mullin, J.B. and White, A.M., *J. Phys. Chem. Solids*, Vol. 36, p. 1041 (1975).
275. White, A.M., Dean, P.J., Tayor, L.L., Clarke, R.C., Ashen, D.J. and Mullin, J.B., *J. Phys. C.*, Vol. 5, p. 1727 (1972).
276. Stillman, G.E., Cook, L.W., Roth, T.J., Low, T.S. and Skromme, B.J., *GaInAsP Alloy Semiconductors*, chap. 6, T.P. Pearsall, ed., New York, John Wiley & Sons (1982).
277. Reynolds, D.C., Bajaj, K.K. and Litton, C.W., *Microscopic Identification of Electronic Defects in Semiconductors*, p. 339, 1985, San Francisco, USA (1985).
278. Lang, D.V., *J. Appl. Phys.*, Vol. 45, p. 3023 (1974).
279. Hewlett-Packard, Palo Alto, CA, USA.
280. Materials Development Corporation, Chatsworth, CA, USA.
281. Polaron, Watford, Hertfordshire, England.
282. Sula Technologies, Lafayette, CA, USA.
283. Lang, D.V., Cho, A.Y., Gossard, A.C., Ilegems, M. and Wiegman, W., *J. Appl. Phys.*, Vol. 47, p. 2558 (1976).
284. Blood, P. and Harris, J.J., *J. Appl. Phys.*, Vol. 56, p. 993 (1984).
285. DeJule, R.Y., Haase, M.A., Stillman, G.E., Palmateer, S.C. and Hwang, J.C.M., *J. Appl. Phys.*, Vol. 57, p. 5287 (1985).
286. Hikosata, K., Mimura, T. and Hiyamizu, S., *Int. Symp. GaAs and Related Compounds*, 1981, Japan (1982).
287. McAfee, S.R., Tsang, W.T. and Lang, D.V., *J. Appl. Phys.*, Vol. 52, p. 6165 (1981).
288. Zhou, B.L., Ploog, K., Gmelin, F., Zheng, X.Q. and Schulz, M., *Appl. Phys.*, Vol. A28, p. 223 (1982).
289. Yamanaka, K., Nanituska, S., Mannoh, M., Yuasa, S.T., Nomura, Y., Mihara, M. and Ishi, M., *J. Vac. Sci. Technol.*, Vol. B2, p. 229 (1984).
290. Sax, N.I., *Dangerous Properties of Industrial Materials*, New York Van Nostrand Rheinhold Co. (1984).
291. Webb, D.R., Sipes, I.G. and Carter, D.E., *Toxicol. Appl. Pharmacol.*, Vol. 76, p. 96 (1984).
292. U.S. Dept. of HEW, *Occupational Exposure to Inorganic Arsenic; New Criteria - 1975*, Order from: Superintendent of Documents, U.S. Gov. Printing Office, Washington, D.C., 20402.
293. Harrison, R.J., *Occupational Medicine*, J. LaDou, ed., Vol. 1, p. 49, Hanley & Belfus, Inc., Philadelphia (1986).
294. Harris, J.J., Joyce, B.A. and Dodson, P.J., *Surf. Sci.*, Vol. 103, p. L90 (1981).
295. Wood, C.E.C., *Sur. Sci.*, Vol. 108, p. L441 (1981).
296. Neave, J.H., Joyce, B.A., Dobson, P.J. and Norton, N., *Appl. Phys.*, Vol. A31, p. 1 (1983).

297. Lewis, B.F., Lee, T.C., Grunthaner, F.J., Madhukar, A., Fernandez, R. and Maserjian, J., *J. Vac. Sci. Technol.*, Vol. B2, p. 419 (1984).
298. Sakamoto, T., Kawai, N.J.J., Nakagawa, T., Ohta, K. and Kojima, T., *Appl. Phys. Lett.*, Vol. 47, p. 617 (1985).
299. Aizaki, N. and Tatsumi, T., 2nd Int. Conf. on Modulated Semiconductor Structures, Collected Papers, p. 282, Kyoto, Japan (1985).
300. Lewis, B.F., Fernandez, R., Madhukar, A. and Grunthaner, F.J., *J. Vac. Sci. Technol.*, Vol. B4, p. 560 (1986).
301. Choi, H.K., Tsaur, B.-Y., Metze, G.M., Turner, G.W. and Fan, J.C.C., *IEEE Elec. Dev. Lett.*, Vol. EDL-5, p. 207 (1984).
302. Windhorn, T.H., Metze, G.M., Tsaur, B.-Y. and Fan, J.C.C., *Appl. Phys. Lett.*, Vol. 45, p. 309 (1984).
303. Wang, W.I., *Appl. Phys. Lett.*, Vol. 44, p. 1149 (1984).
304. Morkoc, H., Peng, C.K., Henderson, T., Kopp, W., Fischer, R., Erickson, L.P., Longerbone, M.D. and Youngman, R.C., *IEEE Elec. Dev. Lett.*, Vol. EDL-6, p. 381 (1985).
305. Fischer, R., Henderson, T., Klem, J., Masseulink, W.T., Kopp, W. and Morkoc, H., *Electron. Lett.*, Vol. 20, p. 945 (1984).
306. Windhorn, T.H. and Metze, G.M., *Appl. Phys. Lett.*, Vol. 47, p. 1031 (1985).
307. Fischer, R., Kopp, W., Morkoc, H., Pion, M., Specht, A., Burkhart, G., Appellman, H., McGougan, D. and Rice, R., *App. Phys. Lett.*, Vol. 48, p. 1360 (1986).
308. Fischer, R., Neuman, D., Zabel, H., Morkoc, H., Choi, C. and Otsuka, N., *Appl. Phys. Lett.*, Vol. 48, p. 1223 (1986).
309. Choi, H.K., Turner, G.W. and Tsaur, B.-Y., *IEEE Elec. Dev. Lett.*, Vol. EDL-7, p. 241 (1986).
310. Bachrach, R.Z. and Krusor, B.S., *J. Vac. Sci. Technol.*, Vol. 18, p. 756 (1981).
311. Ito, T., Shinohara, M. and Imamura, Y., *Jpn. J. Appl. Phys.*, Vol. 23, p. L524 (1984).
312. Wang, Y.H., Liu, W.C., Liao, S.A., Cheng, K.Y. and Chang, C.Y., *Jpn. J. Appl. Phys.*, Vol. 24, p. 628 (1985).
313. Watanabe, N., Fukunaga, T., Kobayashi, K.L.I. and Nakashima, H., *Jpn. J. Appl. Phys.*, Vol. 24, p. L498 (1985).
314. Tatsumi, T., Aizaki, N. and Tsuya, H., *Jpn. J. Appl. Phys.*, Vol. 24, p. L227 (1985).
315. Weng, S.-L., Webb, C., Chai, Y.G. and Bandy, S.G., *Appl. Phys. Lett.*, Vol. 47, p. 391 (1985).
316. Fujiwara, K., Nishikawa, Y., Tokuda, Y. and Nakayama, T., *Appl. Phys. Lett.*, Vol. 48, p. 701 (1986).
317. Fronius, H., Fischer, A. and Ploog, K., *Jpn. J. Appl. Phys.*, Vol. 25, p. L137 (1986).
318. Kirchner, P.D., Woodall, J.M., Freeouf, J.L. and Petit, G.D., *Appl. Phys. Lett.*, Vol. 38, p. 427 (1981).
319. Pettit, G.D., Woodall, J.M., Wright, S.L., Kirchner, P.D. and Freeouf, J.L., *J. Vac. Sci. Technol.*, Vol. B2, p. 241 (1984).
320. Shinohara, M., Ito, T., Wada, K. and Imamura, Y., *Extended Abstracts of the 16th Conference on Solid State Dev. and Materials*, p. 193 (1984).

321. Wood, C.E.C., Rathbun, L., Ohno, H. and DeSimone, D., *J. Cryst. Growth*, Vol. 51, p. 299 (1981).
322. Chai, Y.G. and Chow, R., *Appl. Phys. Lett.*, Vol. 38, p. 796 (1981).
323. Akimoto, K., Dohsen, M., Arai, M. and Watanabe, N., *J. Cryst. Growth*, Vol. 73, p. 117 (1985).
324. Tsang, W.T., *Appl. Phys. Lett.*, Vol. 46, p. 1086 (1985).
325. Tokumitsu, E., Kudou, Y., Konagai, M. and Takahashi, K., *J. Appl. Phys.*, Vol. 55, p. 3163 (1984).
326. Kondo, K., Ishikawa, H., Sasa, S. and Hiyamizu, S., Proc. 12th Symp. GaAs & Related Compounds, Karuizawa, Japan, *1985 Inst. Phys. Conf. Ser.*, Vol. 79, p. 86 (1986).
327. Calawa, A.R., *Appl. Phys. Lett.*, Vol. 33, p. 1020 (1978).
328. Pao, Y.C., Liu, D., Lee, W.S. and Harris, J.S., *Appl. Phys. Lett.*, Vol. 48, p. 1291 (1986).
329. Tokumitsu, E., Kudou, Y., Konagai, M. and Takahashi, K., *Jpn. J. Appl. Phys.*, Vol. 24, p. 1189 (1985).
330. Briones, F., Golmayo, D., Gonzalez, L. and de Miguel, J.L., *J. Vac. Sci. Technol.*, Vol. B3, p. 568 (1985).
331. Morris, F.J. and Fukui, H., *J. Vac. Sci. Technol.*, Vol. 11, p. 506 (1974).
332. Chow, R. and Chai, Y.G., *J. Vac. Sci. Technol.*, Vol. A1, p. 49 (1983).
333. Panish, M.B. and Sumski, S., *J. Appl. Phys.*, Vol. 55, p. 3571 (1984).
334. Temkin, H., Panish, M.B., Petroff, P.M., Hamm, R.A., Vandenberg, J.M. and Sumski, S., *Appl. Phys. Lett.*, Vol. 47, p. 394 (1985).
335. Vandenberg, J.M., Hamm, R.A., Macrander, A.T., Panish, M.B. and Temkin, H., *Appl. Phys. Lett.*, Vol. 48, p. 1153 (1986).
336. Kondo, K., Ishikawa, H., Sasa, S., Sugiyama, Y. and Hiyamizu, S., *Jpn. J. of Appl. Phys*, Vol. 25, p. L52 (1986).
337. Tsang, W.T. and Miller, R.C., *Appl. Phys. Lett.*, Vol. 48, p. 1288 (1986).
338. Kawaguchi, Y., Asahi, H. and Nagai, H., *Jpn. J. Appl. Phys.*, Vol. 24, p. L221 (1985).
339. Kawaguchi, Y., Asahi, H. and Nagai, H., 12th Symp. Int. GaAs and Related Compounds, Karuizawa, Japan, 1985, *Inst. Phys. Conf. Ser.*, Vol. 79, p. 79 (1986).
340. Tsang, W.T., Dayem, A.H., Chiu, T.H., Cunningham, J.E., Schubert, E.F., Ditzenberger, J.A., Shah, J., Zyskind, J.L. and Tabatabaie, N., *Appl. Phys. Lett.*, Vol. 49, p. 170 (1986).
341. Tsang, W.T., *J. Appl. Phys.*, Vol. 58, p. 1415 (1985).
342. Esaki, L. and Tsu, R., *IBM J. Res. Dev.*, Vol. 14, p. 61 (1970).
343. Osbourn, G.C., *J. Appl. Phys.*, Vol. 53, p. 1586 (1982).
344. Lee, G.S., Lo, Y., Lin, Y.F., Bedair, S.M. and Laidig, W.D., *Appl. Phys. Lett.*, Vol. 47, p. 1219 (1985).
345. Bean, J.C., Feldman, L.C., Fiory, A.T., Nakahara, S. and Robinson, I.K., *J. Vac. Sci. Technol.*, Vol. A2, p. 436 (1984).
346. Jorke, H. and Herzog, J.-J., *J. Electrochem. Soc.*, Vol. 133, p. 998 (1986).
347. Faurie, J.P., Sivananthan, S., Chu, X. and Wijewarnasuriya, P.A., *Appl. Phys. Lett.*, Vol. 48, p. 785 (1986).
348. Monfroy, G., Sivananthan, S., Chu, X., Faurie, J.P., Knox, R.D. and Staudenmann, J.L., *Appl. Phys. Lett.*, Vol. 49, p. 152 (1986).

349. Fujii, T., Hiyamizu, S., Yamakoshi, S. and Ishikawa, T., *J. Vac. Sci. Technol.*, Vol. B3, p. 776 (1985).
350. Noda, S., Fujiwara, K. and Nakayama, T., *Appl. Phys. Lett.*, Vol. 47, p. 1205 (1985).
351. Schaff, W.J. and Eastman, L.F., *J. Vac. Sci. Technol.*, Vol. B2, p. 265 (1984).
352. Morioka, M., Mishima, T., Hiruma, K., Katayama, Y. and Shiraki, Y., 12th Int. Symp. on GaAs and Related Compounds, Karuizawa, Japan, 1985, *Inst. Phys. Conf. Ser.*, Vol. 79, p. 121 (1986).
353. Schubert, E.F., Fischer, A., Horikoshi, Y. and Ploog, K., *Appl. Phys. Lett.*, Vol. 47, p. 219 (1985).
354. Vojak, B.A., Zajac, G.W., Chambers, F.A., Meese, J.M., Chumbley, P.E., Kaliski, R.W., Holonyak, N., Jr. and Nam, D.W., *Appl. Phys. Lett.*, Vol. 48, p. 251 (1986).
355. Osbourn, G.C., *J. Vac. Sci. Technol.*, Vol. B2, p. 176 (1984).
356. Temkin, H., Pearsall, T.P., Bean, J.C., Logan, R.A. and Luryi, S., *Appl. Phys. Lett,* Vol. 48, p. 963 (1986).
357. Pearsall, T.P., Temkin, H., Bean, J.C. and Luryi, S., *IEEE Elect. Dev. Lett.*, Vol. EDL-7, p. 330 (1986).
358. Faurie, J.P., Reno, J. and Boukerche, M., *J. Cryst. Growth*, Vol. 72, p. 111 (1985).
359. *Automation in Semiconductor Manufacturing*, SRC Document No. Q85057 (February 1985).
360. Ueda, S., Kamohara, H., Ishikawa, Y., Tamura, H., Katoo, S. and Shiraki, Y., *J. Vac. Sci. Technol.*, Vol. A4, p. 602 (1986).
361. Bean, J.C. and Butcher, P., *Proc. 1st Int. Symp. on Silicon MBE*, J.C. Bean, ed., Electrochem. Soc., p. 429, Pennington, NJ (1985).
362. Harbison, J.P., *J. Vac. Sci. Technol.*, Vol. A4, p. 1033 (1986).
363. Miyauchi, E. and Hashimoto, H., *J. Vac. Sci. Technol.*, Vol. A4, p. 933 (1986).
364. Okamoto, K., Wood, C.E.C. and Eastman, L.F., *Appl. Phys. Lett.*, Vol. 38, p. 636 (1981).
365. Tsang, W.T., *Appl. Phys. Lett.*, Vol. 33, p. 1022 (1978).
366. Stall, R., Wood, C.E.C., Board, K. and Eastman, L.F., *Electron. Lett.*, Vol. 15, p. 800 (1979).
367. Cho, A.Y. and Reinhart, F.K., *Appl. Phys. Lett.*, Vol. 21, p. 355 (1972).
368. Tsang, W.T. and Cho, A.Y., *Appl. Phys. Lett.*, Vol. 32, p. 491 (1978).
369. Tsang, W.T. and Ilegems, M., *Appl. Phys. Lett.*, Vol. 31, p. 301 (1977).
370. Cho, A.Y. and Ballamy, W.C., *J. Appl. Phys.*, Vol. 46, p. 783 (1975).
371. Metze, G.M., Levy, H.M., Woodard, D.W., Wood, C.E.C. and Eastman, L.F., *Appl. Phys. Lett.*, Vol. 37, p. 628 (1980).
372. Shimizu, S., Tsukakoshi, O., Komiya, S. and Mikita, Y., *Jpn. J. Appl. Phys.*, Vol. 24, p. 1130 (1985).
373. Takagi, T., Yamada, I. and Sasaki, A., *Thin Solid Films*, Vol. 45, p. 569 (1977).
374. Yamada, I. and Takagi, T., *Thin Solid Films*, Vol. 80, p. 105 (1981).
375. Yamada, I., Takaoka, H., Usui, H. and Takagi, T., *J. Vac. Sci. Technol.*, Vol. A4, p. 722 (1986).
376. Fraas, L.M., *J. Appl. Phys.*, Vol. 52, p. 6939 (1981).

377. Sugiura, O. and Matsumura, M., *Jpn. J. Appl. Phys.*, Vol. 24, p. L925 (1985).
378. Kondo, N., Kawashima, M. and Sugiura, H., *Jpn. J. Appl. Phys.*, Vol. 24, p. L370 (1985).
379. Kondo, N. and Kawashima, M., 12th Int. Symp. on GaAs and Related Compounds, Karuizawa, Japan, 1985, *Inst. Phys. Conf. Ser.*, Vol. 79, p. 97 (1986).
380. Takahashi, K., 12th Int. Symp. GaAs and Related Compounds, Karuizawa, Japan, 1985, *Inst. Phys. Conf. Ser.*, Vol. 79, p. 73 (1986).
381. Nishizawa, J., Abe, H., Kurabayashi, T. and Sakurai, N., *J. Vac. Sci. Technol.*, Vol. A4, p. 706 (1986).
382. Elsner, G., Hinkel, H., Kempf, J. and Stahl, R., *IBM Techn. Disclosure Bulletin*, Vol. 26, p. 6215 (1984).
383. de Jong, T. and Saris, F.W., *Semiconductor Processing and Equipment Symposium*, Zurich, Switzerland, p. 210 (1983).
384. Larsen, C.A., Chen, C.H., Kitamura, M., Stringfellow, G.B., Brown, D.W. and Robertson, A.J., *Appl. Phys. Lett.*, Vol. 48, p. 1531 (1986).

7

Metal-Organic Chemical Vapor Deposition: Technology and Equipment

J.L. Zilko

1. INTRODUCTION

The growth of thin layers of compound semiconducting materials by the co-pyrolysis of various combinations of organometallic compounds and hydrides, known generically as metal-organic chemical vapor deposition (MO-CVD), has assumed a great deal of technological importance in the fabrication of a number of opto-electronic and high speed electronic devices. State of the art performance of a number of active devices that have been fabricated with this technology has been demonstrated. These devices include lasers,[1] solar cells,[2] phototransistors,[3] photocathodes,[4] field effect transistors[5] and modulation doped field effect transistors.[6] The efficient operation of these devices requires the grown films to have a number of excellent materials properties, including purity, high luminescence efficiency, and/or abrupt interfaces. In addition, this technique has been used to deposit virtually all III-V and II-VI semiconducting compounds and alloys in support of materials studies. The III-V materials GaAs and AlGaAs have been the most extensively studied materials due to their technological importance for lasers, light emitting diodes, photocathodes and high speed field effect transistors, although InP, InGaAs, and InGaAsP are being studied to an increasing extent for light emitting and detecting devices at longer wavelengths. Also, there has been much recent interest in the II-VI materials HgCdTe[7] and ZnSSe[8] for optical devices in the far infrared and visible wavelength ranges, respectively. Finally, there is the promise (largely unfulfilled for many applications) of producing highly uniform layers over large areas.

Much of the appeal of MO-CVD lies in the fact that readily transportable, relatively pure organometallic compounds can be made for most of the elements that are of interest in the epitaxial deposition of doped and undoped compound semiconductors. In addition, a large driving force (i.e., a large free energy change) exists for the pyrolysis of the source chemicals. This means that a wide variety of materials can be grown using this technique that are difficult to grow by other epitaxial techniques. The growth of Al-bearing compounds (difficult by chloride and hydride vapor phase epitaxy due to thermodynamic constraints[9]) and P-bearing compounds (difficult by molecular beam epitaxy due to the high vapor pressure of P[10]) are especially noteworthy. The large free energy change also allows the growth of single crystal semiconductors on non-semiconductor (insulator) substrates[11] (heteroepitaxy) as well as on semiconductor substrates.

In spite of the versatility of MO-CVD and the demonstrations of high quality growth, it is still not considered the epitaxial growth technique of choice for a number of device applications. This is because other techniques, while more limited, presently can or previously have been able to produce material of higher quality (higher purity, higher luminescence efficiency, more abrupt interfaces, larger areas, etc.) or require less care in system design, construction, and use to routinely produce material of equivalent quality. For example, the $AsCl_3/Ga/H_2$ vapor phase epitaxy technique produced high purity GaAs[12] earlier and more reproducibly than has MO-CVD in systems which probably had considerably less leak integrity. On the other hand, the $AsCl_3/Ga/H_2$ technique is not easily able to grow materials containing Al,[9] thus eliminating a large body of devices such as lasers and modulation-doped field effect transistors that can be produced using this technique.

Much of the effort of the last few years has centered around improving the quality of material that can be grown by MO-CVD, particularly those materials properties that are important for device structures. In large part, this effort has resulted in improvements in MO-CVD equipment design and construction. Although a great deal of progress has been made, much work remains to be done, especially in the areas of increasing wafer size and throughput, improving the reproducibility of some of the materials properties, and understanding the basic chemical processes. It is likely that the resolution of all of these issues will result in further equipment modifications, particularly at the reaction chamber. Nevertheless, the technique has reached a sufficient level of maturity so that a review of the basic technology and equipment is appropriate.

In this chapter, we will review MO-CVD technology and equipment with an emphasis on providing a body of knowledge and understanding that will enable the reader to gain practical insight into the various technological processes and options. We assume that the reader has some knowledge of compound semiconductors and devices and of epitaxial growth. Materials and device results will not be discussed in this chapter because of space limitations except to illustrate equipment design and technology principles. For a more detailed discussion of materials and devices, the reader is referred to several recent reviews.[1,13] An excellent, more general review of the process has also recently appeared.[14] Although most of the discussion is applicable to growth of compound semiconductors on both semiconductor and insulator substrates, we will be primarily concerned with the technologically more useful semiconductor substrate growth.

This chapter is organized into four main sections. We first discuss some of the physical and chemical properties of the sources that are used in MO-CVD. Because the sources used for MO-CVD have rather unique physical properties, are generally very toxic and/or pyrophoric, and are chemically very reactive, knowledge of source properties is necessary to understand MO-CVD technology and system design. The discussion of sources will focus on the physical properties of sources used in MO-CVD and source packaging.

The next section deals with deposition conditions and chemistry and reaction mechanisms. Because MO-CVD uses sources that are introduced into a reaction chamber at temperatures around room temperature and are then thermally decomposed at elevated temperatures in a cold wall reactor, large temperature and concentration gradients and nonequilibrium reactant and product concentrations are present during film growth.[15] Thus, materials growth takes place far from thermodynamic equilibrium and system design and growth procedure have a large effect on the film results that are obtained. As a consequence, results from various laboratories that are using nominally the same growth conditions can be very different and sometimes contradictory. Furthermore, different effects are important for the growth of materials from different alloy systems because growth is carried out in different growth regimes. For these reasons, it is presently impossible to write an "equation of state" that describes the MO-CVD process. Nevertheless, we will attempt to give a general framework to the chemistry of deposition and point out effects that control the quality of deposition for several classes of materials.

In the next section, we consider system design and construction. A schematic of a typical MO-CVD system is shown in Figure 1. An MO-CVD system can be thought of as being composed of several functional subsystems. The subsystems include reactant storage, a gas handling manifold, a reaction chamber, and an exhaust. This section of the chapter is organized into several subsections that deal with the generic issues of leak integrity and cleanliness and the gas manifold, reaction chamber, and exhaust subsystems. Reactant storage is not covered because this is a local safety issue and because equipment and use information can be obtained from a number of suppliers. It will once again be emphasized that because materials growth takes place far from thermodynamic equilibrium, seemingly insignificant changes in equipment design can have a large effect on the grown materials.

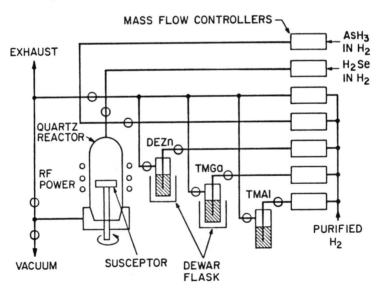

Figure 1: Schematic drawing of a typical MO-CVD system. From Dupuis.[1]

Most of the discussion on system design and construction will be specifically concerned what is recognized as "conventional" MO-CVD which takes place at pressures of ~0.1 to 1 atmosphere in thermally heated, cold wall, open tube flow systems. We will conclude this chapter with a short section on several more exploratory variations of this technique which may provide significant advantages in the future. These variations either operate at significantly lower pressures than conventional MO-CVD or use nonthermal means of decomposing reactants: vacuum MO-CVD,[16] plasma-enhanced MO-CVD,[17] and metal-organic-molecular beam epitaxy (MO-MBE)[18] (also known as "chemical beam epitaxy")[19] and photo-enhanced MO-CVD.[20] These exploratory versions of MO-CVD are all still in the early stages of development.

In performing literature searches in the field of MO-CVD, it should be recognized that a number of names have been used for this growth technique including organometallic chemical vapor deposition (OM-CVD), metal-organic vapor phase epitaxy (MO-VPE), organometallic vapor phase epitaxy (OM-VPE), organometallic pyrolysis (OMP) or metal-alkyl vapor phase epitaxy. Nevertheless, the technique is essentially the same no matter which name is used to describe it. We will use MO-CVD as the name of the technique in this chapter because this is the original name and is the most general term for the process even though most applications require the epitaxial nature of the process. Ludowise gives a short discussion of the merits of the various names for this technique.[14]

2. PHYSICAL AND CHEMICAL PROPERTIES OF SOURCES USED IN MOCVD

Sources that are used in MO-CVD for both major film constituents and dopants are various combinations of organometallic compounds and hydrides. The III-V and II-VI compounds and alloys are usually grown using low molecular weight metal alkyls such as dimethyl cadmium (DMCd - chemical formula: $(CH_3)_2Cd$) or trimethyl gallium (TMGa - chemical formula: $(CH_3)_3Ga$) as the metal (group II or group III) source although organometallic adduct compounds such as trimethyl indium-trimethyl phosphine (TMInTMP - chemical formula: $(CH_3)_3In-P(CH_3)_3$) have been used as the In source for In-containing films. The non-metal (group V and group VI) source is either a hydride such as AsH_3, PH_3, H_2Se, or H_2S or an alkyl such as trimethyl antimony (TMSb) or diethyl tellurium (DETe). The sources are introduced as vapor phase constituents into a reaction chamber at approximately room temperature and are thermally decomposed at elevated temperatures by a hot susceptor and substrate to form the desired film in the reaction chamber. The reaction chamber walls are not deliberately heated (a "cold wall" process) and do not directly influence the chemical reactions that occur in the chamber. The general overall chemical reaction that occurs during the MO-CVD process can be written:

$$(1) \quad R_n M(v) + ER'_n(v) \xrightarrow{\nabla} ME(s) + n\, RR'(v)$$

where R and R' represent a methyl (CH_3) or ethyl (C_2H_5) radical or hydrogen, M is a group II or group III metal, E is a group V or group VI element, and n = 2 or 3 depending on whether II-VI or III-V growth is taking place.

From Equation 1, it is seen that the vapor phase reactants R_nM and ER'_m are thermally decomposed at elevated temperatures to form the nonvolatile product ME which is deposited on the substrate and susceptor and the volatile product RR' which is carried away by the H_2 flush gas to the exhaust. An example would be the reaction of $(CH_3)_3Ga$ and AsH_3 to produce GaAs and CH_4. It should be noted that Equation 1 only describes a simplified overall reaction and ignores any side reactions and intermediate steps. We will consider reaction pathways and side reactions in more detail in the Section 3. The MO-CVD growth of mixed alloys can be described by Equation 1 by substituting two or more appropriate reactant chemicals of the same valence in place of the single metal or nonmetal species. It should also be noted that Equation 1 allows the use of both hydride and organometallic compounds as sources. Virtually all of the possible III-V and II-VI compounds and alloys have been grown by MO-CVD. An extensive list of the materials grown and sources used is given in a review that can be obtained from Alpha Products.[21]

In this section, we will discuss some of the physical properties and chemistry of MO-CVD sources, both organometallic and hydride. Due to space limitations, we will only discuss those properties that are important for the growth of material. Properties that will be discussed include vapor pressure, thermal stability, and source packaging. Growth conditions, material purity and chemical interactions between species will be discussed in the next section on deposition chemistry. Interspersed within this section will be short discussions of the specific sources that are used for the most important materials that have been grown by this technique. More extensive reviews are available in references 14 and 21. Because organometallics and hydrides have rather different physical properties, we will discuss them separately in this section.

2.1 Physical and Chemical Properties of Organometallic Compounds

The organometallic compounds that are used for MO-CVD are generally clear liquids or occasionally white solids around room temperature. They are often pyrophoric or highly flammable and have relatively high vapor pressures in the range of 0.5-100 Torr around room temperature. They can be readily transported as vapor phase species to the reaction chamber by bubbling a suitable carrier such as H_2, He or N_2 through the material as it is held in a container at temperatures near room temperature. The organometallic compounds are generally monomers in the vapor phase except for trimethyl aluminum (TMAl) which is dimeric.[14] Typically, low molecular weight alkyls such as TMGa or DMCd are used for compound semiconductor work because their relatively high vapor pressures allow relatively high growth rates. For In-based compounds and alloys, both adduct sources such as TMIn-TMP and simple alkyls such as TMIn have been used. As a general rule, the low molecular weight compounds tend to have a higher vapor pressure at a given temperature than the higher molecular weight materials. Thus, TMGa has a vapor pressure of 65.4 Torr at $0°C$[21] while triethyl gallium (TEGa) which has also been used to grow GaAs has a vapor pressure of only 18 Torr at the much higher temperature of $48°C$.[21] A list of the vapor pressures as a function of temperature for all the readily obtainable organometallic species is given in references 14 and 21 and is reproduced in Table 1 for the convenience of the reader. Table 1 also gives melting and boiling temperatures and an indication of the chemical reactivity in air for each compound.

Table 1: Physical Properties of Metal Alkyls For MOCVD[21]

Compound Formula, State[a] Reactivity[c]	MP°C	BP°C	Reported Vapor Pressure mm Hg @°C	Vapor Pressure Equation $\log_{10} P(\text{mm Hg}) = B - A/T$ [b]
ALUMINUM				
Diisobutyl aluminum hydride ($C_4H_9)_2$AlH, Liquid, P	-70		1 @ 118 4 @ 140	
Triethyl aluminum ($C_2H_5)_3$Al, Liquid, P	-58	194	4 @ 80	10.784 - 3625/T; (110°C-140°C)
Triisobutyl aluminum ($C_4H_9)_3$Al, Liquid, P	4	130	1 @ 50	
Trimethyl aluminum ($CH_3)_3$Al, Liquid, P	15.4	126	8.4 @ 20	7.3147 - 1534.1/(T - 53); (17°C-100°C)
ANTIMONY				
Triethyl antimony ($C_2H_5)_3$Sb, Liquid, P	-98	160	17 @ 75	
Trimethyl antimony ($CH_3)_3$Sb, Liquid, P	-87.6	80.6		7.7280 - 1709/T
ARSENIC				
Triethyl arsenic ($C_2H_5)_3$As, Liquid, F	-91	140	15.5 @ 37	
Trimethyl arsenic ($CH_3)_3$As, Liquid, F	-87.3	50-52		7.7119 - 1563/T
BERYLLIUM				
Diethyl beryllium ($C_2H_5)_2$Be, Liquid, P	12	194		7.59 - 2200/T
BISMUTH				
Trimethyl bismuth ($CH_3)_3$Bi, Liquid, P	-107.7	107.9		7.6280 - 1816/T
CADMIUM				
Dimethyl cadmium ($CH_3)_2$Cd, Liquid, P	-4.5	105.5	350 @ 80	7.764 - 1850/T
GALLIUM				
Triethyl gallium ($C_2H_5)_3$Ga, Liquid, P	-82.3	143	18 @ 48	8.224 - 2222/T; (50°C-80°C)
Trimethyl gallium ($CH_3)_3$Ga, Liquid, P	-15.8	55.7	65.4 @ 0	8.07 - 1703/T
Trimethyl gallium-Trimethyl arsenic Adduct ($CH_3)_3$Ga·As($CH_3)_3$, Solid, F	24	121		9.114 - 2458/T
GERMANIUM				
Tetramethyl germanium ($CH_3)_4$Ge, Liquid F	-88	43.6	139 @ 0	
INDIUM				
Triethyl indium ($C_2H_5)_3$In, Liquid P	-32	184	1.2 @ 44 3 @ 53 12 @ 83	
Trimethyl indium ($CH_3)_3$In, Solid, P	88.4	133.8	7.2 @ 30 72 @ 70	10.520 - 3014/T
Trimethyl indium-Trimethyl phosphine Adduct ($CH_3)_3$In·P($CH_3)_3$, Solid, F	44.5			6.9534 - 1573/T

(continued)

Table 1: (continued)

Compound Formula, State[a] Reactivity[c]	MP°C	BP°C	Reported Vapor Pressure mm Hg @ °C	Vapor Pressure Equation $LOG_{10}P(mm\ Hg) = B - A/T$ [b]
MAGNESIUM				
Bis(cyclopentadienyl) magnesium, $(C_5H_5)_2Mg$, Solid H	176		0.043 @ 25	25.14 - 2.18 ln T - 4198 T (where ln = natural log)
MERCURY				
Dimethyl mercury $(CH_3)_2Hg$, Liquid, F		96		7.575 - 1750/T (20.5°C-78.7°C)
PHOSPHORUS				
Triethyl phosphine $(C_2H_5)_3P$, Liquid, P	-88	127		7.86 - 2000/T;(18-78.2°C)
Trimethyl phosphine $(CH_3)_3P$, Liquid, P	-85	37.8		7.7329 - 1512/T
SELENIUM				
Diethyl selenide, $(C_2H_5)_2Se$, Liquid, F		108		
SILICON				
Triethyl silane $(C_2H_5)_3SiH$, Liquid, F	-157	107-108		
TELLURIUM				
Diethyl telluride $(C_2H_5)_2Te$, Liquid, F		137-138		7.99-2093/T
Dimethyl telluride $(CH_3)_2Te$, Liquid, F	-10	82	14 @ 30	7.97 - 1865/T
TIN				
Tetraethyl tin $(C_2H_5)_4Sn$, Liquid, F	-112	181		
Tetramethyl tin $(CH_3)_4Sn$, Liquid, F	-53	78	10 @ -21	7.495 - 1620/T;(18°C-78.9°C)
ZINC				
Diethyl zinc $(C_2H_5)_2Zn$, Liquid, F	-28	118	15 @ 29	8.280 - 2190/T
Dimethyl zinc $(CH_3)_2Zn$, Liquid	-42	46	124 @ 0	7.802 - 1560/T

(a) @ 20°C
(b) T in degrees Kelvin
(c) Reactivity In Air At Ambient Conditions; P = Pyrophoric, F = Flammable, H = Hygroscopic

It should be noted that it is generally desirable to use organometallic cylinders at temperatures below ambient in order to eliminate the possibility of condensation of the chemical on the walls of the tubing that leads to the reaction chamber. Of course, if the source has a low vapor pressure, it may become necessary to use a source temperature above room temperature in order to achieve the desired growth rates. In this case, condensation can be prevented by either heating the system tubing to a temperature above the source temperature or by diluting the reactant with additional carrier gas in the system tubing.

The commonly used sources are generally thermally stable around room temperature, although triethyl indium (TEIn) and diethyl zinc (DEZn) have been reported to decompose at low temperatures in the presence of H_2.[21] Thus, the materials are, for the most part, expected to be stable under use conditions even when stored for extended periods of time. The reactants will begin to thermally decompose in the MO-CVD reaction chamber as they encounter the hot susceptor. The temperature at which an organometallic compound will begin to decompose is not particularly well defined. It has been found to be a function of both the surfaces with which the organometallic comes into contact[22] and the gas ambient.[23] Furthermore, workers have obtained rather different results even when using the same nominal experimental conditions.[24,25] Finally, the decomposition will be affected by the residence time of the chemical species near the hot pyrolyzing surface, which implies a flow rate and perhaps a reactor geometry dependence of the thermal decomposition. Generally, however, the reported decomposition temperatures are in the range of 200° to 400°C[22-25] for most of the metal alkyls. Exceptions to this are the P- and As-containing alkyls which decompose at much higher temperatures.[15,26] The high decomposition temperature of the P- alkyls greatly reduces their utility as sources for P in MO-CVD.

2.2 Organometallic Source Packaging

As indicated in Table 1, most of the commonly used organometallic compounds are pyrophoric or at least air and water sensitive and therefore require reliable, hermetic packaging to prevent the material from being contaminated by air and to prevent fires resulting from contact with air. The organometallic compounds are generally shipped from the supplier in the package that will be used for film growth. Thus, the package should be considered an integral part of the source product.

Although early organometallic packaging consisted of commercially available cylinders with pipe thread connections, the packages that are generally used today are welded stainless steel cylinders with bellows or diaphragm valves and vacuum fittings on the inlet and outlet. The use of welded cylinders and bellows or diaphragm valves was motivated by a desire for improved leak integrity and reduced dead volumes. At least one supplier also teflon coats the inside of the cylinders to reduce contamination of the organometallic material by the cylinder walls.

The container is usually in the form of a bubbler (for liquid sources) or a sublimer (for solid sources) in which a carrier gas is passed through the bottom of the material via a dip tube as is pictured in the cross-sectional view of a typical cylinder in Figure 2. The carrier gas then transports the source material into the reactor. If thermodynamic equilibrium is assumed between the condensed source and the vapor in the container above it, then the molar flow (ν) can be written:

$$(2) \qquad \nu = \frac{P_v P_{cyl}}{(P_{cyl} - P_v)} \frac{f_v}{k T_{std}}$$

where ν is the molar flow in mols/min, P_v is the vapor pressure of the organometallic species at bath temperature T_b in atmospheres, P_{cyl} is the total pressure in the organometallic cylinder, f_v is the volume flow rate through the bubbler in ℓ/min, k is the gas constant, and $T_{std} = 273°K$. P_v can be calculated from the data in Table 1. Typical molar flows for organometallic species are in the range of 5×10^{-6} to 5×10^{-5} mols/min.

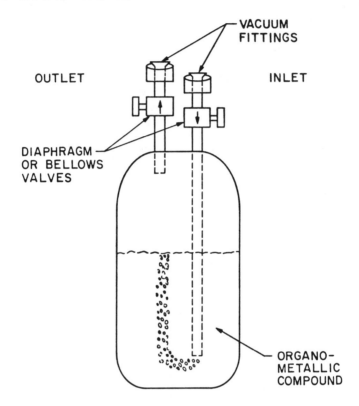

Figure 2: Schematic drawing of an organometallic cylinder.

The approximation of thermal equilibrium between the condensed and vapor phases is a good one for liquid sources such as TMGa. Since most sources are used as liquids (see Table 1), equation 2 is usually a valid description of organometallic molar flows.

Unfortunately, this approximation is expected to be a poor one for solid organometallic sources, the most important of which is trimethyl indium (TMIn). Solid sources are in the form of agglomerated powder and are packaged in the same containers that is used for liquid bubblers (see Figure 2). Because of the lack of bubble formation and the uncertain surface area of the solid, the condensed phase of the source will often not be in equilibrium with the vapor phase, especially at higher carrier gas flows. In this case, the molar flow of reactant will be less than that calculated from equation 2 which was developed assuming thermal equilibrium.[27,28] Mircea et al[28] have measured the time integrated mass flow from a TMIn cylinder at various carrier flows and found that the cylinder deviated from equilibrium at rather low carrier flows. Their curve is reproduced in Figure 3. It might be expected that each cylinder will have a different dependence of the molar flow on the carrier flow and that the dependence must be determined experimentally. In addition, the surface area of the source inside the cylinder can vary as the cylinder is used. Thus, one might expect a long term degradation of the growth rate of material grown with a solid source as the internal source surface area decreases with time. Unfortunately, these expectations have not, to date, been rigorously confirmed.

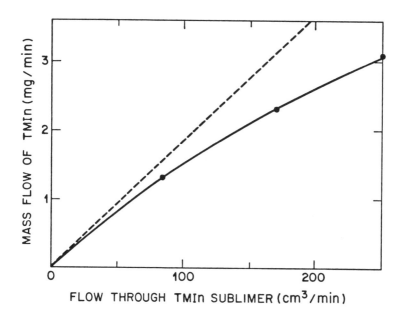

Figure 3: Mass flow of TMIn as a function of the flow through the TMIn sublimer. Sublimer temperature = 25°C. Dashed line represents a linear dependence of mass flow on carrier gas flow. From Mircea et al.[28]

One additional method of packaging that is often used for dopants which require small molar flows is that of diluting the organometallic to several hundred parts per million in a high pressure carrier gas cylinder. This method allows somewhat more control over the low molar flows that are often required for low doping levels since the partial pressure of reactant is set by the cylinder concentration rather than the vapor pressure of the material. One can also use this method to supply reactants for major constituents.[29]

2.3 Hydride Sources and Packaging

In the growth of III-Vs containing As or P and II-VIs containing S or Se, the hydrides AsH_3, PH_3, H_2S and H_2Se are usually used as the sources. This is because they are relatively inexpensive, are available as dilute mixtures in high pressure cylinders and are often more pure than the corresponding organometallic sources. Furthermore, there is evidence that suggests that C contamination, which arises from the use of an organometallic group III source, can be reduced by the use of a hydrogen containing group V species.[16,30] These gases are used in either pure concentrations or, more often, as mixtures diluted to 5 to 10% in H_2. In practice, the choice of cylinder concentration is determined by the flows needed for growth and safety considerations.

The hydrides, AsH_3 and PH_3, are rather thermally stable, generally decomposing at temperatures higher than most organometallics (but lower than the As and P-containing alkyls) and thought to require substrate catalysis for decomposition under many growth conditions.[15] This is especially true for PH_3. Ban[31] measured decomposition efficiencies of AsH_3 and PH_3 in a hot wall reactor and found that under his experimental conditions and at typical GaAs or InP growth temperature of 600°C, 77% of the AsH_3 but only 25% of the PH_3 was decomposed. As expected, the percentage of decomposed PH_3 increased more rapidly than AsH_3 as the temperature was increased so that, for example, at 800°C, 90% of the AsH_3 and 70% of the PH_3 were decomposed. It should be recognized that the data that Ban reported should not be used quantitatively. In a cold wall MO-CVD reactor even less AsH_3 and PH_3 is expected to be decomposed because there will be less time in which the gas is in contact with a hot surface. The poor PH_3 thermal decomposition efficiency and the high vapor pressure of P leads to the use of large PH_3 flows for the growth of P-bearing compounds and alloys. More will be said on this subject in Section 3 of this chapter.

Several groups have attempted to increase the cracking efficiency by pyrolyzing PH_3 in a high temperature furnace prior to introduction into the reaction chamber.[32] Whereas the use of a PH_3 pyrolyzing furnace does appear to decrease the use of PH_3, it is difficult to design a furnace that efficiently pyrolyzes only PH_3 and no other species while avoiding downstream P condensation. Furthermore, it is a system design complication that is apparently not needed since a number of workers have produced high quality InP and InGaAsP without pyrolyzing PH_3 prior to its introduction into the reaction chamber.[28,33,34]

The group VI hydrides thermally decompose at lower temperatures than the group V hydrides with H_2Se decomposing at a lower temperature than H_2S. Although the growth of mixed II-VI alloys containing Se and S is possible at temperatures less than 400°C because the group VI hydrides will decompose at these temperatures, the difference in H_2Se and H_2S decomposition temperatures results in difficulty in compositional control at these low substrate temperatures.[8] More will be said about this subject in Section 3.1.

3. GROWTH CONDITIONS, MECHANISMS AND CHEMISTRY

In this section, deposition chemistry is considered. In spite of a considerable amount of empirical data on growth conditions, there is a great lack of knowledge concerning the mechanism of the growth process. In this section, we will briefly describe the growth conditions that are used for the MO-CVD of compound semiconductors. We will then summarize a framework for considering growth mechanisms that was proposed by Stringfellow.[15] In the discussion of growth mechanisms, we use as examples some of the specific growth conditions that were previously described. We devote an entire subsection to gas phase reactions between sources, a topic that has been a major source of controversy for a number of years.

3.1 Growth Conditions and Materials Purity

The most basic growth parameters that are varied in MO-CVD growth are the growth (susceptor) temperature and the input reactant molar flows. For the growth of III-V's, temperatures ranging from 550° to 900°C have been used successfully, with relatively low melting temperature materials such as GaAs or InP generally grown at the lower end of that range and relatively high melting temperature materials such as GaP and GaN generally grown at the higher end of that range.

Almost all III-V growth is carried out with input III-V ratios between 5 and 400 with GaAs and AlGaAs being the prototypical examples. It has been experimentally found by a number of workers[13-15] that the growth rate is independent of substrate temperature, proportional to the inlet group III molar flow rate and independent of the inlet group V molar flow rate over a wide temperature range. Compilations of some of these data are given in Figures 4[15] and 5.[13] In similar studies, the composition of III-V alloys with mixed group III elements has been found to be proportional to the relative input ratios of the group III constituents. An example for several alloys is shown in Figure 6.[15] These results imply a growth regime in which the growth rate is limited by the gas phase diffusion of group III species through a boundary layer above the substrate.[13-15] The above statements are true in the absence of reactant depletion effects[14,15] which have often plagued the growth of In-containing compounds and alloys. More will be said on this subject in the next two subsections.

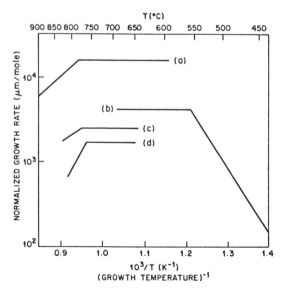

Figure 4: Temperature dependence of the growth rate of GaAs using TMGa. The growth rate is normalized to the inlet TMGa molar flow. Data are from (a) Manasevit and Simpson,[88] (b) Krautle et al,[89] (c) Gottschalch et al,[90] (d) Leys and Veenvliet.[91] From Stringfellow.[15]

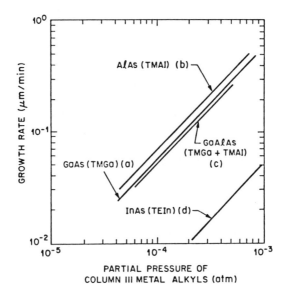

Figure 5: Dependence of the growth rate on the inlet partial pressure of the group III organometallic compounds for a number of III-V's. Data are from (a) Manasevit and Simpson,[88] (b) Coleman et al,[92] (c) Aebi et al,[93] (d) Baliga and Ghandi.[94] From Dapkus.[13]

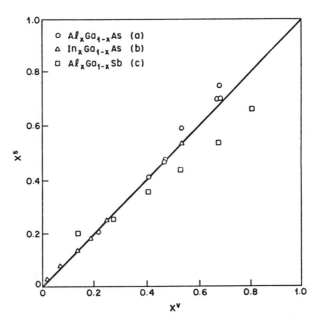

Figure 6: Dependence of the solid composition x^s on the vapor composition x^v for a number of III-V alloys. Data are from (a) Mori and Wantanabe,[95] (b) Ludowise et al,[27] (c) Cooper et al.[96] From Stringfellow.[15]

For the growth of III-V alloys with mixed group V elements, one must consider two rather different cases. For alloys containing As and P, the relative As and P incorporation is determined primarily by the relative thermal stabilities of AsH_3 and PH_3. As previously discussed in section 2.3, PH_3 thermally decomposes much more poorly than AsH_3 at low temperatures. This is reflected in the relative incorporation probabilities of As and P in GaAsP and InAsP alloys as a function of substrate temperature. A compilation of these data is shown in Figure 7.[15]

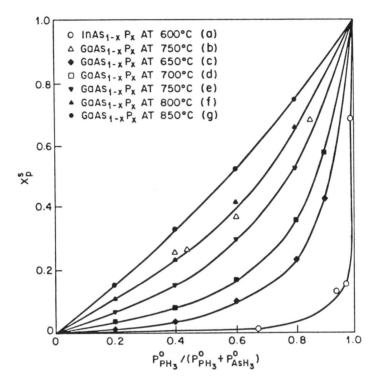

Figure 7: Dependence of the P concentration in the solid on the PH_3 concentration in the vapor. Data are from (a) Fukui and Horikoshi,[97] (b) Ludowise and Dietze,[98] (c), (d), (e), (f), (g) Samualson et al.[99] From Stringfellow.[15]

Surprisingly (in view of the large chemical potential difference between the input reactants and the grown film), the incorporation of Sb into mixed As-Sb alloys is controlled by the relative thermodynamic stabilities of binary antimonides and arsenides. The relative Sb and As incorporation probabilities can be predicted from regular solution theory[13,35] when the input V/III ratio is greater than one. In this case, substrate thermodynamics determine the relative As/Sb elemental incorporation and results in a low Sb incorporation probability. When input V/III ratios less than one are used, the relative Sb incorporation probability increases rather rapidly to one.[15,35] These data are shown in Figure 8.[35]

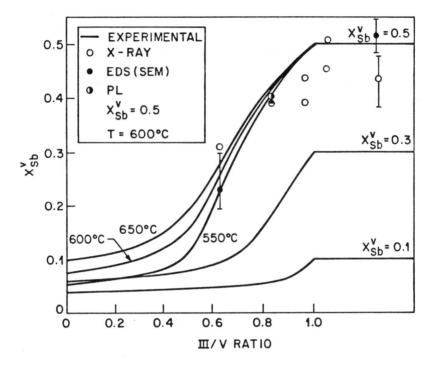

Figure 8: Dependence of the GaSb concentration in the solid on the III-V input ratio for various Sb vapor concentrations, X_{Sb}^v and at various growth temperatures. The solid lines are calculations and the experimental conditions are shown in the box. From Cherng et al.[35]

There has been considerably less work on II-VI's than on III-V's. For the growth of II-VI's, VI/II ratios greater than one are often used[8] although large overpressures of the group II element Hg are needed for the growth of HgCdTe.[7] The VI/II ratio required for good growth of the II-VI's appears to be controlled primarily by the relative thermal decomposition of the sources at the typical growth temperatures of 350° to 450°C. These temperatures (or lower) are highly desirable for the growth of II-VI's because of the large diffusion coefficients in these materials (of particular importance in the growth of CdTeHgCdTe multilayer structures), and because a great decrease in native defect concentration can be realized by growth at low temperatures. As mentioned in Section 2.1, these temperatures are just in the temperature range at which organometallic sources begin to thermally decompose. Thus, the growth of II-VI's is often extremely temperature sensitive. As will be discussed in the last section on future developments, one of the major recent thrusts in the growth of II-VI's is to decouple the source decomposition from the substrate temperature.

The purity of most grown materials is determined, for the most part, by the purity of the starting source material. The purity of the most widely used organometallic compounds (especially TMGa and TMIn) has seen a rather remarkable improvement in the last few years as larger sales volumes have allowed suppliers to improve their synthesis techniques, source analysis, and quality control. The purity of less widely used compounds, however, is still open to question. Most of the effort in improving purity has centered around reducing metallic impurities which can be incorporated as electrically active species. Little work has been done on organic impurities which may or may not play a role in the quality of material grown. The major impurity in hydride sources is often relatively large and highly variable amounts (many tens of ppm) of H_2O which has extremely deleterious effects on Al-bearing materials. This will be discussed in more detail in Section 4.2.

3.2 Growth Mechanisms

Little is known about the details of film growth processes that occur during MO-CVD. Furthermore, only recently have investigations of MO-CVD growth mechamisms been reported. As a result, there is still a great lack of knowledge and some controversy concerning certain aspects of the technique, especially those related to gas phase chemical reactions, incorporation of C and other impurities into the layers, and the respective roles of thermodynamics, surface and gas kinetics, and reaction chamber design in determining the electrical and optical properties of the deposited layers. Stringfellow[15] has recently presented an excellent discussion of the present state of knowledge of growth mechanisms in MO-CVD which we summarize in this subsection.

As Stringfellow points out,[15] the MO-CVD growth process can be divided into four regimes as shown schematically in Figure 9: a reactant input regime, a reactant mixing regime, a boundary layer regime immediately above the substrate, and the growth (substrate) surface itself. Growth complications that can occur in these regimes include gas phase reactions during reactant mixing, reactant diffusion and/or pyrolysis in the boundary layer above the substrate, and thermodynamic or reaction kinetic rejection from the substrate. All of these effects have been seen to one extent or another in MO-CVD and many of the growth conditions discussed in Section 3.1 reflect the presence of these effects.

In MO-CVD, growth conditions are deliberately set up so that a large driving force exists for film deposition. This means that a large difference in free energy [or chemical potential (μ)] exists between the reactant input and film growth regimes. The relative chemical potentials at each point are schematically depicted in Figure 10. It is this large free energy difference that allows the growth of such a wide variety of materials and structures that are often difficult to grow by other techniques. In spite of the large free energy difference, the quality and characteristics of growth are often sensitive to equipment design and construction. This is because the design of the MO-CVD system can modify the thermodynamic path (i.e., the kinetics) between the reactant input and growth regimes. One modified path, representing the possiblity of gas phase chemical reactions depleting a source, is shown schematically in Figure 10 by a dashed line. It should be noted that an alternative pathway that is undesirable for epitaxial growth such as that depicted in Figure 10 may, in many circumstances, be the one that is thermodynamically favored. In that case, optimum growth conditions rely on reaction kinetics to suppress the thermodynamically favored pathway and allow the desired pathway.

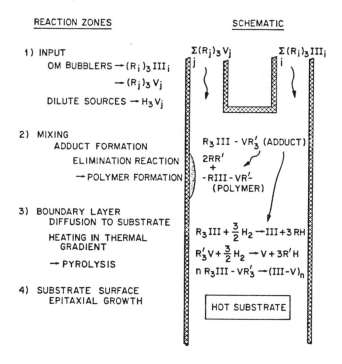

Figure 9: Reaction regimes for the MO-CVD process. From Stringfellow.[15]

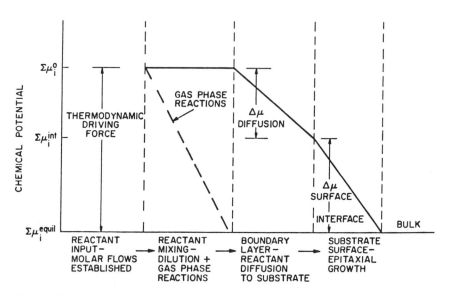

Figure 10: Chemical potentials for the various MO-CVD growth regimes. The sum of the chemical potentials for each component, $\Sigma \mu_i$, decreases as the process proceeds as shown by the solid line. An alternative path, representing gas phase reactions, is shown by a dashed line. From Stringfellow.[15]

3.3 Gas Phase Chemical Reactions

One of the more long-lasting and confusing controversies in the MO-CVD community has concerned the presence or absence of gas phase chemical reactions between hydride and organometallic reactant species prior to coming into contact with the hot susceptor and the effect of these reactions on the quality of the grown films. Lewis acid-Lewis base reactions between group II or III organometallics and group V or VI organometallics or hydrides can occur to form an adduct of the form $R_m M-ER'_n$ where, as before, R and R' represent a methyl or ethyl radical or hydrogen, M is a group II or III metal, E is a group V or VI element and n = 2 or 3 depending on whether III-V or II-VI sources are being used. Although many adducts are stable around room temperature, it has been reported that those containing In[24,26,32,36] and some group II-containing alkyls[8] are thermodynamically unstable around room temperature and decompose to form a low vapor pressure polymer of the form $(-RM-ER'-)_n$ which can condense on the walls of the system tubing or reaction chamber prior to reaching the substrate. Because most work has been done on In compounds, our discussion will focus on these materials.

Organometallic-hydride reactions of the type described above have been cited to explain upstream In depletion that has been observed in the growth of InP and InGaAs using TeIn, TMIn, PH_3, and AsH_3.[24,26,27,32,37,38] The In depletion takes the form of a significantly lower incorporation probability for In than for Ga. In order to reduce the prereactions, it has often been found desirable to design the growth systems to keep the reactants separate until just before the substrate,[39,40] to use adducts such as TMIn-TEP[26] or to use low pressure growth.[32] All of these solutions have certain problems associated with them. Keeping reactants separate until just before they reach the substrate results in a non-uniform gas composition and, therefore, highly non-uniform growth.[27] Adduct sources have extremely low vapor pressures and, therefore, must be used at temperatures significantly higher than room temperature.[41] Low pressure (∼0.1 atm.) growth requires more complicated equipment and sometimes can result in increased reactant consumption, especially of the group V sources since reactant partial pressures are all decreased by a factor of 10.

On the other hand, a number of other workers[33,34,38,42,43] report no In depletion even when using systems in which no special designs have been implemented and employing no special growth conditions. Their conclusion is that, although the gas phase reaction is thermodynamically favored, the reaction kinetics are slow enough so that the reaction only minimally occurs in many real MO-CVD systems. Furthermore, the polymer reaction product is very weakly bound and decomposes into its organometallic and hydride constituents at temperatures $<100°C$.[25,36] The question then becomes: why the discrepancy between various groups concerning In depletion.

Some workers believe that the In depletion effects that many workers had observed were caused by impure In sources.[26] In addition, if sluggish reaction kinetics are preventing significant reactant depletion, then it is likely that small changes in the details of reactor design or growth conditions will alter the reaction kinetics and result in very different results from different laboratories. The metal alkyl-hydride polymer formation also appears to be more severe for TEIn than for TMIn.[38] Thus, many laboratories currently use TMIn as the In source for InP, InGaAs, and InGaAsP growth.

Confusing matters still further, another In-alkyl depletion mechanism has recently been identified. It has been found that under certain growth conditions, the gas immediately upstream of the susceptor can be so hot that the TMIn source can thermally decompose.[44,45] It was found[45] that decreasing the gas temperature (i.e., increasing the temperature gradient between the hot susceptor and the gas flow) by such means as using a low thermal conductivity carrier gas[28,44,45] or increasing the total flow resulted in much decreased depletion of the In source.

On the basis of the above discussion, a few practical suggestions can be offered for the growth of In-containing films. In view of the often contradictory results that have been obtained by various workers, the uncertainties that persist concerning gas phase chemical reactions, and the likelihood that the presence or absence of observable In depletion is reactor dependent, it is probably wise to design systems to minimize the length of tubing in which organometallics and hydrides are mixed, to mix reactants in diluted flows to reduce their concentrations, to use TMIn rather than TEIn as the In source, and to make the temperature gradient between the hot susceptor and the incoming gas stream as sharp as possible.

4. SYSTEM DESIGN AND CONSTRUCTION

In this section, we consider system design and construction. Much of the material discussed in the previous sections will be used in this section to formulate guidelines for the design and construction of an MO-CVD system. Figure 11 is a block diagram of an MO-CVD system in which functional subsystems are defined. Subsystems consist of reactant storage (hydrides, organometallics and H_2), a gas manifold in which flows are controlled and directed to the proper locations, a reaction chamber in which deposition takes place, and an exhaust which can consist of either a pump for low pressure operation or an atmospheric pressure exhaust. In this section, we first consider the generic issues of leak integrity and oxygen gettering techniques. We then consider the design and construction of the functional subsystems in more detail. Our emphasis will be on the gas manifold and reaction chamber subsections because these are the system subsections that primarily effect the quality of the material that is grown in the systems.

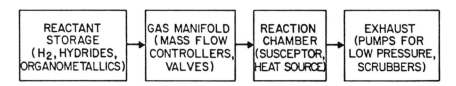

Figure 11: Schematic block diagram of an MO-CVD system showing functional subsections.

4.1 Leak Integrity and Cleanliness

At this point, some comments on the importance of leak integrity and general cleanliness are in order. In all phases of equipment design and construction, it is absolutely critical that leak integrity be established and that components be cleaned and assembled in a clean environment. Not only can system leaks allow the toxic chemicals that are used in the process to escape to the atmosphere, thus presenting a safety hazard, but room air can readily contaminate the reactants and the system. Although a certain amount of O_2 and H_2O contamination can be tolerated in the growth of some compound semiconductors,[46] O_2 and H_2O have extremely deleterious effects on grown material containing Al. In the growth of AlGaAs, for example, altered composition, degraded morphology, deep electronic levels and low photoluminescence intensity have been observed in samples grown in contaminated systems.[47,48] Although oxygen contamination can come from the hydride sources,[49] this source of contamination can be reduced by several gettering techniques[46,48,49,50,51] which will be discussed in more detail in the next subsection. It continues to be important, therefore, to eliminate system leaks as a potential source of O_2 and H_2O.

In addition to eliminating leaks to atmosphere, it is also important to minimize the number of virtual leaks in the system. Virtual leaks serve to trap residual gases and are very difficult to flush, thus providing internal sources of air contamination. In addition, they also serve to trap reactants and thus can contribute to compositional grading, especially at interfaces in grown material.

4.2 Oxygen Gettering Techniques

Because of the extreme sensitivity of AlGaAs properties to trace amounts of O_2 and H_2O and the variability in the quality of gas sources, a considerable amount of work has been done on gettering of these contaminants. First, it should be pointed out that the need for O_2/H_2O gettering in the growth of AlGaAs can be reduced or eliminated by using a growth temperature greater than 780°C.[48] Unfortunately, high temperature growth cannot be used for several important device structures, in particular modulation doped field effect transistors, because the purity of GaAs grown at these temperatures is found to be much worse than material grown at lower temperatures (650°C).[52,53] Thus, other workers have concentrated on removing contaminants from either the gas lines or the reaction chamber.

The first oxygen gettering technique that was reported was the use of graphite baffles. Graphite baffles[46,54,55] are placed in the reaction chamber upstream of the substrate. They getter system oxygen by catalyzing the reaction between TMAl and oxygen to produce Al_2O_3. Unfortunately, TMAl is slowly desorbed from the graphite which results in the formation of nonabrupt junctions.[55] For this reason, this gettering technique is presently not popular.

Two gas line gettering techniques which are compatible with the production of abrupt interfaces are presently used. Efficient gettering of source gases can be obtained by passing the gas through a bubbler containing a room temperature liquid metal alloy consisting of Al-In-Ga.[50] The Al in the melt is oxidized by oxygen containing species and floats to the top of the melt as an Al_2O_3 scum. There are two disadvantages of this technique. First, particle filters must be used in the tubing after the bubbler to prevent small, atomized liquid metal particles from being swept toward the reaction chamber. Second, regeneration of the liquid metal bubblers is somewhat clumsy.

The final gettering technique to be discussed is the use of molecular sieve traps. This is a well-known technique for gas purification by molecular diameter-selective adsorption.[56] Molecular sieve traps have the potential for particle generation so particle filters may be necessary in the gas lines. Regeneration of the traps, however, is easy and is accomplished by baking the traps. Molecular sieve traps can be operated at either room temperature or lower temperatures. Although the gettering action of the traps improves as the temperature decreases, it's probably wise not to decrease the operating temperature lower than the boiling temperature of AsH_3 (−62.5°C at 1 atm pressure).

4.3 Gas Manifold Design

The gas manifold contains the valves, mass flow controllers, and tubing that regulate the flows of all reactants and direct their destinations. It is desired that the flows be accurately controlled with a minimum of fluctuations, particularly when the flows are redirected to a different location. Generally, computer control is used to regulate mass flow controllers and valves, particularly when complex device structures are required. Most of the major reactants require carrier gas flows in the range of 1 to 400 cm^3/min for growth rates in the range of 1 to 10 μm/h. Gas flows can be accurately metered by electronic mass flow controllers. The low reactant flows are then added to a much larger (typically 2 to 10 ℓ/min) dilution flow. The dilution flow usually consists of H_2 or N_2 and acts to increase the speed with which the reactants are swept to the reactor. The dilution gas and the reactant flow can be mixed at either of two locations as illustrated schematically in Figure 12. In Figure 12(a), the dilution gas is being added to each reactant gas stream individually. The diluted gas streams then mix in the tubing or at the reactor inlet. Of course, two or more reactants out of many can be mixed and then added to the dilution flow. This design is used when there is a desire to keep reactants separate to prevent predeposition reactions from occurring within the tubing. In particular, many systems are designed so that organometallics and hydrides are not mixed until just before the reactor inlet.[32] A second feature of this approach is that transient effects associated with flushing of tubing can be reduced or eliminated. For example, a version of this approach has been used for the growth of structures such as modulation doped AlGaAs/GaAs which require extremely sharp interfaces for high quality device performance. By using separate TEGa sources for the GaAs active layer and the AlGaAs layers and by separately introducing the reactants required for each layer, modulation doped structures with high transconductances have been produced.[6] The disadvantage of adding the dilution flow to the reactant individually is that one additional mass flow controller and one additional tee connection must be added to the gas manifold for each reactant used. This adds cost to the system and can take up a considerable amount of space, depending, of course, on the number of species being used.

The reactant flow can also be added to a common dilution flow as illustrated schematically in Figure 12(b). The design is less complicated than that illustrated in Figure 12(a) and is used when there is no concern of premature gas phase reactions such as in the growth of GaAs/AlGaAs laser wafers using TMGa, TMAl, and AsH_3 sources.[1] This design has also been used successfully for the atmospheric pressure growth of structures which require sharp interfaces such as InGaAs/InP heterostructures.[57,58] For sharp interfaces, however, care must be taken to minimize the distance between reactant injection points and to balance the pressure within the manifold. These points will be discussed further later in this section when we discuss "fast switching" valves.

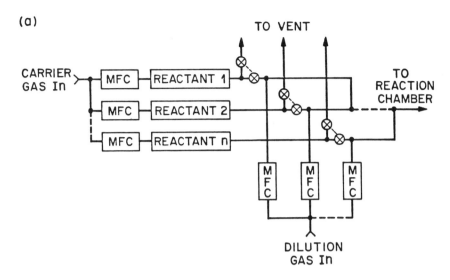

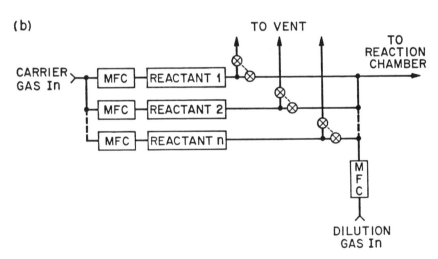

Figure 12: Schematic diagram of reactant and dilution gas mixing schemes. In (a), dilution gas is added to each reactant individually. In (b), reactants are added to a common dilution line. A "vent/run" valve switching configuration is used for both mixing schemes.

While most systems are designed so that reactants are mixed in the tubing prior to introduction into the reaction chamber, some or all reactants can be introduced separately into the reaction chamber if separate dilution flows are provided for each of the inlets.[52,53] This is done when there is a desire to separate reactants that can undergo gas phase reactions. The major disadvantage of separate introduction of reactants is that the gas composition is not uniform across the cross section of the reaction chamber and, therefore, composition or thickness uniformity is difficult to achieve[26] without some form of mixing within the reaction chamber.

It is generally recognized that systems should be configured with a "vent/run" design[59,60] which is shown schematically in Figure 12. In this configuration, reactant flows are established in a line that goes directly to vent prior to their introduction into the growth chamber. The establishment of flows helps to reduce transient flow effects and allows the temperature of the organometallic cylinder to reach a constant steady state value. The reactant flows are then simply switched from the vent line to the line that goes to the reaction chamber ("run" line) in order to initiate growth.

In systems more than a couple of years old, individual vent/run valves were ganged together in the design and construction of systems. As was mentioned above, very sharp interfaces are possible in systems such as these. Recently, however, evidence has begun to mount that interfacial sharpness can be obtained more easily and reproducibly in a vent/run configuration through the use of "fast switching" valves[59,61] and pressure balancing of the vent and system lines.[57,58,61] In essence, a fast switching valve is a number of vent/run valves that are placed together in one valve body. The advantage of this approach is that a number of reactant inputs can be made very close together and in smooth walled tubing, thus reducing dead volumes and increasing the probability of all reactants reaching the reaction chamber at the same time. Furthermore, fast switching valves are compatible with pressure balancing of vent and run lines. Pressure balancing is accomplished through the use of needle valves and reduces or eliminates flow transients through the mass flow controllers. Pressure balancing is facilitated if the vent and run lines are physically identical so that the flow impedance per unit length of tubing are identical. Using the combination of fast switching valves and pressure balancing GaAs/AlGaAs modulation doped FET's[51] and InP/InGaAsP laser structures[57] have been fabricated with excellent device properties.

4.4 Reaction Chamber

The reaction chamber is where the thermal decomposition reaction and deposition occurs. Because the reaction chamber of an MO-CVD system contains the longest time constant for reactant flushing, the abruptness of interfaces is largely controlled by this section. The layer thickness and compositional uniformity that can be achieved is also determined by the reaction chamber. Chambers are cold wall, that is, the susceptor upon which the substrate sits is the only part of the reactor that is deliberately heated. The cold wall design reduces the probability that competing chemical reactions at the walls (such as thermal decomposition) will interfere with film growth. Unfortunately, cold wall designs also favor the establishment of thermally driven convection in the chamber because of the large temperature gradients present. The chamber walls can be either water or ambient-air cooled. For ambient air cooled walls, the wall temperature has been measured to be as high as 200°C when the susceptor temperature is 650°C.[45]

There are two basic geometries of reaction chamber that are used in MO-CVD which are designated "vertical" and "horizontal" and are shown in Figure 13. Both types are made from fused silica because of the chemical inertness of the material and because the optical transparency and electrical insulating properties are compatible with the two main substrate heating methods: infrared (IR) absorption from a suitable IR source, usually quartz lamps,[62] and radio frequency (rf) induction heating.

Both of these heating methods use sources of heat that are usually placed external to the reaction chamber to prevent undesirable interactions between the heating source and the reactant gases. Induction heating uses radio frequency generators in the 200 to 400 kHz and 5 to 20 kW range and has been used for both horizontal and vertical geometry systems. The IR source must be positioned so that fogging of the reaction chamber walls in the path of the radiation from reactant pyrolysis is avoided. In horizontal geometry systems, this means that the lamps are positioned under the susceptor.[62] In vertical systems, the lamps must be positioned inside a hollow susceptor or downstream of the susceptor. The susceptor must be electrically conducting (for rf heating) or optically absorbing (for IR heating). In addition, the susceptor must be chemically inert at the growth temperature so that the film is not contaminated. Usually, the susceptor is fabricated from graphite and is coated with SiC.

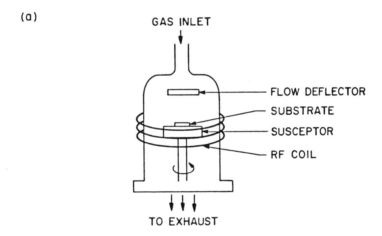

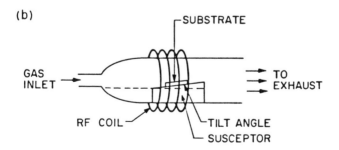

Figure 13: Reaction chamber geometries used in MO-CVD. (a) Vertical geometry, (b) horizontal geometry.

The vertical geometry was the original configuration used by the originators of the MO-CVD.[11] With vertical geometry, the flow is introduced normal to the substrate surface as is shown in Figure 13(a). A flow deflector can be used to spread the flow distribution[10] and produce films with excellent uniformity over moderately large areas (~10-15 cm^2).[63,64] Generally, substrate rotation is used to provide radial thickness uniformity, although reactant introduction in the form of a vortex can be used with a stationary substrate to produce uniform thicknesses. There have been very few flow analyses[65] that have been published for vertical geometry systems and it is difficult to predict hydrodynamic parameters that can be used to improve uniformity. It is generally conceded, however, that vertical geometry systems are susceptible to the recirculation of reactants due to thermally driven convective cell formation upstream of the susceptor. This recirculation can lead to compositional grading at interfaces unless growth is interrupted between layers.[66] However, even with this recirculation, abrupt interfaces such as those required for double heterostructure lasers can be accomplished by using growth interruptions between layers of various different compositions.[1,66]

Horizontal geometry systems, shown in Figure 13(b), are often favored because flow patterns are laminar and can be more easily modeled. In the horizontal geometry, the reactant flow is approximately parallel to the substrate and susceptor surface. Reactant depletion effects cause the thickness of grown material to decrease with distance along the susceptor. For this reason, the susceptor is inclined several degrees to the flow to compensate for this depletion.

Flow analyses of horizontal geometry systems are properly done numerically.[67,68] However, various closed form approximations are very desirable for engineering a reaction chamber design and process.

One reasonably successful approximation used in modeling growth rate variations with distance along the susceptor in horizontal systems has been used by Ghandi and Field.[68,69] Since growth is limited by diffusion of reactant molecules through a boundary layer, at least in III-V's, the reactant flux toward the substrate surface can be written:

(3) $$J = - D N_o/\delta$$

where D is the diffusion coefficient of the reactant molecule within the boundary layer, N_o is the inlet reactant concentration, and δ is the boundary layer thickness. These authors assume a susceptor of 0° and no effect of the far chamber wall to derive a diffusional boundary layer thickness of:

(4) $$\delta = \left(\frac{\pi D x}{v}\right)^{1/2}$$

where x is the distance along the susceptor and v is the horizontal gas velocity. Figure 14 shows the results of their simple calculation for GaAs growth and compares the calculated growth rate variation (which assumes no tilt) with the measured growth rate variation and a numerical calculation (both of which used a 7° tilt). As can be seen from Figure 14, horizontal geometry system are often subject to thickness variations along their length.

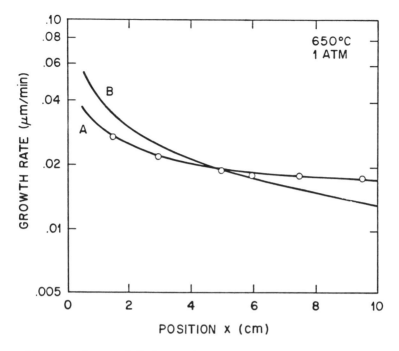

Figure 14: Growth rate of GaAs as a function of distance along susceptor. (A) Numerical calculation (tilt angle = 7°). (B) From equations 3 and 4 (tilt angle = 0°). From Ghandi and Field.[69]

Ludowise[14] briefly summarizes a detailed flow model that was developed by Berkman et al.[70] The decrease in layer thickness with distance along the susceptor that is due to reactant depletion can be calculated from this model. The susceptor tilt angle, θ, that is required to compensate for reactant depletion is calculated[14] from:

$$(5) \qquad \sin\theta = \left(2D/b_o v\right)\left(T_a/T_o\right)^{0.88}$$

where D is the diffusion coefficient of the reactant molecule at the temperature of the gas, T_o, far from the hot susceptor, T_a is the average temperature of the gas above the susceptor,[69] b_o is the boundary layer thickness, and v is the horizontal flow velocity upstream of the susceptor. The tilt angle is usually in the range of 3° to 5°. As Figure 14 shows, however, susceptor tilt does not completely compensate for the reactant depletion.

Komeno et al[71] have succeeded in growing very uniform GaAs layers in a horizontal system with a 0° tilt by using combination of substrate revolution and rotation during growth. They report thickness uniformity of ±1.5% over three 50 mm diameter substrates.

Although horizontal and vertical geometry systems have been extensively used for research and developmental MO-CVD systems, these geometries have inherent capacity limitations. Following the lead of silicon technology, attempts have been made at producing reaction chambers that can grow large areas and multiple wafers. These chambers use a vertical chamber with a barrel susceptor as is shown in Figure 15. With the barrel geometry susceptor, the reactants are introduced so that they flow approximately parallel to the susceptor surface in a manner similar to the horizontal susceptor. However, the flow patterns and thickness and compositional uniformity of a barrel geometry reactor are expected to be more complicated than a horizontal geometry system. This is due to the increased size of the system and components and because of increased flow disturbance when the inlet flow encounters the susceptor. Thickness uniformity of ±10% and doping uniformity of ±50% over an area of 480 cm^2 (sixty 2 cm x 4 cm wafers) has been achieved for GaAs/AlGaAs solar cell material[71] in a "production" barrel reactor. It appears, however, that these results have been obtained by empirical parameter adjustment and at a great expense of time and effort. It is clear that a great deal of work remains to be done on reaction chamber design and analysis for MO-CVD to become a dominant manufacturing technology for many applications.

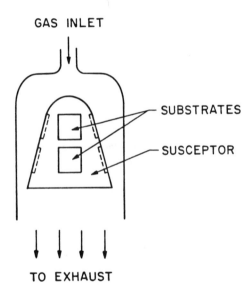

Figure 15: Schematic of barrel geometry growth chamber.

4.5 Exhaust and Low Pressure MO-CVD

The discussions in this section are generally applicable to atmospheric pressure growth. For atmospheric pressure systems, exhaust considerations are relatively straightforward. It is expected,[14,67,68,70] however, that uniformity will be easier to achieve over larger wafer areas at reduced pressures because of reduced boundary layer thickness and reduced reactant depletion.[68] Indeed, the multiple wafer growth that was discussed in the last subsection was done at 400 Torr.[72]

In the next few paragraphs, we will describe some of the additional issues that need to be considered for low pressure growth.

There are three ranges of pressure that can be identified for growth at less than atmospheric pressure: low pressure MO-CVD (LP-MO-CVD), vacuum MO-CVD, and MO-molecular beam epitaxy (also called Chemical Beam Epitaxy). The latter two techniques are still rather exploratory and will be discussed in Section 5 of this chapter on future developments.

LP-MO-CVD resembles atmospheric pressure MO-CVD most closely in that the typical pressure range of 76 to 100 Torr is within the viscous flow regime, and many gas phase collisions are expected to occur prior to the thermal decomposition reaction. Thus, flow can be modeled from hydrodynamic considerations. As such, most of the gas manifold and reaction chamber considerations that were discussed for atmospheric pressure are still valid for LP-MO-CVD. The reduction in operating pressure is accomplished through the use of a vacuum pump which is specially prepared for chemical service and which can be obtained commercially from several vendors. Because of the particulate material that is generated in the reactor, a pump oil filter is usually required. In addition, an inert gas ballast should be used during operation to prevent dissolution of toxic reactants in the hot pump oil. The pump must be throttled in order to control the pressure in the chamber. The throttling can be accomplished by one of two methods. A butterfly throttle valve can be placed in the vacuum line which is controlled by a feedback loop to the reaction chamber pressure. The other alternative is to introduce an additional flow of inert carrier to the pump in order to increase the pressure in the reaction chamber. Once again, this should be automatically controlled by a feedback loop to the reaction chamber pressure measurement. Finally, a method for pressure reduction is usually provided between the chamber and the source cylinders. The pressure reduction can be through a manual or automatic needle valve. It should be noted that at least one laboratory[73] maintains low pressure all the way to the organometallic sources.

5. FUTURE DEVELOPMENTS

Several more exploratory methods of producing films using organometallic sources are beginning to attract attention. These methods fall into two categories: growth using thermal decomposition of sources at much lower pressures and growth using nonthermal means to decompose sources. In this final section, we discuss both of these categories.

The motivation for MO-CVD growth at pressures substantially lower than 0.1 atm is to reduce or eliminate hydrodynamic effects from the growth process. Hydrodynamics, particularly the presence of a thick boundary layer, can lead to poor reactant utilization in a diffusion-limited growth process because reactants that do not diffuse through the boundary layer to the substrate surface are swept away to the exhaust and are lost to deposition. In addition, the spacial variation in boundary layer thickness leads to thickness nonuniformities. Furthermore, reactant flushing from the reaction chamber is very rapid at reduced pressures so that very abrupt interfaces are expected. Finally, if the pressure is reduced far enough, the flow regime is molecular rather than viscous, or Knudsen, and the mean free path of gas molecules is on the order of the dimensions of the reaction chamber. In the molecular flow regime, line-of-sight deposition and the attendant possibility of shadow masked growth, similar to that seen in molecular beam epitaxial growth,[74] is operative.

There are two low pressure exploratory MO-CVD techniques that can be differentiated on the basis of the operating vacuum (and, thus, the flow regime): vacuum MO-CVD and metal-organic molecular beam epitaxy (MO-MBE), also known as chemical beam epitaxy (CBE). Fraas and co-workers[16,75,76,77] have developed vacuum MO-CVD for the production of GaAsP/GaAsSb solar cell material. For this application, production cost reduction is extremely important and can be achieved by efficient reactant utilization. A schematic of the vacuum MO-CVD reaction chamber is shown in Figure 16. Fraas et al use an internally mounted hot wall chamber with GaAs substrates mounted over holes in the wall. The hot wall promotes the efficient thermal decomposition of AsH_3. The vacuum is established through the use of a turbomolecular pump and operating pressures are in the range of 5×10^{-3} Torr. In this pressure range, hydrodynamic effects are mostly eliminated, but multiple gas phase collisions are, nevertheless, expected. Using this system, high efficiency solar cell material has been produced.

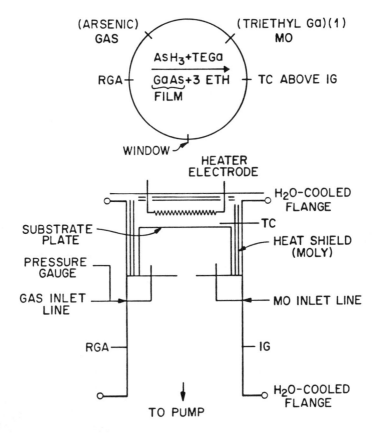

Figure 16: Schematic diagram of vacuum MO-CVD system. From Fraas et al.[75]

MO-MBE[18,78] or CBE[19] is, in reality, MBE using metal alkyl and hydride sources. As such, this version of the technique operates at much lower pressures (5×10^{-5} to 5×10^{-4} Torr) than does vacuum MO-CVD. Important characteristics of MBE[74] such as abrupt interfaces, thickness control, and the use of mechanical shadow masking are also expected to be present with MO-MBE. A schematic of an MO-MBE system is shown in Figure 17. It is a conventional MBE chamber which is equipped with leak valves for the organometallic compounds in place of furnaces for elemental sources. The metal alkyls are cracked at the substrate surface. It has been found, however, that As and P sources, both organometallic and hydride, must be precracked prior to introduction into the chamber because of the poor cracking efficiency of these chemicals. Recent results[79,80] on producing laser and detector material with this technique have been quite encouraging and much excitement exists around the world.

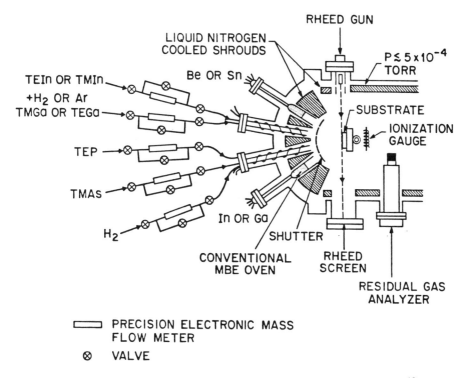

Figure 17: Schematic diagram of an MO-MBE system. From Tsang.[19]

The two final exploratory techniques, plasma-enhanced MO-CVD and photo-enhanced MO-CVD are being developed to decouple source decomposition from deposition, processes that occur essentially simultaneously in conventional MO-CVD. Although these variations can include growth under high vacuum conditions, their distinguishing feature is the nonthermal decomposition of sources. Nonthermal source decomposition allows film growth at temperatures below the source decomposition temperature, a feature that is especially important for the growth of II-VI's because of the high bulk diffusion coefficients[7] and the high native defect densities[8] in these alloys. In addition, the photo-enhanced version also allows selective patterned growth on the substrate surface with a focused light source.

Plasma-enhanced MO-CVD uses plasmas which can be formed by either dc[17,81] or capacitively coupled radio frequency[82-84] bias to enhance the reactivity of source molecules which allows growth at low substrate temperatures. A schematic of one system is shown in Figure 18.[17] Growth generally takes place at pressures ~1.0 Torr. Care must be taken in the chamber design to eliminate the exposure of the substrate to the plasma to eliminate the possibility of radiation damage. If that is done, material with good properties can be grown at low temperatures. For example, Pande and Seabaugh[17] produced single crystal GaAs at temperatures under 400°C and Mino et al[84] produced ZnSe electroluminescent cells by plasma-enhanced MO-CVD.

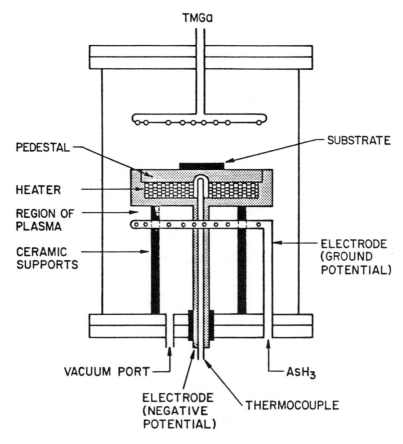

Figure 18: Schematic diagram of a plasma MO-CVD system. From Pande and Seabaug.[17]

Photo-enhanced MO-CVD has gotten recent attention for the growth of the II-VI compounds and alloys at low temperatures and for potential direct writing applications in the III-V materials. Irvine and co-workers[20,85] used the system shown schematically in Figure 19 to deposit HgCdTe using ultraviolet photolytically enhanced MO-CVD at atmospheric pressure. Epitaxial growth was achieved at temperatures ~250°C. It was found that using an inert carrier gas rather than H_2 reduced gas phase nucleation and promoted epitaxy. In similar experiments, Ando et al[86] grew single crystal ZeSe at temperatures above 450°C and polycrystalline ZnSe at temperatures as low as 150°C using photo-enhanced MO-CVD.

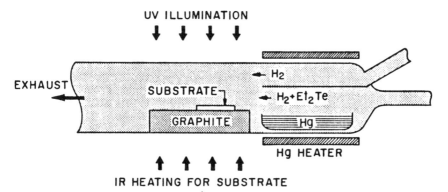

Figure 19: Schematic diagram of a photo-enhanced MO-CVD system. From Irvine et al.[20]

Photo-enhanced MO-CVD of InP has recently been accomplished using excimer laser (ultraviolet) radiation.[87] The use of a laser opens the possibility of direct writing of semiconductor structures on substrates. Although still very much in its infancy, the photo-enhanced MO-CVD may lead to great improvements in the kinds of structures that can be grown.

Acknowledgements

The author would like to thank a number of present and former colleagues at AT&T Bell Labs for helping him to gain whatever understanding of the MO-CVD process he possesses. Included in this list are D. L. Van Haren, J. Long, R. Karlicek, P-Y Lu, R.D. Dupuis, and V.M. Donnelly of AT&T Bell Labs and N.E. Schumaker, R.A. Stall, and W.R. Wagner of Encore. The author would also like to thank P-Y. Lu for his comments on the manuscript.

REFERENCES

1. Dupuis, R.D., *J. Cryst. Growth.*, Vol. 55, p. 255 (1981).
2. Hamaker, H.C., Ford, C.W., Wertken, J.G., Virshup, G.F. and Kaminar, N.R., *Appl. Phys. Lett.*, Vol. 47, p. 762 (1985).
3. Milano, R.A., Windhorn, T.H., Anderson, E.R., Stillman, G.E., Dupuis, R.D. and Dapkus, P.D., *Appl. Phys. Lett.*, Vol. 34, p. 562 (1979).
4. Allenson, M. and Bass, S.J., *Appl. Phys. Lett.*, Vol. 28, p. 113 (1976).
5. Nakanisi, T., Udagawa, T., Tanaka, A. and Kamei, K., *J. Cryst. Growth*, Vol. 55, p. 255 (1981).

6. Kobayashi, N., Fukui, T. and Tsubaki, K., *Jpn. J. Appl. Phys.*, Vol. 25, p. 1176 (1984).
7. Mullin, J.B., Irvine, J.C. and Tunnicliffe, J., *J. Cryst. Growth*, Vol. 68, p. 214 (1984).
8. Cockayne, B. and Wright, P.J., *J. Cryst. Growth*, Vol. 68, p. 223 (1984).
9. Stringfellow, G.B., *Ann. Rev. Mat. Sci.*, Vol. 8, p. 73 (1978).
10. Dapkus, P.D., *J. Cryst. Growth*, Vol. 68, p. 345 (1984).
11. Manasevit, H.M., *J. Cryst. Growth*, Vol. 55, p. 1 (1981).
12. Dilorenzo, J.V., *J. Cryst. Growth*, Vol. 17, p. 189 (1972).
13. Dapkus, D.P., *Ann. Rev. Mater. Sci.* Vol. 12, p. 243 (1982).
14. Ludowise, M., *J. Appl. Phys.*, Vol. 58, p. R31 (1985).
15. Stringfellow, G.B., *J. Cryst. Growth*, Vol. 68, p. 111 (1984).
16. Fraas, L.M., McLeod, P.S., Partain, L.D., Cohen, M.J. and Cape, J.A., *J. Electron. Mat.*, Vol. 15, p. 175 (1986).
17. Pande, K.P. and Seabaugh, A.C., *J. Electrochem. Soc.*, Vol. 131, p. 1357 (1984).
18. Kawaguchi, Y., Asahi, H. and Nagai, H., *Jpn. J. Appl. Phys.*, Vol. 23, p. L737 (1984).
19. Tsang, W.T., *J. Vac. Sci. Technol.*, Vol. B3, p. 666 (1985).
20. Irvine, S.J.C., Giess, J., Mullin, J.B., Blackmore, G.W. and Dosser, O.D., *J. Vac. Sci. Technol.*, Vol. B3, p. 1450 (1985).
21. *Organometallics for Vapor Phase Epitaxy*, Alpha Products, 152 Andover Street, Danvers, Massachusetts 01923.
22. Haigh, J. and O'Brien, S., *J. Cryst. Growth*, Vol. 68, p. 550 (1984).
23. Yoshida, M., Watanabe, H. and Uesugi, F., *J. Electrochem. Soc.*, Vol. 132, p. 677 (1985).
24. Cheng, C.H., Jones, K.A. and Motyl, K.M., *J. Electron. Mat.*, Vol. 13, p. 703 (1984).
25. Karlicek, R., Long, J.A. and Donnelly, V.M., *J. Cryst. Growth*, Vol. 68, p. 123 (1984).
26. Moss, R.H., *J. Cryst. Growth* Vol. 68, p. 78 (1984).
27. Ludowise, M.J., Cooper, C.B., III and Saxena, R.R., *J. Electron. Mat.*, Vol. 10, p. 1051 (1981).
28. Mircea, A., Azoulay, R., Dugrand, L., Mellet, R., Rao, K. and Sacilott, M., *J. Electron. Mat.*, Vol. 13, p. 603 (1984).
29. Yoshikawa, A., Yamaga, S. and Tanaka, K., *Jpn. J. Appl. Phys.*, Vol. 23, p. L388 (1984).
30. Kuech, T.F. and Venhoff, E., *J. Cryst. Growth*, Vol. 68, p. 148 (1984).
31. Ban, V.S., *J. Cryst. Growth*, Vol. 17, p. 19 (1972).
32. Duchemin, J.P., Hirtz, J.P., Razeghi, M., Bonnet, M. and Hersee, S.D., *J. Cryst. Growth*, Vol. 55, p. 64 (1981).
33. Hsu, C.C., Cohen, R.M. and Stringfellow, G.B., *J. Cryst. Growth*, Vol. 63, p. 8 (1983).
34. Zu, L.D., Chan, K.T. and Ballantyne, J.M., *Appl. Phys. Lett.*, Vol. 47, p. 47 (1985).
35. Cherng, M.J., Cohen, R.M. and Stringfellow, G.B., *J. Electron. Mat.*, Vol. 13, p. 799 (1984).
36. Didchenko, R., Alix, J.E. and Toeniskoetter, R.H., *J. Inorg. Nucl. Chem.*, Vol. 14, p. 35 (1960).

37. Whiteley, J.S. and Ghandi, S.K., *J. Electrochem. Soc.*, Vol. 129, p. 383 (1982).
38. Kuo, C.P., Cohen, R.M. and Stringfellow, G.B., *J. Cryst. Growth*, Vol. 64, p. 461 (1983).
39. Fukui, T. and Norikoshi, Y., *Jpn. J. Appl. Phys.*, Vol. 19, p. L395 (1980).
40. Ogura, M., Inoue, K., Ban, Y., Uno, T., Morisaki, M. and Hase, N., *Jpn. J. Appl. Phys.*, Vol. 21, p. L548 (1982).
41. Long, J.A., Riggs, V.G. and Johnston, W.D., Jr., *J. Cryst. Growth*, Vol. 69, p. 10 (1984).
42. Bass, S.J. and Young, M.L., *J. Cryst. Growth*, Vol. 68, p. 311 (1984).
43. Carey, K.W., *Appl. Phys. Lett.*, Vol. 46, p. 89 (1985).
44. Koppitz, M., Vestavik, O., Pletschen, W., Mircea, A., Heyen, M. and Richter, W., *J. Cryst. Growth*, Vol. 68, p. 136 (1984).
45. Zilko, J.L., Van Haren, D.L., Lu, P.Y., Schumaker, N.E. and Leung, S.Y., *J. Electron. Mat.*, Vol. 14, p. 563 (1985).
46. Kisker, D.W., Miller, J.N. and Stringfellow, G.B., *Appl. Phys. Lett.*, Vol. 40, p. 614 (1982).
47. Terao, H. and Sunakawa, H., *J. Cryst. Growth*, Vol. 68, p. 157 (1984).
48. Tsai, M.J., Tashima, M.M. and Moon, R.L., *J. Electron. Mat.*, Vol. 13, p. 437 (1984).
49. Fraas, L.M., Cape, J.A., McLeod, P.S. and Partain, L.D., *J. Vac. Sci. Technol.*, Vol. B3, p. 921 (1985).
50. Shealy, J.R., Kreismanis, V.G., Wagner, D.K. and Woodal, J.M., *Appl. Phys. Lett.*, Vol. 42, p. 83 (1983).
51. Bhat, R., Chan, W.K., Kastalsky, A., Koza, M.A. and Davisson, P.S., *Appl. Phys. Lett.*, Vol. 47, p. 1344 (1985).
52. Wantanabe, M., Tanaka, A., Nakanisi, T. and Zohta, Y., *Jpn. J. Appl. Phys.*, Vol. 20, p. L429 (1981).
53. Aebi, V., Bandy, S., Nishimoto, C. and Zdasink, G., *Appl. Phys. Lett.*, Vol. 44, p. 1056 (1984).
54. Stringfellow, G.B. and Hom, G., *Appl. Phys. Lett.*, Vol. 34, p. 794 (1979).
55. Kisker, D.W. and Stevenson, D.A., *J. Electron. Mat.*, Vol. 12, p. 459 (1983).
56. Breck, D.W., *Zeolite Molecular Sieves*, John Wiley and Sons, New York (1974).
57. Nelson, A.W., Moss, R.H., Regnault, J.C., Spurdens, P.C. and Wong, S., *Electron. Lett.*, Vol. 21, p. 331 (1985).
58. Mircea, A., Mellet, R., Rose, D., Robein, D., Thibierge, H., Leroux, G., Daste, P., Godfrey, S., Ossart, P. and Pounet, A.M., Electronic Materials Conference, Boulder, Colorado (1985).
59. Roberts, J., Mason, N.J. and Robinson, M., *J. Cryst. Growth*, Vol. 68, p. 422 (1984).
60. Thrush, E.J., Wale-Evans, G., Whiteaway, E.A., Lamb, B.L., Wright, D.R., Chew, N.G., Cullis, A.G. and Griffiths, R.J.M., *Electron. Mat.*, Vol. 13, p. 969 (1984).
61. Griffiths, R.J.M., Chew, N.G., Cullis, A.G. and Joyce, G.C., *Electron. Lett.*, Vol. 19, p. 989 (1983).
62. Beneking, H., Escobosa, A. and Kräute, H., *J. Electron. Mat.*, Vol. 10, p. 473 (1981).
63. Dupuis, R.D. and Dapkus, P.D., *IEEE J. Quant. Electron.*, Vol. QE-15, p. 128 (1979).

64. Kanber, H., Zielinski, T. and Whelan, J.M., *J. Electron. Mat.*, Vol. 14, p. 769 (1985).
65. Houtman, C., Graves, D.B. and Jensen, K.F., *J. Electrochem. Soc.*, Vol. 133, p. 961 (1986).
66. Dupuis, R.D., Miller, R.C. and Petroff, P.M., *J. Cryst. Growth*, Vol. 68, p. 398 (1984).
67. Juza, J. and Cermak, J., *J. Electrochem. Soc.*, Vol. 129, p. 1627 (1982).
68. Field, R.J. and Ghandi, S.K., *J. Cryst. Growth*, Vol. 69, p. 581 (1984).
69. Ghandi, S.K. and Field, R.J., *J. Cryst. Growth*, Vol. 69, p. 619 (1984).
70. Berkman, S., Ban, V.S. and Goldsmith, N., *Heterojunction Semiconductors for Electronic Devices*, p. 264, edited by G.W. Cullen and C.C. Wang, Springer, New York (1977).
71. Komeno, J., Tanaka, H., Itoh, H., Ohori, T., Kasai, K. and Shibatomi, A., Electronic Materials Conference, Boulder, Colorado (1985).
72. Tandon, J.L. and Yeh, Y.C.M., *J. Electrochem. Soc.*, Vol. 132, p. 662 (1985).
73. Shealy, J.R., Wicks, G.W., Ohno, H. and Eastman, L.F., *Jpn. J. Appl. Phys.*, Vol. 22, p. L639 (1983).
74. Cho, A.Y., *Thin Solid Films*, Vol. 100, p. 291 (1983).
75. Fraas, L.M., *J. Appl. Phys.*, Vol. 52, p. 6939 (1981).
76. Fraas, L.M., McLeod, P.S., Cape, J.A. and Partain, L.D., *J. Cryst. Growth*, Vol. 68, p. 490 (1984).
77. Fraas, L.M., McLeod, P.S., Partain, L.D. and Cape, J.A., *J. Vac. Sci. Technol.*, Vol. B4, p. 22 (1986).
78. Pütz, N., Veuhoff, E., Heinecke, H., Heyen, M., Luth, H. and Balk, P., *J. Vac. Sci. Technol.*, Vol. B3, p. 671 (1985).
79. Tsang, W.T., *Appl. Phys. Lett.*, Vol. 48, p. 511 (1986).
80. Tsang, W.T. and Campbell, J.C., *Appl. Phys. Lett.*, Vol. 48, p. 1416 (1986).
81. Pande, K.P., Nair, V.K.R. and Gutierrez, D., *J. Appl. Phys.*, Vol. 54, p. 5436 (1983).
82. Shiosaki, T., Yamamoto, T., Yagi, M. and Kawabata, A., *Appl. Phys. Lett.*, Vol. 39, p. 399 (1981).
83. Shimizu, M., Matsueda, Y., Shiosaki, T. and Kawabata, A., *J. Cryst. Growth*, Vol. 71, p. 209 (1985).
84. Mino, N., Kobayaski, M., Konagai, M., Takahaski, K., *Jpn. J. Appl. Phys.*, Vol. 24, p. L383 (1985).
85. Irvine, S.J.C., Giess, J., Mullin, J.B., Blackmore, G.W. and Dosser, O.D., *Materl. Lett.*, Vol. 3, p. 290 (1985).
86. Ando, H., Inuzuka, H., Konagi, M. and Takahashi, K., *J. Appl. Phys.*, Vol. 58, p. 802 (1985).
87. Donnelly, V.M., Geva, M., Long, J. and Karlicek, R.F., *Appl. Phys. Lett.*, Vol. 44, p. 951 (1984).
88. Manasevit, H.M. and Simpson, W.I., *J. Electrochem. Soc.*, Vol. 116, p. 1968 (1969).
89. Krautle, H., Roehle, H., Escobosa, A. and Beneking, H., *J. Electron. Mat.*, Vol. 12, p. 215 (1983).
90. Gottschalch, V., Petzke, W.H. and Butter, E., *Kristall. Tech.*, Vol. 9, p. 209 (1974).
91. Leys, M.R. and Veenvliet, H., *J. Cryst. Growth*, Vol. 55, p. 145 (1981).

92. Coleman, J.J., Dapkus, P.D., Holonyak, N., Jr., and Laidig, W.D., *Appl. Phys. Lett.*, Vol. 38, p. 894 (1981).
93. Aebi, V., Cooper, C., Moon, R.L. and Saxena, R.R., *J. Cryst. Growth*, Vol. 55, p. 517 (1981).
94. Baliga, B.J. and Ghandi, S.K., *J. Electrochem. Soc.*, Vol. 122, p. 683 (1975).
95. Mori, Y. and Wantanabe, N., *J. Appl. Phys.*, Vol. 52, p. 2792 (1981).
96. Cooper, C.B., Saxena, R.R. and Ludowise, M.J., *Electron. Lett.*, Vol. 16, p. 892 (1980).
97. Fukui, T. and Horikoshi, Y., *Jpn. J. Appl. Phys.*, Vol. 19, p. L53 (1980).
98. Ludowise, M.J. and Dietze, W.T., *J. Electron. Mat.*, Vol. 11, p. 59 (1982).
99. Samualson, L., Omling, P., Titze, H. and Grimmeiss, H.R., *J. Physique*, Vol. 43, p. C5-323 (1982).

8

Photochemical Vapor Deposition

Russell L. Abber

INTRODUCTION

In conventional, or thermally activated, chemical vapor deposition (CVD), reactive gases are passed over a heated surface. The thermal energy derived from this surface drives the reactions, depositing a film. As in any heterogeneous reaction system, the gas phase species can combine in the gas phase, on the surface, with the surface, or in any combination thereof.

CVD is commonly used in semiconductor processing to deposit thin films such as metals, dielectrics, and semiconductors. Dopants, which are controlled impurities, are added to these films in order to obtain the desired electrical properties. The dopants can either be added during the deposition process or after the film is grown.

For this application, the major drawback of thermally activated CVD are the problems associated with high temperature processing. Depending upon the desired properties of the film and the precursors, temperatures can range as high as 1200°C. At temperatures in excess of 800°C, dopants can not only diffuse out of the top surface, but also between buried layers, altering the impurity profiles which are essential for device applications. In addition, any impurities in the reactor, the susceptor, or the backside of the wafer can diffuse into the growing film.

Hillocks, or small nodules, will form in aluminum films above 450°C due to the thermal mismatch between aluminum and silicon. Temperatures above this level will also cause a reaction between the aluminum and silicon, degrading the thin film interface.

High temperatures pose problems in compound semiconductor processing due to the difference in volatility among the various species in a particular layer. Furthermore, high temperatures can physically damage a wafer, generating defects in the devices.

In recent years, the problems associated with high temperature processing have led to the development of alternative, low temperature processing. Plasma-enhanced CVD (PECVD) is one such technique. Although plasma processing is widely used to etch a variety of films, there are some problems that have limited the adoption of plasma to film deposition. Perhaps the most difficult problem that PECVD poses is the radiation induced damage to the film. PECVD relies on an RF source, such as microwaves, to ionize the reactive gases. In this manner, the reactants then possess enough energy to react and form a film. However, these reactants are highly energized, and their collisions with the surface can transfer enough energy to damage the film. Although many of these damages can be annealed out, this eliminates the advantages of low temperature processing, since annealing is done at elevated temperatures. The highly energized reactants are also more likely to react with the reactor vessel, introducing new sources of contamination.

The kinetics of PECVD are extremely complex, primarily due to the large number of species at various energy levels. For a detailed discussion of PECVD, see Chapter 4.

Photo-CVD is a relatively new, low temperature, nonsurface damaging method of depositing thin films. Photons from lamps, lasers, or plasmas provide energy to dissociate chemical species. Unlike high energy PECVD particles, however, the dissociated reactants are electrically neutral, free radicals. Because a given molecule will absorb light only at specific wavelengths, the choice of photons will dictate which chemical bonds will be broken. The breaking of a specific bond allows for a high degree of selectivity and a greater control of the deposition process.

This chapter will focus on the application of Photo-CVD to thin film deposition, discussing three vital aspects: first, an overview of Photo-CVD theory; second, a review of thin film deposition processes to which Photo-CVD has been applied; lastly, a discussion of two commercially available reactors for Photo-CVD and some important points for reactor construction.

THEORY

Photo-CVD processes can be divided into two broad categories, but, in practice, the distinction is rarely made. Photo-activated CVD refers to a set of reactions under certain conditions that, without the addition of photons, either would not occur or would occur so slowly as to be imperceptible. If, by the addition of light, the reaction rate is increased, or the morphology of the film is modified, then the reactions are photo-enhanced.

For the purpose of this chapter, Photo-CVD is defined as the use of photons to dissociate molecules or excite atoms in the gas phase or on a surface to form a thin film, without significantly increasing the thermal energy of the reactants and/or surface. In general, this implies that the photons are of high energy, usually in the ultraviolet (UV) and that the surface does not absorb the photons.

The primary sources for UV light are low pressure mercury discharge lamps and excimer lasers. In addition, xenon lamps and gas discharges may also serve as sources of UV light. Typical wavelengths and intensities of a discharge lamp and pertinent excimer lasers are presented in Table 1.

Table 1: Wavelength and Intensity of The Major Light Sources

Light Source	Wavelength (nm)	Intensity W/cm^2
Mercury discharge lamp	254	10-100 x 10^{-3}
	185	2-25 x 10^{-3}
Excimer lasers		
F$_2$	157	3-10
ArF	193	3-10
KrCl	222	3-10
KrF	248	3-10
XeCl	308	3-10
XeF	350	3-10

Before considering the effect of light on CVD, one must first understand the steps the reactive gases go through before leaving the reactor. CVD can be described as a heterogeneous reaction system: one in which gases react either with or on a solid surface to form solid and gaseous products. Outlined below are the steps that could occur in a heterogeneous reaction system.

(1) Mixture of the reactants in the gas phase. They may react to form intermediate species or, in a process called gas phase nucleation, the reaction will go to completion, creating solid particles in the gas phase.

(2) Transport of the gaseous species from the bulk fluid to the surface.

(3) Adsorption of these species at the surface.

(4) Reaction among the adsorbed species and/or the surface.

(5) Diffusion of the products along the surface.

(6) Desorption of adsorbed products.

(7) Transport of the products away from the surface into the bulk fluid.

The addition of light to the heterogeneous reaction system can increase the rate of any or all of these steps with the exception of transporting the species to and from the bulk. In order to control and optimize the deposition process, it is necessary to understand the effect of light on the sequence of reactions. Furthermore, it is important to understand the effects of temperature, total pressure, partial pressure of the reactants, light intensity, and other variables on the rate limiting step, and thus on the whole deposition process.

Although each set of reactions is different, it is possible to generalize about the effects of some of the variables. Most Photo-CVD processes are independent of or weakly dependent on, the surface temperature. Increasing the partial pressure of the rate limiting species will increase the deposition rate while lowering the total pressure tends to decrease the deposition rate. (However, operating at higher pressures has a few disadvantages. At atmospheric pressure, it becomes difficult to control environmental contamination and the use of toxic and pyrophoric gases should be avoided.)

If the light aids in the surface reactions or increases the mobility of the surface species, then the surface must be irradiated. On the other hand, if the light energy does not affect these processes, then whether the light irradiates the surface or not is a matter of convenience, not necessity.

When the photo-dissociation proceeds through a single photon pathway, the deposition rate is directly proportional to the intensity of the photons. In multiphoton reactions, the same dependency is observed, but it is weaker.

It should be remembered that the rate limiting reaction can be changed by varying the deposition parameters. For example, at low light intensities, the deposition rate will increase with light intensity on a one-to-one basis. Once the system is saturated with UV light, an increase in light intensity will have no affect on the deposition process.

In some systems, it is either undesirable or impossible for the reactants to absorb the available radiation. In these cases, a sensitizer must be used to decompose the reactants. A sensitizer will only aid in the gas phase reactions - no effect will be observed if the deposition proceeds through a surface decomposition.

The most common sensitizer is mercury, and it is generally used with a mercury discharge lamp. The mercury absorbs the emitted photons and is excited to a higher energy state. The excited atom decomposes one or all of the reactants by transferring its energy in a gas phase collision. Deposition rates tend to be higher with a sensitizer than without for identical conditions. However, the increase in deposition rate is offset by the concern of possible mercury incorporation into the film and the health hazards of handling mercury.

REVIEW OF PHOTO-CVD APPLICATIONS

A wide variety of films have been grown using Photo-CVD.[1-83] In selecting a material to be deposited, a number of criteria must be used. The most important of these, and one that applies to all forms of CVD, not just Photo-CVD, is that the reactants must be volatile. Some method must exist for the reactants to enter the deposition chamber in the vapor phase. Liquids and solids may be vaporized, or a carrier gas bubbled through or passed over them. A more in-depth discussion on handling liquid and solid sources is presented later in this chapter.

Once a set of possible reactants has been selected, the suitability of each reactant to Photo-CVD must be determined. The available light energy should be absorbed by the reactants. Failing this, the wavelength of the light should be changed, or a sensitizer added.

The next step in designing a Photo-CVD process is the design and construction of a reactor. This is the focus of the last section of the chapter.

Although a detailed discussion of each Photo-CVD application is beyond the scope of this chapter, a brief review of thin films deposited by Photo-CVD is provided in the following pages.

Silicon

It should come as no surprise that more Photo-CVD work has been performed with silicon than with any other material. Epitaxial and amorphous silicon have been deposited with mercury discharge lamps, and lasers have been used to grow amorphous and polycrystalline films.

Frieser[1] used hexachlorodisilane to deposit epitaxial films on silicon surfaces, achieving single crystal growth at temperatures as low as 700°C. Without the UV light, polycrystalline films at half the deposition rate were grown. Frieser theorized that the silicon diatom was important in achieving epitaxial growth.

Kumagawa et al[2] found that the increased deposition rates only occurred when the light irradiated the surface. Minimal decrease in the threshold temperature (the lowest temperature at which single crystal growth will occur) was observed when silicon tetrachloride was used as the source of silicon. Yamazaki et al,[3] publishing 16 years after the previous works, demonstrated epitaxial growth at 630°C, using disilane. When silane was used as a reactant, the mercury xenon lamp had no effect on the deposition rate, but the activation energy of photo-dissociated disilane was 14 kcal/mol below the activation energy for the thermal decomposition.

One important problem associated with the films grown by the above research groups, is the high temperature (1050° to 1200°C) cleaning step necessary for epitaxial growth.[1-3] Ishitani et al[4] were able to achieve single crystal films with a precleaning step at 730°C. Using dichlorosilane, they deposited epitaxial silicon at 650° to 730°C.

The most promising work to date was performed by Nishida et al.[5] By mixing difluorosilane with disilane, they achieved single crystal growth at 100° to 300°C. Their cleaning procedure consisted mainly of a 10^{-5} Torr bakeout. The fluorine and fluoride radicals play an important role not only in removing the native oxide on the initial surface, but also at later stages of the deposition. No hydrogen or fluorine were detected in the films by infrared measurements.[5]

Low temperature epitaxy is a promising technique, because of the sharp dopant profiles that can be obtained. Very little work has been done on the doping of Photo-CVD epitaxial silicon. Until the effects of doping have been determined and devices have been made with the dopants, the suitability of this process for VLSI applications will remain unknown.

Hydrogenated amorphous silicon (a-Si:H) is widely used in the formation of silicon solar cells. By doping with boron and phosphorus, p-i-n junctions can be formed. Mishima et al[6] obtained a-Si:H films with disilane and mercury discharge lamps at temperatures below 300°C. Typical deposition rates were 1.5 nm/min. The deposition rate was independent of substrate temperature for undoped and phosphorus doped films, while it was thermally activated and dramatically increased by boron doping.[7,9] Substitutional doping and increasing the deposition temperature decreased the amount of hydrogen in the films.

Tarui et al,[8] using what they called a 185 nm lamp, achieved deposition rates as high as 60 nm/min for a mercury sensitized silane process. The 185 nm lamp is a normal, low pressure mercury discharge lamp, made from synthetic quartz (Suprasil), rather than fused quartz, so that the lower wavelength light is not absorbed by the lamp. Saitoh et al[10] used the mercury photo-sensitization of silane to grow films below 300°C. Unlike the deposition of epitaxial silicon films, a-Si:H deposition is limited by the gas phase reactions.[10]

When mercury was added to a disilane and hydrogen mixture, the deposition rate was increased by 300 to 400%,[11] as compared to the direct photo-decomposition.[6]

Tarui et al[12] utilized a novel reactor design to deposit films from silane without the use of a mercury catalyst. The lamp, an argon or mercury doped argon discharge, was incorporated directly into the reactor. More details of this reactor will be presented later.

Ashida et al[13] describe the effects of doping on a thermally activated (500°C) disilane process to deposit a-Si:H. Phosphine doping decreased the deposition rate due to a catalytic poisoning of the surface. On the other hand, the addition of diborane increased the reaction rate by catalytically activating the surface. A similar phenomenon was noted for the Photo-CVD of disilane. Mishima et al[16] stated that phosphine decreased the deposition rate and diborane increased the rate. Furthermore, for identical gas ratios, the incorporation of phosphorus was higher for the Photo-CVD process than the thermal process. The reverse effect was observed for boron doping. In addition to the aforementioned mechanism,[13] Mishima et al[16] attribute the influence of the dopants on deposition rates and dopant incorporation to the absorption cross sections in the 190 to 200 nm region. The cross section of diborane is an order of magnitude lower than that of disilane, while the phosphine cross section is one order of magnitude higher than disilane. Therefore, given equivalent gas ratios, one could expect higher phosphorus incorporation in the Photo-CVD films than the thermal films.

Decreasing the hydrogen content of a-Si:H by substitutional doping lowers the optical band gap, causing a problem in silicon solar cell fabrication.[13] Inoue et al[14] added acetylene and dimethylsilane to form a wide optical band gap hydrogenated amorphous silicon carbide a-SiC:H). They used this a-SiC:H to form solar cells with an efficiency of 8.29%.[15] Although this a-SiC:H has a wide band gap, its dark conductivity is quite low—1.0×10^{-6} Ohm^{-1}cm^{-1}.[15] Pure methylsilanes [$SiH_n(CH_3)_{4-n}$, n = 1, 3] and acetylene were used in follow-up work to deposit a fairly conductive (7.0×10^{-5} Ohm^{-1} cm^{-1}) a-SiC:H, which gave a solar cell efficiency of 9.46%.[17] This efficiency was obtained when monomethylsilane was used as the carbon source in a-SiC:H. The optical band gap of the amorphous layers was higher, at equivalent dark conductivities, for the direct photo-dissociation as opposed to the mercury sensitized dissociation.[17]

Nishida et al[18] continued their solar cell fabrication with the Photo-CVD of microcrystalline a-Si:H. Highly conductive (dark conductivities were: n-type, 20 Ohm^{-1}cm^{-1}, p-type, 1 Ohm^{-1}cm^{-1}) and wide optical band gap films were grown from disilane, diborane, phosphine, and hydrogen at 200°C.

In order to reduce the cross contamination of doped layers and the amount of oxygen and water from the loading and unloading procedure, a four chamber reactor, three for deposition and one for loading, was designed.[19] Solar cells made in this reactor had a high efficiency of 9.64%.

Trisilane has a higher absorption of 185 nm light than disilane, so its deposition rate should be higher in a direct photo-dissociation. This was verified by Kumata et al.[20]

Polycrystalline silicon films (with average grain sizes up to 500 nm) were deposited on silicon dioxide using ArF and KrF excimer laser to dissociate silane.[21] The high intensity of the laser (1 to 20 MW/cm^2) may account for the film being polycrystalline rather than amorphous.

a-Si:H films were also deposited with the same wavelength lasers from disilane.[22] The optical band gap in the laser deposited films was higher than that of plasma-CVD or mercury sensitized Photo-CVD films. Yamada et al[23] also used an

ArF laser to deposit a-Si:H from disilane at deposition rates as high as 300 nm/min.

In the above works, the laser beam was either normal to the surface[22,23] or 60° above the horizontal.[21] In contrast, Taguchi et al[24] used a focused ArF laser parallel to the surface and about 1 mm above it, to photo-dissociate disilane. Deposition rates ranged from 2 to 5 nm/min. In a similar system, but with the beam normal to the surface, Zarnani et al[25] controlled the hydrogen content of the film by varying the substrate temperature.

Dielectrics and Insulators

The Photo-CVD process which forms PHOTOX™ (PHOTOX is a trademark of Hughes Aircraft Company) is very promising for VLSI applications. Peters[26,27] first described the PHOTOX process in 1981. Peters[26,27,29] and Kim et al[28] proposed the following set of reactions to describe the process.

Direct Photolysis

(1) $N_2O + h\nu$ (184.9 nm) $\rightarrow N_2 + O(^1D)$

(2) $Hg_0 + h\nu$ (253.7, 184.9 nm) $\rightarrow Hg^*$ (photoexcited)

(3) $Hg^* + N_2O \rightarrow N_2 + O(^3P) + Hg_0$

(4) SiH_4 (gas) $\rightleftharpoons SiH_4$ (adsorbed)

(5) SiH_4 (adsorbed) $\rightarrow$ Si (deposit) $+ 2H_2$ (adsorbed)

(6) O (gas) $\rightleftharpoons$ O (adsorbed)

(7) SiH_4 (adsorbed) $+ xO$ (adsorbed) $\rightarrow SiO_x$ (deposit)

(8) H_2 (adsorbed) $\rightleftharpoons H_2$ (gas)

(9) 2O (adsorbed) $\rightarrow O_2$ (adsorbed) $\rightleftharpoons O_2$ (gas)

If no mercury is present, oxygen atoms are formed through reaction 1. Reactions 2 and 3 dominate in the presence of mercury. Silane is transparent to the radiation from a mercury discharge lamp, so it will not be photo-decomposed. The silicon oxide is formed by reaction 7, where x can vary from zero to two, depending on the ratio of nitrous oxide to silane. If there is no nitrous oxide present, reaction 5 will dominate, forming pure silicon. The recombination of atomic oxygen, reaction 9, is the rate limiting step. As the rate of this reaction increases, the amount of oxygen available for reaction 5 decreases, retarding the formation of PHOTOX.

Although deposition rates for the mercury sensitized processes tend to be higher than direct photolysis rates, some concern has been expressed about the possible incorporation of mercury in the film. The mercury, if present as a free metal, would decrease the dielectric strength of the insulating film. Employing secondary ion mass spectroscopy (SIMS), no mercury was detected in the films above the established limit of sensitivity (100 ppm).[28]

Mishima et al[30,31] deposited silicon dioxide from a disilane and oxygen mixture. The UV directly dissociates the disilane, which in turn reacts with oxygen. At a temperature of 150°C, the Photo-CVD deposition rate is about 40 times larger than the rate for thermal CVD of a disilane and oxygen mixture. Some formation of ozone is likely occurring, accelerating the deposition rate.

In the silane-oxygen system, the reaction definitely proceeds through the formation of ozone since silane is not dissociated by the photons from mercury discharge lamps. Okuyama et al[33] used a deuterium lamp, which emitted VUV light above 160 nm, to dissociate silane without a sensitizer. Tarui et al[35] and Takahashi and Tabe[36] used a mercury discharge lamp to deposit silicon dioxide from a silane-oxygen mixture, also without a sensitizer. The activation energy of the silane-oxygen system is 0.22 eV[35] which is less than half the disilane-oxygen system activation energy of 0.53 eV.[31] The deposition rates for both of these systems are comparable within a temperature range of 100° to 300°C. Mercury photo-sensitization of a silane and nitrous oxide mixture was studied by other researchers.[32,34]

In general, the properties of PHOTOX are commensurate with thermal oxides, although slightly less dense.[34] The PHOTOX and thermal oxide etch rates in buffered hydrofluoric acid, however, are not the same, due to the differences in density and hydrogen content. The etch rate of the undensified PHOTOX is eight times greater than that of a thermal oxide. In contrast, when the film is annealed in oxygen at 925°C, the etch rates are identical.[34]

Likewise lasers can be used to deposit silicon dioxide films.[37-39] Boyer et al[37,38] achieved deposition rates of 300 nm/min by photo-dissociating a silane and nitrous oxide mixture with an ArF excimer laser. Although the deposition rate was independent of substrate temperature, film quality was improved at temperatures above 200°C.[38] Tate et al[39] deposited a ternary mixture of germanium-silicon-oxygen using an ArF laser to dissociate silane and germanium mixed with nitrous oxide.

Silicon nitride is another important insulator for VLSI that can be readily deposited by Photo-CVD. In a typical system, ammonia and silane are photodissociated by a mercury sensitizer with a low pressure mercury discharge lamp.[40,41] Photo-nitride has been deposited at temperatures as low as 100°C, at a rate of about 6.5 nm/min. Without the mercury photo-sensitizer, deposition rates were only slightly lower - 4 nm/min.[42]

Other oxides and nitrides that have been deposited by Photo-CVD include aluminum oxide,[37,43] aluminum oxynitride $[(Al_2O_3)_{1-c}(AlN)_c]$,[44] phosphorus nitride,[45] and zinc oxide.[37,46]

Metals

Aluminum is the most important metal currently used in semiconductor processing. Although alternatives to the use of aluminum are being explored, it is unlikely that aluminum will ever be completely replaced. Deutsch et al[47] demonstrated that small spots of aluminum could be grown from trimethylaluminum (TMA) using UV lasers. They subsequently extended the work to direct writing of aluminum films.[48,49] The UV laser is used to dissociate the adsorbed aluminum alkyl wherever the laser beam scans. Later, thermal CVD techniques are used to continue the growth only where the film had previously been prenucleated. Tsao and Ehrlich[50] used a laser to write patterns onto a silicon dioxide surface with triisobutylaluminum (TIBAL). Very thin films, less than 5 nm, were deposited by the prenucleation step. These surfaces catalyzed the deposition of aluminum films, so that growth occurred only in the prenucleated pattern.

Higashi and Fleming[54] used an unfocused laser and a mask, rather than a focused scanning laser to achieve similar results. Other metals, such as cadmium, zinc, tin, bismuth, germanium, silicon, and boron, were also deposited with this method.[47-49]

Higashi and Rothberg[51,53] studied the mechanism of TMA photo-decomposition on sapphire substrates. Comparing deposition results using two lasers, ArF and KrF, they found the shorter wavelength ArF laser was more effective at removing the methyl groups and nucleating purer films. Furthermore, Motooka et al[52] demonstrated that the photo-decomposition of TMA proceeds through the absorption of a single photon and that CH fragments are formed by the chemical reactions among the free radicals.

Calloway et al[55] used incoherent vacuum ultraviolet (VUV), of wavelength 105-190 nm, to deposit TMA at room temperature. Due to the high energy of the photons, TMA was dissociated not only into free radicals, but also into ionized fragments. These fragments were attracted to the surface by a -200V bias on the substrate to increase the deposition rate.

Refractory metals and their silicides are gaining in importance as the electrical and mechanical requirements for VLSI exceed the limits of aluminum. Solanki et al[56] demonstrated that tungsten, molybdenum, and chromium hexacarbonyls could be used to deposit the metal films by UV light. The work was extended from small spots ($10^{-4} cm^2$)[56] to large areas (>5 cm^2).[57] Deposition rates were as high as 150 nm/min at room temperature.[57] Metallic films resulted if the laser beam was normal to the surface, but if the beam grazed the surface, only black particulate films were formed. The metallic films contained less than 1% carbon and less than 7% oxygen.[57]

Additionally, using titanium chloride, Tsao et al[58] were able to deposit titanium films on $LiNbO_3$. They found evidence that the rate limiting step was the photolysis of the adsorbed titanium chloride surface layer.

Tungsten films were deposited on silicon and silicon dioxide by UV lasers using tungsten hexafluoride.[59] Deutsch and Rathman[59] compared these films with those grown thermally. The tungsten deposition rate for Photo-CVD was significantly greater than for thermal CVD, between 200 and 300°C. The presence of UV light was essential for deposition of tungsten onto silicon dioxide surfaces below 440°C; otherwise, no deposition occurred. Unlike deposition from the carbonyl,[57] these films contained less than 1% of both carbon and oxygen.

Gupta et al[60] deposited titanium silicide films from a titanium chloride and silane mixture with an excimer laser. The UV laser photo-dissociated the titanium chloride. When a carbon dioxide laser was used to heat the silane independently of the substrate, greater control of the process was possible.[61]

Creighton[62] and Flynn et al[65] used hexacarbonyls to deposit molybdenum,[62,65] chromium,[65] and tungsten,[65] but in all cases the carbon and oxygen levels were too high to be of practical use.

Toyama et al[63,64] used a xenon lamp to increase the deposition rate of tungsten from tungsten hexafluoride and hydrogen. They noted that the rate increased in proportion to irradiation time for exposures of 10-100 minutes. Although the experimental data have not yet been published, they proposed a few models to account for the increased deposition rate and its time dependence.

Cadmium,[47,67] copper,[66] and chromium[57] have also been deposited from UV light sources. The volatile source for copper deposition was bis(1,1,1,5,5,5-hexafluoro-2,4-pentanedionate) copper (II), CuHF, which is a solid at room temperature.[66]

Compound Semiconductors

The use of compound semiconductors has increased over the past few years, and continued growth is forecast for the future. Because the deposition of many compound semiconductors are extremely sensitive to high process temperatures, Photo-CVD provides an excellent alternative to conventional processing. Putz et al[68] employed a low pressure mercury discharge lamp to increase the deposition rate and improve the morphology of gallium arsenide films grown from trimethylgallium (TMG) and arsine. As would be expected, the UV light had the strongest affect when the reaction was surface kinetically controlled. This work was continued using triethylgallium (TEG) as a gallium source, which allowed the deposition temperature to be decreased by 150°C.[71] However, the UV light had a stronger affect on films grown from TMG than TEG.

Aoyagi et al[69] showed that light from an Ar laser (514.5 nm) had little affect on the deposition rate above 750°C where presumably the deposition is transport controlled. Nishizawa et al[70] used excimer lasers to demonstrate that the effect of UV light was wavelength dependent. When the light source was a KrCl, XeCl, or XeF laser, no change in the deposition rate was observed, but it increased sharply when a KrF laser was used.

Nishizawa et al[72,73] applied UV light to a molecular layer epitaxy (MLE) process. Arsine and TMG or TEG were alternately introduced into the chamber where they combined on the surface one monolayer at a time. The chamber was evacuated prior to the introduction of each gas. When the growing surface was irradiated with an high pressure mercury discharge lamp, an Ar laser, a frequency doubled Ar laser,[72] or excimer lasers,[73] the morphology and electrical properties were improved. The improvement appeared to be caused by an increase in the migration length of the adsorbed gallium complex.

Donnelly et al[74,75] used an ArF laser to photo-dissociate $(CH_3)_3InP(CH_3)_3$ and $P(CH_3)_3$ to form indium phosphide and indium oxide. Focusing the laser promoted multiphoton absorption and increased the incorporation of carbon in the film. However, the latter could be suppressed by keeping the beam normal to the surface.[74]

Zinc, selenium, and nonstoichiometric zinc-selenide were deposited by the photo-dissociation of dimethylzinc and dimethylselenide with a mercury xenon arc lamp at room temperature.[76] The only significant impurity in the film was carbon. A low pressure mercury discharge lamp was also used to dissociate diethylzinc and dimethylselenide to form zinc-selenide.[77] Without the irradiation of UV light, the deposition rate rapidly decreased below 400°C, whereas, with irradiation, growth occurred down to 200°C.[77]

Epitaxial mercury-telluride films were grown from diethyltelluride and mercury vapor using a mercury discharge lamp.[78] High mercury vapor pressures (>7 Torr) were needed for adequate growth.

Photo-CVD can also be used to deposit sulfides with a mercury discharge lamp. In this system, carbonyl sulfide is dissociated either directly or with a mercury catalyst.[79]

Miscellaneous

Kitahoma et al[80] explored the possibility of depositing diamond films with UV light. An ArF excimer laser was used to dissociate acetylene diluted with hydrogen. When the unfocused laser beam was aimed at the surface, a brown-colored amorphous film, probably of high order organic polymers, was formed. If the laser was focused 1° to 5° above the horizontal, however, diamond lines were formed along previously made scratches in the substrate.

Germanium films have also been deposited by laser Photo-CVD.[81-83]

PHOTO-CVD EQUIPMENT

Commercial Equipment

Two commercial Photo-CVD reactors are available as of Fall, 1986. One, based on mercury discharge lamps, is manufactured by Tylan Corp., Carson, CA; while the other is a laser-CVD system manufactured by Nippon Tylan Corp., Tokyo.

The photon-enhanced reactor consists of twin deposition chambers linked by a single gas delivery and exhaust system (Figures 1 and 2). The reactor was primarily developed for pilot line and R&D fabrication of advanced semiconductor devices, but its versatility enables it to be used for many low temperature, low pressure, Photo-CVD research applications. A substrate heater provides temperature control from 50° to 250°C, and all process gases are mass-flow controlled. Both automatic and manual process controls are provided. Thickness uniformities, film characteristics, and equipment issues for the PHOTOX deposition process are shown in Table 2.

Figure 1: Photochemical CVD reactor.

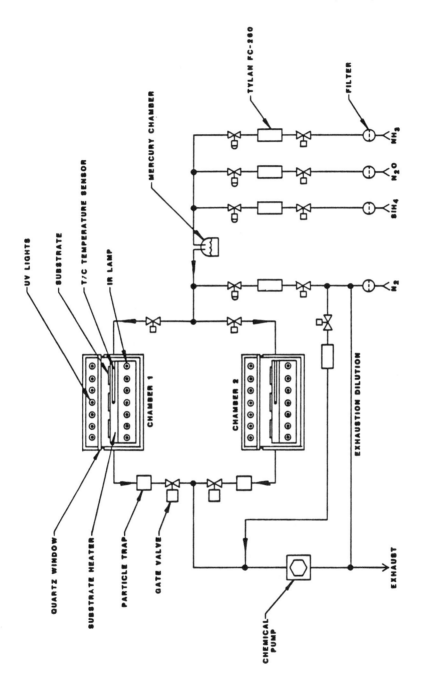

Figure 2: Flow diagram for photochemical CVD reactor.

Table 2: PHOTOX Properties Deposited in Photon Enhanced Reactor

Thickness Uniformity (100 mm wafer)[1]

Across substrate	±4.5%
Within run	±3.5%
Repeatability	±3.5

Film Characteristics

Density (g/cm^3)	2.10–2.27
Index of refraction	1.46–1.50
Dielectric constant	3.8–3.9
Dielectric strength[2]	0.5–1.5 × 10^6 V/cm
Particulate density[3]	<4/cm^2
Deposition rate[4]	20–60 nm/min
Total cycle time for 150 nm film	25–35 min
Process temperature	50°–200°C
Deposition pressure	1.15–1.50 Torr
Film adhesion (tension)	>6.9 × 10^7 Pa

Equipment Characteristics

UV grid lamp lifetime	500–1,000 hours
Mercury reservoir temperature	25°–65°C
Mercury vapor in pump exhaust	2–25 mg/m^3

Notes:
1) Uniformity calculations are based on 1 σ deviation.
2) For 120–650 nm thickness.
3) Added particles per square centimeter (0.5 micron and larger).
4) Dependent on mercury reservoir temperature, gas flow rates, reactor pressure, and substrate temperature.

The Laser CVD System, shown in Figures 3 and 4, is an excimer laser Photo-CVD system. It consists of a reactor chamber with a mechanical x/y scanner and heated substrate holder, an excimer laser, and a controller for process sequence, x/y scanning, gas distribution, temperature, and pressure. The x/y scanner is accurate to 10 microns over 100 mm with a scanning speed of 0.5 to 60 mm/sec. The chamber can handle wafers up to 150 mm, and it takes about one hour to scan a 100 mm wafer.

Reactor Design

There are two major design considerations which must be contemplated when constructing a reactor. First, because most Photo-CVD reactions are insensitive to temperature, deposition occurs on the reactor walls as well as on the substrate. One of these walls is the window through which light energy passes. The transmission of this window degrades during the deposition process due to film growth on the inside surface. The other major design consideration is the reactant handling system. Because it is not always possible to choose gaseous

Photochemical Vapor Deposition 283

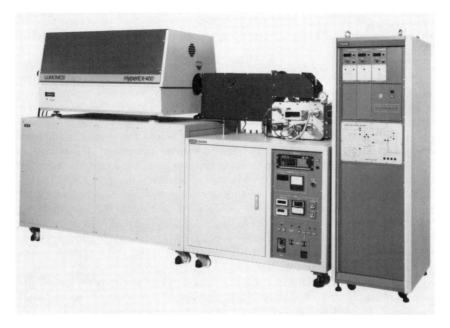

Figure 3: Laser enhanced photo-chemical reactor.

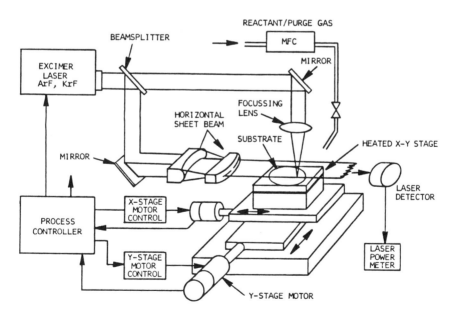

Figure 4: Schematic of laser enhanced CVD reactor.

precursors, the choice of reactants is limited to species that absorb the available photons. Hence, liquid and solid sources, which require special handling, must be used.

When considering designs for limiting growth on the window in a given process, the combination of deposition time and film thickness should be carefully examined. If the film can be deposited before the window film growth significantly attenuates the light through absorption, then the method employed to limit window deposition was successful. However, when growth on the window reduces the reaction rate below a critical value before the process is complete, further steps must be taken to decrease unwanted deposition. It has been assumed that for most applications it would be unacceptable to open the chamber to atmosphere once film growth begins. Aside from the introduction of particulate and contamination to the film in mid-layer, it would add significantly to the processing time. This is because the extreme toxicity and hazards of typical Photo-CVD reactive gases (i.e., silane, diborane, phosphine, mercury, etc.) require the chamber to be thoroughly purged prior to opening. Furthermore, once the reactor has been exposed to the environment, the adsorbed oxygen and water must be outgassed.

For a research system, the window can be cleaned after each run, if necessary. However, this procedure would complicate a production reactor, and a method for preventing window deposition must be developed.

There are two basic methods used to prevent window deposition: dilution of the reactive gases near the window and the use of physical or chemical barriers to prevent film growth on the window surface. Any combination of these methods can be employed.

Passing an inert purge gas across the window surface will prevent the bulk transport of the gases to the window. However, the reactive gases will still diffuse across the boundary layer, especially at lower pressures. Designing nozzles to aim the gases onto the substrate, and presumably away from the window, would aid in slowing the transport of the gases to unwanted surfaces. But, no matter how much the gases are diluted, the window will still be coated, albeit at a slower rate.

Locating the window far from the substrate would likely further decrease the transport of gases to the window. However, since the intensity of the light source decreases with distance, the deposition rate would fall off if the distance between substrate and window were to be increased.

Physical barriers can also be employed to prevent the reactive gases from getting to the window. Numasawa et al[42] used a 150 mm diameter, 2 mm thick, Suprasil disk as a separator plate to prevent window deposition. The disk consisted of 1,200 holes, each of which was 0.6 mm in diameter, and spaced 4 mm apart. An inert gas was passed between the separator plate and an optical window, effectively diluting the reactive gases.

Peters and Gebhart[84] employed a mobile, transparent barrier to keep the gases away from the window. The barrier material was continuously rolled past the window so that any deposition occurred on the moving barrier. The transparent film could be replaced each time the chamber was opened or as otherwise needed. In practice, this method did not work well enough for a production environment. The reactive gases were still able to diffuse around the barrier and deposit onto the window, and the polyvinylidene fluoride material that was used in some experiments[84] presented a significant outgassing problem.

Baron et al[85] modified the design so as to achieve a more effective barrier. A mobile teflon film physically separated the deposition and window areas, unlike the Peters and Gebhart[84] film which just covered the window. The window area was partially isolated from rest of the chamber, with its own gas inlet and outlet. This allowed the upper portion of the chamber to be both, kept at a higher pressure, minimizing the diffusion of the reactants into the area, and to be flushed with an inert gas and a deposition inhibitor, reducing film growth on the window. If outgassing of the barrier mater does not present a problem, this scheme would be suitable to a production environment, since the deposition time is limited only by the amount of teflon film and its movement rate past the window.

Another common barrier for reducing window deposition is a coating of a low vapor pressure oil, such as Fomblin (a perfluorinated fluid) on the inside surface of the window. The oil must be transparent to the light and have a vapor pressure low enough to ensure against film contamination. The oil acts as an effective barrier by delaying the nucleation of the film on the window. Care must be taken to apply the oil uniformly and of just the right thickness. If the film is too thin, growth will occur on the window, and if it is too thick, the oil can drip into the chamber and possibly onto the substrate.

The Fomblin film works only for a limited time, so it is not suitable for a production system. Scapple et al[86,87] were able to improve upon the manual application of oil to allow for long run times. Fomblin was continuously applied to the top of a cylindrical window and allowed to drip down the exposed surface, collecting at the bottom in a small pan. The oil was spread uniformly across the surface by the continuous action of wiper blades rotating around the cylinder. Using this method, the window can be kept free of unwanted growth indefinitely.

A novel means of eliminating attenuation of light by the growth of a film on the window is to remove the window entirely. Tarui et al[12] did this in what they called a unified reactor, using a discharge above the reactive gases to provide UV light. All of the photons from the discharge are available for the reaction since there is no window to cut off the light. Unfortunately, no special attempt was made to isolate the discharge gases from the deposition gases, and mixing occurred.

Many of the film precursors are liquids and, to a lesser extent, solids, at room temperature. As mentioned previously, some method must be used to convert these liquids and solids to a volatile species for transport into the reactor. Liquid sources can be introduced into the deposition chamber by boiling them or by bubbling an inert carrier gas through them. If the liquid is boiled off, then it can be treated as a gas, and passed through a mass flow controller. The flow rate of a liquid vapor from the bubbler can be controlled by varying either the bubbler temperature or the mass flow rate of a carrier gas. It is recommended that for dense liquids, such as mercury, the gases be passed over, rather than bubbled through, the vessel. Otherwise, the pressure drop across the bubbler will be too high. A solid can be introduced into the deposition chamber by melting it, and then treating it in the same manner as a liquid, by subliming it directly into the chamber, or, if the vapor pressure is high enough, by passing a carrier gas over the solid.

If the liquid or solid is heated above room temperature to allow it to enter the reactor, then care must be taken to prevent recondensation. Liquid droplets, to say nothing about solid particles, can interrupt the flow rate and make mass-flow control impossible. Therefore, the gas delivery system downstream of the source vessel should be heated 5° to 10°C hotter than the bubbler. Condensation problems could also be avoided by minimizing the length of tubing and number of valves downstream of the container. In addition, the coldest point in the deposition chamber should be at least as hot as the manifold. Otherwise, condensation will occur in the reactor, possibly contaminating the film with liquid droplets. These droplets may also serve as impurities in subsequent films deposited in the chamber. Safety hazards are also reduced if the liquid or solid do not condense in the chamber. If condensation does occur, the reactor should be thoroughly purged until the reactants are completely evaporated.

SUMMARY

Photo-CVD is a promising technique for depositing thin films at low temperatures. Photolysis of the reactive gases can occur at the surface or in the gas phase using lamps or lasers. A wide variety of materials have been successfully deposited by this method, and many of these materials have application in VLSI processing. Photo-CVD films exhibit higher purities and lower defect densities than PECVD films without the problems associated with high temperature processing such as substrate damage, dopant diffusion, and interface destruction.

Although most of the discussion in this chapter has focused on semiconductor applications, Photo-CVD can also be used to deposit coating on any temperature sensitive substrate. This includes hard coatings on magnetic discs and optical filters on plastics.

Two commercial systems for Photo-CVD are currently on the market. The PVD-1000 employs low pressure mercury discharge lamps to provide photons for thin film deposition, while the GS-4300B uses excimer lasers.

Two important design considerations, one for limiting film growth on the window and the other for reactant handling, are presented for the reader who wishes to design a custom system.

Acknowledgements

I would like to thank F. Gebhart, J. Hall, J. Linder, and H. Rogers for their helpful discussions. A special thanks to L. Roth and F. Rivero for their timely comments and advice.

REFERENCES

1. Frieser, R.G., *J. Electrochem. Soc.*, Vol. 115(4), p. 401 (1968).
2. Kumagawa, M., Sunami, H., Terasaki, T., and Nishizawa, J., *Jpn. J. Appl. Phys.*, Vol. 7(11), p. 1332 (1968).
3. Yamazaki, T., Ito, T., and Ishikawa, H., *Technical Digest of 1984 Symposium on VLSI Technology*, San Diego, p. 56 (1984).
4. Ishitani, A., Kanamori, M., and Tsuya, H., *J. Appl. Phys.*, Vol. 57(8), p. 2956 (1985).
5. Nishida, S., Shiimoto, T., Yamada, A., Karasawa, S., Konagai, M., and Takahashi, K., *Appl. Phys. Lett.*, Vol. 49(2), p. 79 (1986).

6. Mishima, Y., Hirose, M., Osaka, Y., Nagamine, K., Ashida, Y., Kitagawa, N., and Isogaya, K., *Jpn. J. Appl. Phys.*, Vol. 22(1), p. L46 (1983).
7. Mishima, Y., Ashida, Y., and Hirose, M., *J. Non-Crystalline Solids*, Vol. 59 & 60, p. 707 (1983).
8. Tarui, Y., Sorimachi, K., Fujii, K., and Aota, K., *J. Non-Crystalline Solids*, Vol. 59 & 60, p. 711 (1983).
9. Kazahaya, T., Hirose, M., Mishima, Y., and Ashida, Y., *Technical Digest of the International PVSEC-1*, Kobe, p. 449 (1983).
10. Saitoh, T., Muramatsu, S., Shimida, T., and Migitaka, M., *Appl. Phys. Lett.*, Vol. 42(8), p. 678 (1983).
11. Inoue, T., Konagai, M., and Takahashi, K., *Appl. Phys. Lett.*, Vol. 43(8), p. 774 (1983).
12. Tarui, Y., Aota, K., Kamisako, K., Suzuki, S., and Hiramoto, T., *Ext. Abstr. of the 16th (1984 International) Conference on Solid State Devices and Materials*, Kobe, p. 429 (1984).
13. Ashida, Y., Mishima, Y., Hirose, M., and Osaka, Y., *J. Appl. Phys.*, Vol. 55(6), p. 1425 (1984).
14. Inoue, T., Tanaka, T., Konagai, M., and Takahashi, K., *Appl. Phys. Lett.*, Vol. 44(9), p. 871 (1984).
15. Tanaka, T., Kim, W.Y., Konagai, M., and Takahashi, K., *Appl. Phys. Lett.*, Vol. 45(8), p. 865 (1984).
16. Mishima, Y., Hirose, M., Osaka, Y., and Ashida, Y., *J. Appl. Phys.*, Vol. 56(10), p. 2803 (1984).
17. Yamada, A., Keene, J., Konagai, M., and Takahashi, K., *Appl. Phys. Lett.*, Vol. 46(3), p. 272 (1985).
18. Nishida, S., Tasaki, H., Konagai, M., and Takahashi, K., *J. Appl. Phys.*, Vol. 58(4), p. 1427 (1985).
19. Takei, H., Tanaka, T., Kim, W.Y., Konagai, M., and Takahashi, K., *J. Appl. Phys.*, Vol. 58(9), p. 3664 (1985).
20. Kumata, K., Itoh, U., Toyoshima, Y., Tanaka, N., Anzai, H., and Matsuda, A., *Appl. Phys. Lett.*, Vol. 48(20), p. 1380 (1986).
21. Andreatta, R.W., Abele, C.C., Osmundsen, J.F., Eden, J.G., Lubben, D., and Greene, J.E., *Appl. Phys. Lett.*, Vol. 40(2), p. 183 (1982).
22. Yoshikawa, A., and Yamaga, S., *Jpn. J. Appl. Phys.*, Vol. 23(2), p. L91
23. Yamada, A., Konagai, M., and Takahashi, K., *Jpn. J. Appl. Phys.*, Vol. 24(12), p. 1586 (1985).
24. Taguchi, T., Morikawa, M., Hiratsuka, Y., and Toyoda, K., *Appl. Phys. Lett.*, Vol. 49(15), p. 971 (1986).
25. Zarnani, H., Demiryont, H., and Collins, G.J., *J. Appl. Phys.*, Vol. 60(7), p. 2523 (1986).
26. Peters, J.W., U.S. Patent 4,371,587, February 1, 1983, assigned to Hughes Aircraft Company.
27. Peters, J.W., U.S. Patent 4,419,385, December 6, 1983, assigned to Hughes Aircraft Company.
28. Kim, H.M., Tai, S.-S, Groves, S.L., Schuegraf, K.K., in: Proceedings of the Eighth International Conference on Chemical Vapor Deposition (J.M. Blocher, Jr., G.E. Vuillard, and G. Wahl, eds.), Electrochemical Society, Pennington, NJ, p. 258 (1981).
29. Peters, J.W., *International Electron Devices Meeting*, IEDM-81, p. 240 (1981).

30. Mishima, Y., Ashida, Y., Hirose, M., and Osaka, Y., *Ext. Abstr. of the 15th Conference on Solid State Devices and Materials*, Tokyo, p. 121 (1983).
31. Mishima, Y., Hirose, M., Osaka, Y., and Ashida, Y., *J. Appl. Phys.*, Vol. 55(4), p. 1234 (1984).
32. Hayafuji, N., Nagahama, K., Ito, H., Murotani, T., and Fujikawa, K., *Ext. Abstr. of the 16th (1984 International) Conference on Solid State Devices and Materials*, Kobe, p. 663 (1984).
33. Okuyama, M., Toyoda, Y., and Hamakawa, Y., *Jpn. J. Appl. Phys.*, Vol. 23(2), p. L97 (1984).
34. Chen, J.Y., Henderson, R.C., Hall, J.T., and Peters, J.W., *J. Electrochem. Soc.*, Vol. 131(9), p. 2146 (1984).
35. Tarui, Y., Hidaka, J., and Aota, K., *Jpn. J. Appl. Phys.*, Vol. 23(11), p. L827 (1984).
36. Takahashi, J., and Tabe, M., *Jpn. J. Appl. Phys.*, Vol. 24(3), p. 274 (1985).
37. Boyer, P.K., Roche, G.A., Ritchie, W.H., and Collins, G.J., *Appl. Phys. Lett.*, Vol. 40(8), p. 716 (1982).
38. Emery, K., Boyer, P.K., Thompson, L.R., Solanki, R., Zarnani, H., and Collins, G.J., *SPIE*, Vol. 459, p. 9 (1984).
39. Tate, A., Jinguji, K., Yamada, T., and Takato, N., *J. Appl. Phys.*, Vol. 59(3), p. 932 (1986).
40. Peters, J.W., Gebhart, F.L. and Hall, T.C., *International J. Hybrid Microelectronics*, Vol. 3(2), p. 59 (1980).
41. Peters, J.W., Gebhart, F.L., and Hall, T.C., *Solid State Technol.*, Vol. 23(9), p. 121 (1980).
42. Numasawa, Y., Yamazaki, K., and Hamano, K., *J. Electronic Mater.*, Vol. 15(1), p. 27 (1986).
43. Solanki, R., Ritchie, W.H., and Collins, G.J., *Appl. Phys. Lett.*, Vol. 43(5), p. 454 (1983).
44. Demiryont, H., Thompson, L.R., and Collins, G.J., *J. Appl. Phys.*, Vol. 59(9), p. 3235 (1986).
45. Hirota, Y., and Mikami, O., *Electronic Lett.*, Vol. 21, p. 77 (1985).
46. Solanki, R., and Collins, G.J., *Appl. Phys. Lett.*, Vol. 42(8), p. 662 (1983).
47. Deutsch, T.F., Ehrlich, D.J., and Osgood, R.M., Jr., *Appl. Phys. Lett.*, Vol. 35(2), p. 175 (1979).
48. Ehrlich, D.J., Osgood, R.M., Jr., and Deutsch, T.F., *Appl. Phys. Lett.*, Vol. 38(11), p. 948 (1981).
49. Ehrlich, D.J., and Osgood, R.M., Jr., *Thin Solid Films*, Vol. 90, p. 287 (1982).
50. Tsao, J.Y., and Ehrlich, D.J., *Appl. Phys. Lett.*, Vol. 45(6), p. 617 (1984).
51. Higashi, G.S., and Rothberg, L.J., *J. Vac. Sci. Technol. B*, Vol. 3(5), p. 1460 (1985).
52. Motooka, T., Gorbatkin, S., Lubben, D., and Greene, J.E., *J. Appl. Phys.*, Vol. 58(11), p. 4397 (1985).
53. Higashi, G.S., and Rothberg, L.J., *Appl. Phys. Lett.*, Vol. 47(12), p. 1288 (1985).
54. Higashi, G.S., and Fleming, C.G., *Appl. Phys. Lett.*, Vol. 48(16), p. 1051 (1986).

55. Calloway, A.R., Galantowicz, T.A., and Fenner, W.R., *J. Vac. Sci. Technol. A*, Vol. 1(2), p. 534 (1983).
56. Solanki, R., Boyer, P.K., Mahan, J.E., and Collins, G.J., *Appl. Phys. Lett.*, Vol. 38(7), p. 572 (1981).
57. Solanki, R., Boyer, P.K., and Collins, G.J., *Appl. Phys. Lett.*, Vol. 41(11), p. 1048 (1982).
58. Tsao, J.Y., Becker, R.A., Ehrlich, D.J., and Leonberger, F.J., *Appl. Phys. Lett.*, Vol. 42(7), p. 559 (1983).
59. Deutsch, T.F., and Rathman, D.D., *Appl. Phys. Lett.*, Vol. 45(6), p. 623 (1984).
60. Gupta, A., West, G.A., and Beeson, K.W., *J. Appl. Phys.*, Vol. 58(9), p. 3573 (1985).
61. West, G.A., Beeson, K.W., and Gupta, A., *J. Vac. Sci. Technol. A*, Vol. 3(6), p. 2278 (1985).
62. Creighton, J.R., *J. Appl. Phys.*, Vol. 59(2), p. 410 (1986).
63. Toyama, M., Itoh, H., and Moriya, T., *Jpn. J. Appl. Phys.*, Vol. 25(5), p. 679 (1986).
64. Toyama, M., Itoh, H., and Moriya, T., *Jpn. J. Appl. Phys.*, Vol. 25(5), p. 688 (1986).
65. Flynn, D.K., Steinfeld, J.I., and Sethi, D.S., *J. Appl. Phys.*, Vol. 59(11), p. 3914 (1986).
66. Jones, C.R., Houle, F.A., Kovac, C.A., and Baum, T.H., *Appl. Phys. Lett.*, Vol. 46(1), p. 97 (1985).
67. Wood, T.H., White, J.C., and Thacker, B.A., *Mat. Res. Soc. Sym. Proc.*, Vol. 17, p. 35 (1983).
68. Putz, N., Heinecke, H., Veuhoff, E., Arens, G., Heyen, M., Luth, H., and Balk, P., *J. Crystal Growth*, Vol. 68, p. 1984 (1984).
69. Aoyagi, Y., Masuda, S., Namba, S., and Doi, A., *Appl. Phys. Lett.*, Vol. 47(2), p. 95 (1985).
70. Nishizawa, J., Yokubun, Y., Shimawaki, H., and Koike, M., *J. Electrochem. Soc.*, Vol. 132(8), p. 1939 (1985).
71. Balk, P., Heinecke, H., Putz, N., Plass, C., and Luth, H., *J. Vac. Sci. Technol. A*, Vol. 4(3), p. 711 (1986).
72. Nishizawa, J., Abe, H., and Kurabayashi, T., *J. Electrochem. Soc.*, Vol. 132(5), p. 1197 (1985).
73. Nishizawa, J., Abe, H., Kurabayashi, T., and Sakurai, N., *J. Vac. Sci. Technol. A*, Vol. 4(3), p. 706 (1986).
74. Donnelly, V.M., Geva, M., Long, J., and Karlicek, R.F., *Appl. Phys. Lett.*, Vol. 44(10), p. 951 (1984).
75. Donnelly, V.M., and Karlicek, R.F., *SPIE*, Vol. 476, p. 102 (1984).
76. Johnson, W.E., and Schlie, L.A., *Appl. Phys. Lett.*, Vol. 40(9), p. 798 (1982).
77. Ando, H., Inuzuka, H., Konagai, M., and Takahashi, K., *J. Appl. Phys.*, Vol. 58(2), p. 802 (1985).
78. Irvine, S.J.C., Mullin, J.B., and Tunnicliffe, J., *J. Crystal Growth*, Vol. 68, p. 188 (1984).
79. Peters, J.W., U.S. Patent 4,447,469, May 8, 1984, assigned to Hughes Aircraft Company.

80. Kitahama, K., Hirato, K., Nakamatsu, H., and Kawai, S., *Appl. Phys. Lett.*, Vol. 49(11), p. 634 (1986).
81. Eden, J.G., Greene, J.E., Osmundsen, J.F., Lubben, D., Abele, C.C., Gorbatkin, S., and Desai, H.D., *Mat. Res. Soc. Sym. Proc.*, Vol. 17, p. 185 (1983).
82. Osmundsen, J.F., Abele, C.C., and Eden, J.G., *J. Appl. Phys.*, Vol. 57(8), p. 2921 (1985).
83. Stanley, A.E., Johnson, R.A., Turner, J.B., and Roberts, A.H., *Appl. Spectroscopy*, Vol. 40(3), p. 374 (1986).
84. Peters, J.W. and Gebhart, F.L., U.S. Patent 4,265,932, May 5, 1981, assigned to Hughes Aircraft Company.
85. Baron, B.N., Rocheleau, R.E., and Hegedus, S.S., SERI report SERI/STR-211-2967 (June 1986).
86. Scapple, R.Y., Peters, J.W., Linder, J.F., and Yee, E.M., U.S. Patent 4,597,986, July 1, 1986, assigned to Hughes Aircraft Company.
87. Scapple, R.Y., Peters, J.W., Linder, J.F., and Yee, E.M., U.S. Patent 4,615,294, October 7, 1986, assigned to Hughes Aircraft Company.

9
Introduction to Sputtering

Brian Chapman and Stefano Mangano

PRINCIPLE AND IMPLEMENTATION OF SPUTTERING

Introduction

In this chapter, we shall be looking at the technique of sputtering as a means of depositing thin layers of material, or *thin films* as they are commonly known. The thicknesses of these films are usually from a few angstroms (1 Å = 10^{-10} m) or a few monolayers of material, up to a micron (i.e., 1 micrometer or 10,000 Å) or so. Thin films are of interest not only because they are thin, but also because the ratio of the surface area to the bulk volume is so high that the surface properties become very important; this is not usually true for bulk materials.

We shall focus on the application of sputtering to *deposition*, since this is the main subject of this volume. However, we should note also that sputtering is used also for removing or *etching* a surface – in both physical and physically-assisted chemical processes, as we shall see below.

The purpose of this chapter is to give the reader some familiarity with the subject of sputtering. It is not meant to be a state-of-the-art survey, but is instead an introduction that will enable one to determine whether this may be an appropriate technology for a particular application, and also that will familiarize the reader with some of the benefits and limitations of the technology as well as with the language that is employed in the technical literature – an often daunting hurdle. Much more information is available in a number of reviews (e.g., Chapman[1]), and technical papers appear regularly in a number of journals, including the *Journal of Applied Physics* and the *Journal of Vacuum Science and Technology*, both published by the American Institute of Physics, New York.

Most people would regard sputtering as a fairly new technology. In the semi-conductor industry, for example, the technique has been used extensively only for the past fifteen years; and in other industries the technology is finding application only today or even tomorrow. So it would come as a surprise to many to find that the process was discovered more than 130 years ago by W.R. Grove.[2]

Grove was studying what we have subsequently come to call neon lights or the forerunners of fluorescent light tubes. He observed that a dark coloration formed on the inside of the tube adjacent to the electrode and correctly surmised that somehow the material of the electrode was being transferred to the inside wall of the glass tube. He had discovered sputtering! Grove's equipment (Figure 1) is an inspiration to us all. If we look closely we can see the electrical power supply, the reaction vessel, the electrodes, the vacuum pump (manually operated!), and even the gas supply - a leather bladder! A modern system duplicates each one of these components of the original system. Modern technology has made enormous strides in the materials that we now have available to implement these technologies. However, man has never been short of ideas and it is inspiring to see how successful Grove and his contemporaries were in spite of great adversities.

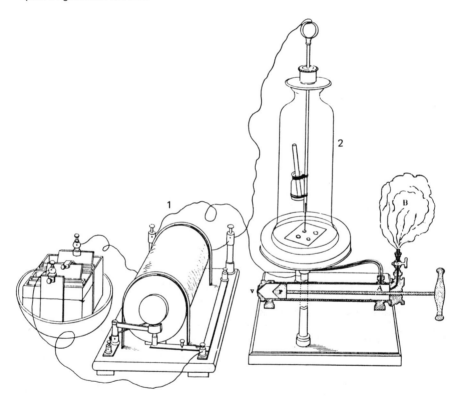

Drawing of apparatus used by Grove for his first sputtering studies. 1 is the coil apparatus used for generating a high voltage; the contact breaker is shown in front. 2 is the vacuum vessel containing an electrode and a polished silver plate, on which deposits were observed. The air pump used to evacuate the vessel is (A), and a bladder (B) contained a gas sample used for studies of discharges in different gases.

Figure 1: The sputtering system developed by Grove (1852).

Although the initial introduction came early, the subsequent development came very slowly, and this again relates primarily to the lack of available materials. However, it was in this gaseous discharge environment that, in the latter half of the 19th century, many of the fundamental discoveries of modern science were made (Table 1). So we have an enormous debt to the environment that sputtering uses.

Table 1: Basic Discoveries in the Gaseous Electronics Environment[3]

Date	Concept	Originator
1600	Electricity	Gilbert
1742	Sparks	Desaguliers
1808	Diffusion	Dalton
1808	Arc (discharge)	Davy
1817	Mobility	Faraday
1821	Arc (name)	Davy
1834	Cathode and anode	Faraday
1834	Ions	Faraday
1848	Striations	Abria
1860	Mean free path	Maxwell
1876	Cathode rays	Goldstein
1879	Fourth state of matter	Crookes
1880	Paschen curve	la Rue and Müller
1889	Maxwell–Boltzmann distribution	Nernst
1891	Electron (charge)	Stoney
1895	X rays	Röntgen
1897	[Cyclotron] frequency	Lodge
1898	Ionization	Crookes
1899	Transport equations	Townsend
1899	Energy gain equations	Lorentz
1901	Townsend coefficients	Townsend
1905	Diffusion of charged particles	Einstein
1906	Electron (particle)	Lorentz
1906	[Plasma] frequency	Rayleigh
1914	Ambipolar diffusion	Seeliger
1921	Ramsauer effect	Ramsauer
1925	Debye length	Debye and Hückel
1928	Plasma	Langmuir
1935	Velocity distribution functions	Allis

It was in the 1920's that the pioneering work of Irwin Langmuir[4,5] laid the foundation for modern plasma physics and led us to a much better understanding of the gas discharge environment in which sputtering operates. Shortly afterwards, aided by the introduction of modern vacuum pumping technology, the process began to find commercial application including the use by RCA Corporation to deposit thin films of metal used in the manufacturing of their phonograph records.

What Is Sputtering?

Sputtering is a process operating on an atomic or molecular scale whereby an atom or molecule of a surface is ejected when the surface is struck by a fast incident particle (Figure 2). The process of sputtering is reasonably well understood. The momentum of the incident atom is transferred to the atoms in the target material and this momentum transfer can often lead to the ejection of a surface atom – the sputtering process. The process is often likened to a game of atomic snooker since the scattering process employs similar mechanics.

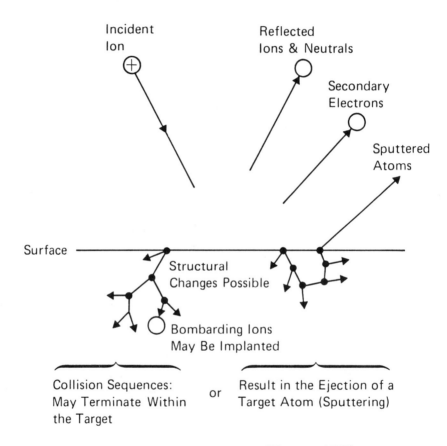

Figure 2: The sputtering process (Chapman 1980).

In order for the sputtering process to be efficient, the incident particle also has to be of atomic dimensions. A very small particle, such as an electron, does not carry enough momentum to be effective, whereas a large particle, such as a brick, is so massive that it cannot interact with individual atoms or molecules in the surface. We therefore end up using atomic sized particles as incident particles or "bullets".

Introduction to Sputtering

One measure of the sputtering process is that of *sputtering yield*. This is defined as the number of atoms sputtered from the surface per incident particle. This number is typically around one, but is understandably a strong function of the momentum of the incoming particle.

Applications of Sputtering

The sputtering process removes atoms of the bombarded surface. This target surface is therefore being eroded or *sputter etched* by the process, which is therefore a technique for removing material in this way. The technique can also be used to generate surface *topography*, or surface patterns, by selectively protecting the surface of the target material during the sputter etch. We shall deal with this further below.

In the sputter etching case, one is not too concerned about the material that is removed. If, however, we place a suitable *substrate* (i.e., a "workpiece" with a surface to be coated) in the flight of the sputtered atom, then there is a good possibility that the sputtered atom will strike and condense on the substrate surface. By the continuation of this process, we therefore have a technique of sputter deposition for coating substrates. Since the atoms arrive individually, it is an appropriate technique for growing films of atomic thicknesses or small multiples thereof. Similarly, films can be grown molecule-by-molecule for compound target materials that sputter as molecules.

Finally, the process can be made to control the topography or roughness of the surface being bombarded – either smoothing ("*planarizing*") the surface, or sometimes enhancing its roughness for specific structures – e.g., to produce the cone-like structures[6] that absorb light effectively and are used as solar collectors.

Sources of Sputtering 'Bullets'

We have established that the sputtering bullets need to be of atomic dimensions. Gases are convenient sources of such bullets. In order to ensure that there is no chemical reaction between the bombarding particle and the material, an inert or noble gas is commonly used to prevent chemical activity. Argon is almost always used because of its general availability in fairly pure form.

As far as we know it doesn't matter whether the energetic particles are charged or uncharged, providing they have enough momentum to carry out the sputtering process. However, it is very easy to accelerate a charged particle with an electric field and very difficult to accelerate a neutral, and so one normally uses *ions*. An ion (and usually it's an argon ion) is an atom from which an electron has been removed leaving it with a net positive charge.

Ion Beam Sputtering

There is a piece of equipment known as an *ion beam gun* or ion beam source, that has the specific task of producing ions. One approach to sputtering would therefore be to take one of these ion beam sources and use it to produce ions that traverse the workspace to strike the target and there cause sputtering (Figure 3). This does require that the ion does reach the target without being diverted by collisions with other gas atoms, and consequently the workspace is evacuated. For practical reasons, this ion beam configuration is not often used for sputter deposition, but it is a common arrangement for sputter etching.

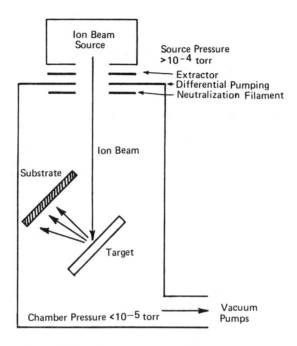

Figure 3: Ion beam sputtering (Chapman 1980).

Ions From Plasmas

Inside the ion beam sources, ions are normally produced by collisions between electrons and atoms. An atom consists of a nucleus containing positively charged protons surrounded by an equal number of (negative) electrons, and so is electrically neutral overall. If an electron collides with this atom and is travelling fast enough, there's a possibility that an electron will be torn out of the atom by the electron impact. This will leave the atom with a net charge – it has become an ion (Figure 4). The threshold for this *ionization process* is usually around 10 to 15 electron volts. An electron volt, usually written as eV, is a convenient measure of energy; it has the value of $1.6 \; 10^{-12}$ ergs.

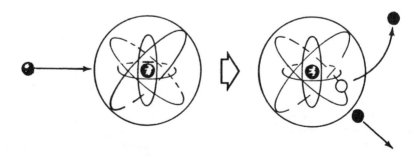

Figure 4: The process of ionization (Chapman 1980).

Note that in the ionization process a second electron has been produced, and this can be used to generate more ions and more electrons, etc. We have begun to form a *plasma* – a gas consisting of a substantial and equal number of electrons and positive ions.

This plasma usually exists within ion beam sources too, but it's more convenient and usual to generate the plasma within the main vacuum chamber so that the ions can be produced close to the material to be sputtered. We then have a *sputtering system*.

Glow Discharge DC Sputtering

In the simplest case, the system can be operated with DC current. A schematic of such a simple system is shown in Figure 5. An electric field is applied between two electrodes. The electric field causes electrons to be accelerated. These electrons collide with neutral atoms, causing ionization and the generation of ions and more electrons as we have seen above. These new electrons are then accelerated by the electric field so that the process becomes continuous.

The ions in their turn will be accelerated by the electric field in the direction opposite to the electrons, so that they will come to bombard the negative electrode, or cathode. This cathode will be sputtered if the energy of the bombarding ions is large enough. The substrate surface to be coated could be placed anywhere in the system, but it's often convenient to place it on the positive electrode (anode) facing the sputtering target, so that it intercepts a large proportion of the material that is sputtered from the target.

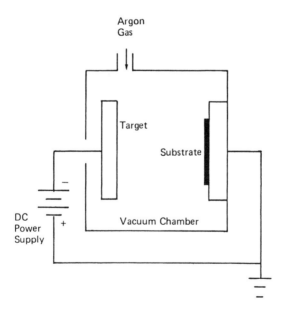

Figure 5: A simple DC sputtering system (Chapman 1980).

Practical DC Sputtering Systems

As one might expect, the number of atoms or molecules sputtered from the target depends on the number or flux of ions that strike the surface, and also on their energy. We have already defined the sputtering yield as the average number of atoms that are sputtered from the surface per incident ion. This sputtering yield is, not surprisingly, a function of ion energy and is typically vanishingly small for ion energies below 20 eV. It then increases monotonically and almost linearly up to about 1 keV. A typical dependence is shown in Figure 6.

The sputtering yield continues to increase, but less than linearly, up to energies of a few tens of keV. It then starts to decrease as the incident ions tend to penetrate into the sputtering target, rather than cause sputtering. We have then moved into the region of *ion implantation*, which is used beneficially in other applications.

From an energy efficiency point of view, it is best to work with ions having energies of a few hundred up to a few keV, and this is the region where most sputtering systems operate. The electrical power supplies need to have corresponding voltages, typically in the range 500 to 5,000 V.

The total rate of material being sputtered depends directly on the impacting ion rate or flux - the number of ions per unit area per unit time. For the process to be practical, (and this is usually defined in terms of having deposition rates of one angstrom/sec or more), then ion beam fluxes need to be of the order of a few tenths of a ma/cm^2.

The pressure that is used for sputtering is a compromise. Ideally the pressure should be extremely low, so that material can easily travel through vacuum from the sputtering target to the substrate. However, we have also seen that collisions between electrons and neutral gas atoms are used to maintain the plasma. This requires a substantial pressure to occur reasonably efficiently. The compromise turns out to be at pressures in the range of 1 millitorr to 100 millitorr, where atmospheric pressure equals 760 Torr.

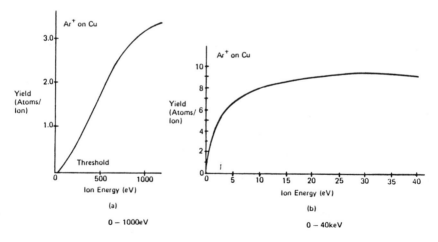

Figure 6: The sputtering yield of argon ions on copper (Carter & Colligon 1968).[7]

CHALLENGES IN SPUTTER DEPOSITION

The technique as we have described it so far is capable of producing reasonable deposited thin films. However, the power of the technology is due to the face that it has been able to respond positively to many of the challenges encountered in thin film science, where greater film control and stringent performance and properties requirements are frequently encountered. These challenges are primarily in the areas of:

(a) *Film purity* – although required film properties are sometimes produced by the controlled incorporation of known impurities,

(b) *Film properties* – since properties different from and superior to those of the bulk can often be produced,

(c) *Step coverage* or the ability to cover complex shapes – with the usual requirement that the film thickness is constant no matter what the topography of the surface (this is known as *conformal coverage*),

(d) *Throughput* of the system (i.e., the number of substrates that can be processed through the system in a given time) – for obvious economic reasons in manufacturing,

(e) *Uniformity of deposition*, both in *film thickness* and *film quality* – again for pragmatic reasons,

(f) *Target integrity* – so that the sputtering target has a useful and practical lifespan, and retains the same properties throughout this life,

(g) *Process repeatability*, so that the same results can be achieved day in, day out, particularly in a manufacturing environment.

HIGH RATE SPUTTERING

For quite a few applications, sputter deposition rates of one angstrom per second are not considered to be fast enough, either for economic reasons or for film structure reasons (since there is a relationship between film microstructure, film purity, and deposition rates). For these applications, a higher rate sputter deposition process is required, and this is now commonly available through the use of magnetically enhanced plasma discharges generically known as *magnetron discharges*. Using this approach, deposition rates up to a few microns per minute are available, and are limited primarily by our ability to prevent the target from melting due to the energy input from the ion bombardment!

DC Magnetrons

The concept of magnetron enhancement was introduced in the 1930s primarily by Penning and his contemporaries[8] and then was developed substantially by Penfold and Thornton[9] in the early 70s before the development of similar sources for other specific applications (e.g., Chapin[10], and Wright & Beardow[11]).

In sputtering discharges, the sputtering rate depends directly on the ion flux, and this in turn depends on the density of ions in the plasma. The major limitation on ion density is the recombination of ions with electrons. This commonly happens on the walls of the vacuum vessel. The use of a magnetic field is primarily to trap electrons close to the sputtering target, so as both to prevent them from escaping to the walls where they will cause ion loss by recombination, but also to cause them to create ions by electron impact close to the sputtering target where they are most useful.

The interaction between the magnetic field **B** and an electron with vector velocity **v** is to create a force **F** on the electron, where **F = e v × B**, so that **F** is in a direction orthogonal to both the magnetic field and electron velocity vectors (Figure 7). As a result, there is a strong tendency for the electron to travel in an epicycloidal motion close to the target surface where both the magnetic field strength and the electron velocity are large. By contrast, the relatively large mass and low velocities of the ions produces much less trapping of the ions, although this can still be a significant effect.

By this means of magnetic enhancement, ion fluxes can be increased to several tens of milliamps per square centimeter, with corresponding increases in deposition rates.

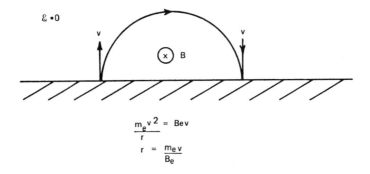

Figure 7: The influence of a magnetic field on electron motion (Chapman 1980).

SPUTTERING OF INSULATORS

In the example that we gave above in Glow Discharge DC Sputtering, it was implicit that the sputtering target could conduct the DC current that was being applied. This would not be true for insulating materials, which therefore present something of a problem. Some amount of resistance is tolerable, provided that the resistance of the targets is no more than that of the discharge itself. Typical conditions of 1,000 volts and 1 ma/cm^2 represent a discharge resistance of 1 Mohm/cm^2 of target. Materials such as silicon can therefore be DC sputtered, but at rather low rates. Another technique, however, enables us to sputter insulators more effectively. This is the technique of *RF sputtering*.

RF Sputtering

If we apply a negative potential to an insulating target, then that target will

Introduction to Sputtering 301

initially be bombarded by positive ions. However, since the target is insulating, it will then charge up positively until it repels any further positive ion bombardment.

When this issue was first confronted it was quickly suggested that the solution would be to use alternating current, so that the target would be alternately bombarded by positive ions and then by negative electrons so as to neutralize the charge. However, the practical implementation of this approach was delayed until it was appreciated by Wehner[12] that frequencies of the order of a Megahertz or more would be required for practical sputtering systems where the targets are reasonably thick (a few mm) to promote their longevity in usage. Such a conclusion can be arrived at by viewing the insulating target as the dielectric of a series capacitor, and considering its capacitance and the typical ion fluxes used in sputtering systems.

Having reached this conclusion, the technique was then soon demonstrated. Although the specific frequency is not critical, it is common to use generators operating at 13.56 MHz and multiplies thereof, since these frequencies are most consistent with the direction of the Federal Communications Commission (FCC) and their international counterparts. Frequencies in this range could otherwise have a major adverse effect on communications systems.

The introduction of RF sputtering greatly extended the range of utility of the sputter deposition technique and this has subsequently led to the usage of the technology for an enormous range of materials. Sputtering is one of the most universal processes since very many materials can be sputtered and sputter deposited.

RF Magnetrons

Just as the DC magnetron was used to enhance the rate of deposition from conducting targets, so also RF magnetrons can be used to enhance the rates of sputter deposition from insulating targets. The degree of enhancement is usually not so great for RF magnetrons as it is for DC magnetrons. This is partly because the fluctuating electric field leads to a less efficient confinement of the electrons, and partly because the electrically insulating nature of many targets usually implies their reduced thermal conductivity, and hence limits the energy input into the target before it degrades or fractures.

REACTIVE PROCESSES

So far we have discussed physical processes where the mechanism is of momentum exchange from the bombarding ion to the sputtered atom or molecule. The plasma environment also turns out to be a useful one for carrying out chemical processes. This is made possible by the process of *dissociation* of a molecule. We have already seen that an electron striking a ground state atom can cause the atom to become ionized by tearing off an electron. When an electron strikes a molecule, which by definition consists of two or more atoms, then the result can be that the molecule is fractured into two or more fragments, each fragment consisting of one or more atoms. This is the process of *electron impact dissociation*.

Most molecules exist because their component atoms would individually be chemically reactive enough that they would satisfy their chemical bonding requirements by grouping together into a molecule. Conversely, when we break up or dissociate a molecule, then we usually enhance the chemical reactivity of the resulting component fragments. In this way, the plasma can be used to create a chemically active environment.

This is particularly true in two plasma processes closely related to sputtering; plasma etching and plasma deposition. These processes are primarily chemical in nature. By contrast, sputtering is primarily a physical process, although we can sometimes use some chemical activity to enhance the sputtering.

Reactive Sputter Deposition

We have commented earlier in this chapter that argon is usually chosen as the sputtering gas so as to avoid any chemical activity between the depositing material and the gaseous environment. In *reactive sputter deposition* we often try to do the opposite. This technique was initially employed before RF sputtering was introduced, since it offered a means of producing insulating films. For example, a tantalum nitride film could be deposited by sputter depositing tantalum in the presence of nitrogen or another nitrogen-containing gas such as ammonia (NH_3) – for various reasons compounds are often more effective than elemental gases.

The technique is not quite as simple as it sounds. The reaction, in this case between tantalum and nitrogen, could take place at any or all of three locations:

(a) At the target.

(b) In the gas phase.

(c) On the substrate surface.

If the reaction takes place at the target, the result is usually to produce an insulating film that would terminate the process if a DC discharge were being employed, or would often slow it down in an RF discharge.

Reaction in the gas phase often will lead to the further agglomeration or nucleation of the resulting molecules so that the material arrives at the substrate in large particles or "powder", producing a friable coating of limited utility.

We normally wish to encourage the reaction on the substrate surface. In order to accomplish this effectively, the process must be well controlled. This can sometimes be facilitated by appropriate tools such as a mass spectrometer to control the gaseous environment in the reactor.

Control of Stoichiometry

Sometimes an insulating target is used to deposit an insulating film. However, the composition of the film may not be quite the same as the composition of the target. We have referred to the sputtering process as the ejection of a single atom or single molecule which is then transported to the substrate. The actual situation may differ somewhat from that. Although in some cases clusters of atoms or molecules may be ejected from the target, it's also true that partial molecules may be ejected from the target (Coburn et al[13]). This is particularly likely if one of the elements of a compound target happens to be a gas. For example, the RF sputtering of a film from a silicon nitride target may often turn out to be deficient in nitrogen since the molecular species that is ejected from the target may lose some of its nitrogen atoms. We refer to the *"stoichiometry"* or composition of the deposited film being different from that of the target.

Reactive sputtering offers us a means of controlling the stoichiometry of the film. To pursue the example quoted above, a silicon nitride target could be RF sputtered in a gas containing nitrogen so as to restore the stoichiometry of the resulting film. This turns out to be a very useful technique.

On the other hand, the nonstoichiometric films may have superior properties to those having the bulk composition, so that this is not necessarily a disadvantage. For stoichiometric and nonstoichiometric films, as for the reactively sputter deposited films discussed above, control of the process is an important feature.

BIAS SPUTTERING

Properties of Bias Sputtered Films

We have seen that sputtering results from the bombardment of a target with high energy ions. It is also possible to arrange for the growing film to be ion bombarded, by applying some electrical power to the substrate. If the magnitude of this bombardment is too great, then the film will be sputtered away from the substrate as fast as it arrives so that no net deposition occurs. However, for lower bombardment energies, net deposition and some very beneficial results can be obtained. This technique of *bias sputtering*, using an electrical bias applied to the substrate to control the amount of ion bombardment there, has been extremely useful in controlling the properties of the deposited film. The mechanism for this change in film properties appears to be multifold.

If we examine the fluxes of particles bombarding the substrate surface, we find that the flux of sputtered atoms that we wish to deposit is very much smaller than the flux of atoms of the sputtering gas. There might also be a significant flux of contaminant atoms; for example, a flux of oxygen atoms and molecules from a leak in the vacuum system. Not surprisingly, some of these sputtering gas and contaminant atoms may become incorporated in the film.

If the growing film is subjected to mild ion bombardment, with energies of a few eV or tens of eV, then a "scrubbing" action will take place so as to remove any loosely bonded material, such as the chemically inert sputtering gas. Some frequently observed effects of bias sputtering appear to be due to removal of this trapped sputtering gas.

Other changes occur that are less well understood. For example, it has been observed that ion bombardment changes the initial formation stages of the thin film – i.e., the nucleation and growth phases – although the reasons for this are not clear. There are also other structural changes that occur due to this ion bombardment.

Almost any film property can be modified by the application of bias. An example is shown in Figure 8.

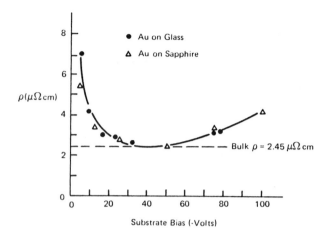

Figure 8: The variation of resistivity of 6000 Angstroms dc sputtered gold films versus rf substrate bias (Vossen and O'Neill 1968).[14]

Topography Control With Bias

Although it is not obvious, it is also found that bias sputtering can be used to control the *topography* or surface contours of a deposited film. One sometimes has need to coat an uneven surface so as to have a film of constant thickness overall; this is known as *conformal* coverage. On the other hand, it is sometimes necessary to coat a rough surface so that the surface of the resulting film is substantially flat; this is known as a *planarizing* process. The bias technique can be used to advantage in both these cases.

The basic idea is to cause enough ion bombardment at the growing film that it is resputtered. Although the net deposition rate is reduced as a result, this resputtering also causes a redistribution of the deposited material so as to cause material to be coated, for example, onto sidewalls of "pits" or "trenches" that would otherwise not be subject to direct deposition by the incoming sputtered atoms. These processes of conformal coverage or planarization are often aided by the use of high substrate temperatures so as to encourage the diffusion of condensing sputtered atoms, enabling them to move around on the substrate surface. The difference between bias used for conformal coverage and bias used for planarizing is usually a matter of degree.

DC or RF Bias?

With insulating substrates the applied bias has to be at high frequencies for similar reasons to those advanced in the description of RF sputtering above. Lower frequencies could be used as the insulator thickness, and hence its ability to store charge, decreases. This is particularly pertinent when one has to deal with insulating films on conducting substrates. However, in practice 13.56 MHz is almost always used when AC is required.

For conducting films on conducting substrates, it is possible to use DC bias, sometimes to advantage. However, this is often difficult to accomplish reproducibly in practice, because a large number of conducting materials form a native oxide or other insulating compound on their surfaces. This prevents the making of a good DC electrical contact between the substrate and the substrate holder. One could remove the insulating film with an etch prior to the deposition step, but this is an additional process step that is usually unwelcome, particularly in a manufacturing process. As a result, RF bias is more usually employed.

SPUTTER DEPOSITION EQUIPMENT

In the days of Grove, sputtering was a laboratory curiosity and systems were produced only by those scientists who were prepared to design and build all of the necessary components. We have earlier referred to the enormous difficulties with which Grove and his contemporaries were faced due to the unavailability of many or most of the components and materials that were required. Today, by contrast, we have an array of companies supplying components and complete systems. For each specific system component, e.g., vacuum pump, pressure gauge, power supply, there are a number of suppliers. The resulting competition forces companies to continuously improve their existing products and to introduce new products in order to provide growth and to respond to market needs. It is this action of pushing and pulling that has allowed system manufacturers to progress from the Grove apparatus to today's fully automated systems.

It would appear, with these multiple choices for components, to be a simple task to design and build a sputtering system, but this is not the case. Part of the reason can be found in the magnitude of the challenges resulting from industries' needs to deposit thin films for highly specialized and ever-demanding applications. As an example, there is a need in semiconductor processing to deposit very thin films into so-called "vias", where a film has to be deposited into a tiny hole in the substrate; the film is typically a micron thick, but so also is the hole a micron deep and a micron wide. This challenge is discussed in more detail in a later section (Future of Sputtering: Sputtering for Step Coverage). Such challenges are stretching the imaginations of sputtering system designers to find increasingly sophisticated solutions.

Variety of Equipment

There are many different types of systems available to meet the very broad range of applications for which sputtering is now used. At one end of the spectrum are systems for coating the very tiny features of an integrated circuit or of a micromechanical device. At the other end are huge systems that are used for coating large sheets of architectural glass that are typically 5 meters by 3 meters, or so-called "roll coaters" that are used to sputter deposit films onto rolls of material that may be meters wide and thousands of meters long.

Further differentiation depends on the useage of the system. Manufacturing systems are highly automated to ensure consistency of performance and to minimize operator intervention. By contrast, research and development systems sacrifice automation in order to achieve flexibility and to ensure that exactly the desired process is being run without the "assistance" of a central controller (which sometimes "edits" our processes, with all good intent!).

The semiconductor industry is probably the most demanding in regard to sputtering process performance. We will describe such a system as an illustration of sophisticated sputtering equipment.

Semiconductor Deposition Equipment

For sputter deposition in semiconductor processing, the main application is for the deposition of aluminum alloys. Several different approaches are used by different equipment manufacturers. The first division is into *single wafer* or *batch* systems, in which the wafers are processed either one wafer at a time or several at a time. Further divisions are:

- static
- planar rotating
- in-line

In practice, there are so many variations on a theme that the divisions are rather blurred. Each of these approaches has its advantages and disadvantages.

Before discussing the differences between these approaches, we can look at some common features. Aluminum is a reactive metal and is likely to produce oxidized films unless great care is taken to avoid this. One aspect of the avoidance is to isolate the process chamber from the environment with a load lock (Figure 9), just like those used in space vehicles (hopefully without an ill-intentioned control computer!). In principle, one can eliminate oxygen and other contaminants from the process chamber by pumping long enough, but this would not lead to the productivity required in manufacturing equipment.

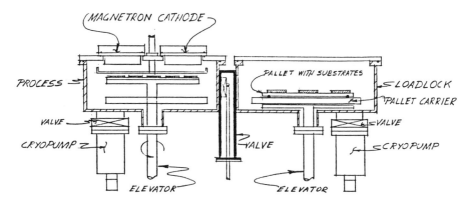

Figure 9: Loadlock System.

Traditionally most sputtering equipment has sputtered *downwards* onto a horizontal substrate surface. Since semiconductor devices are so sensitive to very small amounts of contamination, very extensive measures are taken to avoid contamination of the wafer surface. This is one of the most evident and necessary aspects of the fabrication process. As stringency has increased and as materials have changed, so a concern has arisen about contaminant material falling from the sputtering target surface onto the wafer. This is more likely with friable materials such as titanium-tungsten and any target that is pressed from powder. To minimize this, a good deal of equipment is now built in a *side sputtering* configuration, with the faces of the target and the wafer vertical, and sputter deposi-

Introduction to Sputtering 307

tion taking place in a sideways mode. This isn't fail-safe however, since very tiny particles in the submicron range are likely to become electrically charged and then electrostatic forces will overcome those of gravity.

Yet another common concern relates to the expense of producing the extremely clean environment of the semiconductor processing fab. In order to achieve the most cost-effective utilization of this environment, process equipment is often built in a *through the wall* configuration in which the load and unload sections of the system are in the fab, but the majority of the hardware is in a less clean secondary area known as a *chase*; sometimes the rotary mechanical pumps are even in a tertiary area so as not to contaminate the secondary area.

The differing aspects of systems will be discussed below.

Static Systems. These are inherently single wafer systems. During deposition the substrate is stationary, usually a few cm from the target (Figure 10). The prime motivation here is to achieve maximum deposition rate and minimum contamination from background gases. As well as reducing the electrical conductivity of the aluminum films, oxidation is known to reduce the quality of step coverage. This probably happens by reducing the *mobility* of the deposited sputtered atoms in the period after they arrive on the substrate, until the time when they finally come to rest and become incorporated into the growing film.

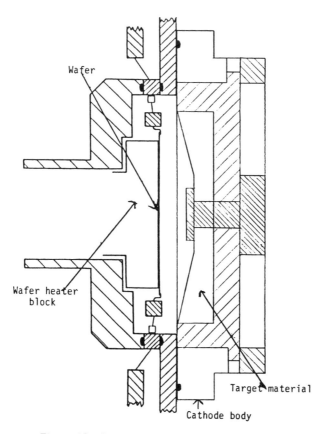

Figure 10: Static Deposition Sputtering System.

308 Thin-Film Deposition Processes and Techniques

The major disadvantage with static systems is that by definition they cannot use substrate-target relative motion to achieve uniformity of film thickness and of step coverage, as is used in the dynamic systems below.

Process steps are performed in individual chambers with the substrate stationary (or sometimes rotating for step coverage improvement, although this is no longer truly static). Multilayer films can be deposited by moving the substrate between process stations.

Planar Rotation Systems. These are batch systems with the substrates loaded on a planar pallet. The batch size in these systems is usually related to wafer size, in order to limit the total system size.

During deposition the pallet rotates under a stationary target. As a result, these systems are often referred to as *multipass* systems. The effective deposition rate and hence the film thickness per pass can be varied by changing the pallet rotation speed. This can be used to influence the film properties.

The process chamber top plate is often equipped with several targets. This can be used either to achieve higher deposition rates or, using targets of different materials, to achieve either *layered* structures or pseudo-homogeneous mixtures of materials, depending largely on the rotation speed. Simultaneous deposition from two or more targets is known as *cosputtering*.

In-Line System. This system configuration is a hybrid of the single wafer approach of the static system and the batch approach of the dynamic system. In an in-line system, the substrates stream one at a time (or sometimes several abreast) past the sputtering target, and sometimes past several targets in series (Figure 11). This approach offers some of the advantages of the other two approaches.

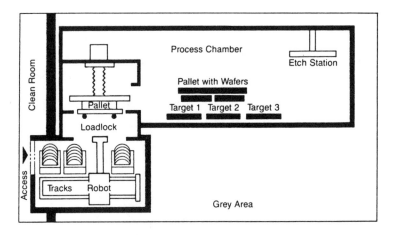

Figure 11: MRC Model 603 Side-Sputtering System (In-Line System).

Modules of a Sputter Deposition System

Now that we are familiar with the major system concepts, let us look at the various modules that make up a system by following a wafer through the deposition process. Again the example is for aluminum alloy films; other films may require more or fewer steps.

The substrate loading is done by a microprocessor-controlled mechanism. The wafer is picked up from its rear surface by a vacuum pick-up or by the edge with a mechanical device. The goal is to generate the fewest particles and break the fewest wafers.

The substrates are transferred to the *load lock*. As we have discussed above, the primary function for this lock is to act as a buffer between the process chamber and the atmosphere. The lock is first evacuated by a mechanical pump to a pressure of 500 to 100 millitorr. Then, a high vacuum pump such as a turbo or cryo pump is used to bring the pressure down to a much lower level in order to minimize the amount of air and other contaminants that will be introduced into the process chamber.

The load lock is also utilized to perform several "preconditioning" steps that would be undesirable in the main process chamber. Typical preconditioning steps are:

> Heating of the wafer to remove adsorbed water from its surface. This water will otherwise interfere with the *nucleation* or formation of the film on the substrate and will also cause oxidation of the growing film since the water will vaporize in the main process chamber and then be dissociated by the plasma into its constituent hydrogen and oxygen atoms.
>
> This heating process may also be the first step of raising the wafer temperature to a particular level in order to control the deposition process in the main chamber.
>
> Another preconditioning step may be the removal of a thin layer of material from the surface of the substrate. This is often done for substrate materials that oxidize easily, for example, in order to remove the surface oxide layer so that the subsequently deposited film can make intimate contact with the substrate itself. This removal process is usually carried out by the ion beam etching or plasma etching processes discussed elsewhere in this chapter.

Upon completion of the previous steps the substrates are transported to the *main process chamber*, which is usually mechanically isolated from the load lock during the deposition process.

> Several more steps are required prior to commencing deposition. First, the chamber is pumped to a low base pressure to minimize the amount of background contamination. For aluminum, pressures of the order of 10^{-8} Torr are achieved using a cryo pump. Some systems are also equipped with other types of pumps; for examples, a Meissner pump is sometimes used for the pumping of water vapor generated during the process, and/or a titanium sublimation pump known as a "getter" to pump the large volumes of H_2 generated by the dissociation of water in the discharge, and the H_2 from the stainless steel used extensively in vacuum equipment, as well as other active gases.
>
> Simultaneously the wafer is brought to process temperature, typically 400°C for aluminum. Temperature is a major factor in con-

trolling the microstructure and stoichiometry of the growing film, as well as other properties such as step coverage. Substrate heating is achieved by flowing hot gas under the substrate, by heater filaments embedded into the substrate holder, or by radiation heating.

There may then be an additional etch step to remove contamination or a compound layer, such as an oxide layer, from the substrate surface if this has not been adequately achieved in the load lock.

Now, we're almost ready to begin the sputtering process. Argon gas is introduced into the chamber to bring the pressure to a predetermined process pressure of typically several millitorr. The maximum flow of argon that can be handled by the vacuum pumps is normally used, again to minimize contamination. To facilitate the task of the vacuum pumps, a *throttle valve* in series with the pump is actuated. This establishes a pressure differential between the process chamber and the pump, allowing the pump to operate under high efficiency conditions determined by its pumping speed versus pressure relationship.

This next step should not be necessary in a good load-locked system, but in systems where the target surface can become contaminated between runs, this target surface must next be cleaned. Cleaning is carried out by rotating a shutter to a position shielding the substrate from the target, and then sputtering the target onto the shutter for an appropriate period of time to remove the surface layer. The shutter is then moved away to allow deposition onto the substrate proper to take place. It is much better to avoid this target cleaning step in the main process chamber where possible since there is a tendency for contaminated material to get around the shutter onto the substrate.

Sputter deposition onto the substrate now takes place. During this time, bias is often applied to the substrate to further control film properties, as we have discussed above. This is clearly a critical step of the process and one where there is considerable latitude for improvement. Fortunately, there is a growing interest in process control for process assurance and enhanced productivity by increasing yield, a subject in which the authors are much involved.

Finally, upon completion of the deposition, the substrate is moved into the next process chamber or into the exit load lock before being unloaded into a substrate holder or cassette.

The system and process steps described above are based on sputtering systems found in the semiconductor industry for the deposition of aluminum alloys. As sputtering evolves, it is finding application in many other industries. In some of these other industries and other applications, quite different system requirements will exist. In many cases, the relevant equipment supply is fledgling or nonexistent. Fortunately, there are specialized companies such as our own that are able to supply processes and systems for these applications. As a result, sputtering applications that did not exist quite recently are now adolescent or mature; compact discs, thin film magnetic recording heads and discs are just examples.

ETCHING

Sputter Etching

Although this chapter is primarily about the use of sputtering for the *deposition* of thin films, it is worth briefly describing the use of the technique for the removal or *etching* of substrate surfaces.

At the beginning of this chapter, we described sputtering as a process whereby an atom or molecule of a target surface is ejected when that surface is struck by a fast incident particle. Subsequently, our attention has been focussed on the application of the phenomenon to the formation of thin films by the deposition of the sputtered target material onto a substrate. However, we should also recognize that the target material has been removed from the target – i.e., the target has been etched.

Although materials can often be more simply etched by conventional wet chemical techniques, plasma processes have the interesting advantage that they can be used to produce *directional etching*.

In order to understand the utility of this process property, it is useful to compare the profiles that are obtained when a surface is selectively etched by the wet and dry techniques (Figure 12). The selectiveness of the etch is accomplished by protecting the surface of the substrate where etching is *not* required, by using a *patterning mask*. In wet chemical etching, the substrate is then placed in a liquid etchant. Since the substrate material is usually isotropic in nature, the etch generally proceeds from the substrate area exposed through the patterning mask in all directions at equal speed. This results in the classic partially spherical shape that is commonly observed in this type of etching. Note particularly that there is an *undercut* beneath the mask so that the top of the etched hole is wider than the corresponding hole in the patterning mask. Although this is not an issue for etching large features, it becomes of increasing disadvantage when the width of the etch approaches its depth.

By contrast, the electric fields that exist in a sputtering system cause the sputtering ions to arrive substantially perpendicular to the substrate surface. As a result, the etch tends to be very directional, often having vertical sidewalls, in contrast to wet etching.

Patterning By Sputter Etching

Sputter etching can be used to generate patterns on a surface by selectively etching the surface using a protective masking material, which is itself patterned, usually by a lithographic technique akin to photography. The masking material may need to be quite substantial for deep etches or it may be a thin film of light-sensitive polymer known as photoresist, which corresponds to the emulsion that it used in photographic film. In this way, very fine dimensions down to a micron or so can sometimes be achieved. However, there are some practical difficulties facing successful implementation of this technique.

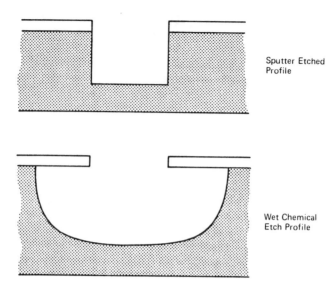

Figure 12: Etch profiles produced by sputter etching and by wet chemical etching (Chapman 1980).

Glow Discharge Etching

A sputter etching system could be as simple as the sputter deposition system described earlier (Figure 5), since etching occurs at the target when deposition occurs on the substrate. However, although this system is simple, there are some limitations: in the comparatively high pressures used by in situ glow discharges, some amount of collision and deflection takes place between the accelerated ions and the gas phase atoms so that some sputtering ions arrive at the substrate at nonperpendicular incidence. This tends to create some loss of the inherent directionality of the process.

In addition, the atoms and molecules sputter ejected from the target do not always eventually escape from the target and are sometimes redeposited there after being deflected, again by collisions with other atoms in the gas phase.

Ion Beam Etching

As a result, one often reverts to the ion beam approach described at the beginning of this chapter (Figure 3), in which ions from an ion beam source are used to bombard a substrate surface situated in a high vacuum environment — typically at pressures of 10^{-6} Torr and below. Under these circumstances, there are essentially no collisions in the gas phase so that ions arrive in a parallel beam at the substrate. In addition, material sputtered from the target is often able to escape from the surface without making the collisions that would result in redeposition.

Limitations of Sputter Etching

Even in this low pressure environment, there continue to be limitations with this technique. One is that the patterning mask also tends to be eroded – this is one of the disadvantages of sputtering being such a universal process. This mask erosion usually has the effect of changing the width of the etch and the shape of the sidewall as well as limiting the depth that can be etched with the mask intact. Additionally, there are also some other limitations commonly known as *"facetting"* and *"trenching"*, which tend to produce uncontrolled and usually unwelcome etch profiles. Finally, sputtered material still recondenses on sidewalls and other surface topography, where there is a further problem with material being efficiently forward sputtered into the valleys on the surface.

These limitations combine to be fairly serious, particularly when one further considers the rather low rates with which materials can be sputter etched. Magnetron systems are not commonly used to enhance sputter etching rates, partially because of the lack of uniformity of etching in such systems (although this has begun to be cured in some magnetron systems) and partly because of the large power inputs which heat the substrate, often limiting the nature of the masking material that can be used.

Consequently, sputter etching is not as popular as was initially expected and is usually employed only when the plasma chemical etch techniques described below cannot be employed. But this latter situation is frequently encountered, and the almost universal nature of the sputtering process means that it will continue to be employed for some applications.

Plasma Etching

Earlier in this chapter, we looked at the technique of reactive sputter deposition whereby the depositing material combined with a gas phase atom to form a compound thin solid film. We have also seen that there are some limitations to the sputter etching process because the sputter ejected material tends to recondense in areas where it is unwelcome. Plasma etching offers a solution to this dilemma by forming a compound product, in like manner to reactive sputter deposition. However, in the etching case the resulting compound is gaseous so that it is pumped away by the vacuum system instead of recondensing within the vacuum process chamber. This process is inherently chemical in nature since we have then transformed a solid material into a gaseous material by a chemical reaction within the system.

As an example, silicon surfaces can be plasma etched in a fluorine-containing gas discharge in very similar hardware to that used for sputtering. The atomic fluorine created by dissociation in the discharge combines with the solid silicon to form silicon tetrafluoride, SiF_4, which is a gas and can be pumped away. As with sputter etching, the accompanying ion bombardment can create a directional etch.

To create some confusion, this process is also known as *reactive sputter etching* and as *reactive ion etching*. Actually they are all variations of the same approach, and the divisions that some erect between these techniques are very largely arbitrary.

Patterning By Plasma Etching

The last decade has seen a tremendous growth in the use of plasma chemical etching to etch surfaces. This technique is particularly valuable because it

314   Thin-Film Deposition Processes and Techniques

can achieve the directional etching that we have described above for the sputter etching process. It can often accomplish this without the disadvantages noted above, and also often at quite high rates. The development of plasma etching has been significant in our ability to define the tiny geometries that are utilized in modern integrated circuit fabrication.

There is, however, a significant limitation of plasma etching which is that the process is by no means universal in the sense that convenient gaseous compounds often do not exist for many of the materials of interest.

Patterning By Lift-Off

As a result of these limitations of sputter etching and plasma etching, there is another technology known as *"lift-off"* for defining patterns, and by contrast this is achieved in the deposition of the film of interest.

The deposition is usually by evaporation, which is another atomic deposition process in which atoms are "boiled off" from a heated source material. More recently, sputter deposition has also been increasingly used, mostly because of sputtering's inherent superiority in producing films of controlled composition.

In lift-off, the substrate is first prepared by generating a patterned mask and then deliberately undercutting the underlying structure, normally by a plasma etching step in a nondirectional mode (Figure 13a).

The directional nature of the subsequent film deposition ensures that the undercut sidewalls do not get coated (Figure 13b).

The undercut material is now etched away together with the masking layer and some of the deposited film, leaving the base substrate patterned as required (Figure 13c).

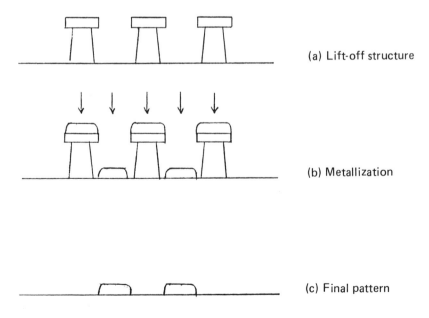

(a) Lift-off structure

(b) Metallization

(c) Final pattern

Figure 13: Patterning by lift-off.

Although the end of the useful life of this technique has been predicted for the last ten years due to its assumed incompatibility with very tiny geometries, the technique of lift-off continues to be used. It has recently been pronounced as in its last phase of implementation - once again!

FUTURE OF SPUTTERING

The almost universal nature of the sputtering process ensures that the technology will continue to be utilized in many fields of activity for years to come.

There are, however, some shortcomings: one of these is in deposition rate. This is particularly true when it comes to the deposition of some thicker insulating materials, for which DC magnetron sputtering cannot be used. This imposes such severe economic restraints that the technique is often limited in its application.

Another shortcoming appears when one considers the uses of sputtering for covering surfaces having complex topographies. We have already referred to this earlier, and this is a limitation facing the application of sputtering in integrated circuit technology today and in the future.

This limitation is best appreciated by considering what would happen if atoms were arriving from a point source onto a substrate having vertically walled "steps" or "pits" in its surface (Figure 14). This is a situation encountered, for example, in I.C. fabrication where one often wishes to deposit an aluminum alloy film over a silicon wafer coated with silicon dioxide, patterned so as to have vertical walls.

The horizontal surfaces on the wafer subtend a large angle at the point source so that the material is deposited quite effectively on these surfaces. By contrast, some of the vertical walls on the wafer are "shadowed" and subtend little or no angle at the point source so that deposition rates on those walls are very small. A metal film deposited on such a surface in this way, would tend to be discontinuous or at least extremely unreliable in its electrical continuity.

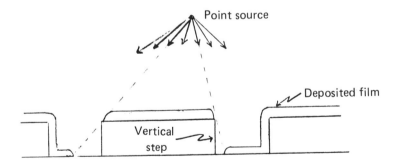

Figure 14: Step coverage problems.

Sputtering For Step Coverage

A sputtering target is not a point source, but is by contrast an extended source. This is a considerable advantage for *step coverage* (although it is a disadvantage when trying to achieve the lift-off structures described above, where the technique of *evaporation*, from essentially a point source, is used to great advantage).

However, the use of an extended source is not by itself adequate to take care of these issues of step coverage. The reason is that the vertical walls still do not see the same extended source that the horizontal surfaces of the substrate would see.

Indeed, the horizontal "plateaux" on the substrate see a much greater source area than the horizontal surfaces in the "valleys", an increasing problem as the valleys become steeper and narrower. We have referred earlier to the use of bias sputtering to help to overcome this problem. By judicious application of ion bombardment at the substrate, some of the atoms are resputtered from the horizontal valleys in the steps onto the sidewalls. This process can be suplemented by the use of substrate heating so as to make the depositing atoms very mobile, and hence encouraging their transfer between the horizontal and vertical areas during the deposition process. By a combination of such techniques, a much improved step coverage can be achieved. This has been particularly noticeable in the last year or so as the demands of submicron technology have become evident in the semiconductor industry. These same technology needs have spurred the development of planarizing processes in similar applications.

Sputtering or CVD?

However, there remains a concern that sputtering will cease to be a useful deposition technique when film geometries in I.C. fabrication go to small fractions of a micron. Chemical vapor deposition (CVD) is often hailed as the replacement for sputtering because of its inherent ability to produce a conformal coverage – a uniform film thickness over complex topographies. CVD employs a gas reactant, which dissociates or decomposes, often on the substrate surface and often as a result of thermal processes, to deposit a thin solid film. For example, silicon films can be deposited from silane (SiH_4) gas.

Because of the needs of the semiconductor industry, sputtering has come into competition with CVD for the deposition of aluminum alloy films – alloyed with silicon or with silicon and copper – over micron-sized steps. Although sputtering or evaporation dominate this application today, much effort has been made to develop a CVD process. Unfortunately, it has been found extremely difficult to find a suitable source material for this application. Some success has been obtained with organometallic sources, primarily TIBAL (tri-isobutylaluminum). However, the surface roughness of the films obtained to date have been marginal, and there are difficulties in incorporating the other elements of the alloy.

Sputter-Assisted Processes

It remains true that CVD has some advantages over sputtering for some applications. However, rather than argue over which is the preferable technique, it is more fruitful to consider how the films may be used in combination. For example, there has been recent success in depositing a film by CVD under conditions where there is simultaneous ion bombardment of the growing film to control its properties and topography. Purdes[15] has shown some examples of this approach.

Other sputter-assisted processes has been developed over the years, particularly for specific applications. Mattox[16] has introduced the technology of *ion plating* whereby material is evaporated onto a substrate subjected to high energy ion bombardment. The resulting forward sputtering of the depositing film *into* the substrate eliminates the interface and produces a very durable film, of particular interest for some metallurgical applications. A related technique is that of *Activated Reactive Evaporation* where evaporated material passes through a reactive plasma to form a compound deposited film.[17]

Conclusions

As was stated above, the almost universal nature of the sputtering process ensures that the technology will continue to be utilized in many fields of activity for years to come.

This has been a brief and inevitably incomplete review of this powerful technology. The reader is encouraged to consult the more detailed literature.

Acknowledgments

Acknowledgments are due to authors and publishers for their kind approval to reproduce figures from their publications, including to John Wiley & Sons, New York, for permission to reproduce figures from Chapman.[12]

REFERENCES

1. Chapman, B.N., *Glow Discharge Processes: Sputtering & Plasma Etching*, John Wiley & Sons, New York (1980).
2. Grove, W.R., *Philos. Trans. Faraday Soc.*, 87 (1852).
3. Brown, S.C., *Gaseous Electronics*, ed. J.W. McGowan and P.K. John, North-Holland, Amsterdam (1974).
4. Langmuir, I., *General Electric Review*, Vol. 26, p. 731 (1923).
5. Langmuir, I. and Mott-Smith, H., *ibid*, Vol. 27, p. 449 et seq (1924).
6. Tarng, M.L. and Wehner, G.K., *J. Appl. Phys.*, Vol. 43, p. 2268 (1972).
7. Carter, G. and Colligon, J.S., *Ion Bombardment of Solids*, Elsevier (1968).
8. Penning, F.M., *Physica*, 873 (1936).
9. Penfold, A.S. and Thornton, J.A., U.S. Patents 3,884,793 (1975); 3,995,187, 4,030,996, 4,031,424 and 4,041,353 (1977).
10. Chapin, J.S., *Research/Development*, p. 37 (January 1974).
11. Wright, M. and Beardow, T., *J. Vac. Sci. Technol.*, A 4(3), p. 388 (1986).
12. Wehner, G.K., *Adv. in Electronics and Electron Physics*, VII, 253 (1955).
13. Coburn, J.W., Taglauer, E., and Kay, E., *Jpn. J. Appl. Phys.*, Suppl. 2, Pt. 1, p. 501 (1974).
14. Vossen, J.L. and O'Neill, J.J., Jr., *RCA Review*, Vol. 29, p. 566 (1968).
15. Purdes, A.J., *Proc. Electrochem. Soc.*, New Orleans (1984).
16. Mattox, D.M., *J. Vac. Sci. Tech.*, Vol. 4, p. 209 (1973).
17. Bunshah, R.F., *Science and Technology of Surface Coating*, ed. B.N. Chapman and J.C. Anderson, Academic Press, London and New York (1974).

10

Laser and Electron Beam Assisted Processing

Cameron A. Moore, Zeng-qi Yu, Lance R. Thompson, and George J. Collins

INTRODUCTION

The fabrication of submicron microelectronic devices in semiconducting crystals requires inducing a wide variety of physical and chemical processes. Epitaxy, oxidation, lithography, etching, ion implantation, and deposition all must be carried out with a high degree of process control and spatial uniformity. As device and feature sizes shrink below one micron, the inherent limitations of traditional furnace-based thermal processing methods become manifest. Foremost, we must reduce the total time-temperature cycling required to complete fabrication steps.[1] This reduction is motivated by the need to minimize thermal diffusion and its associated redistribution of as-implanted doping profiles, to reduce thermal defect creation and migration in the crystalline substrate, and to minimize thermally-induced substrate warpage as required by submicron lithography. When fabricating submicron feature sizes, there is also a need to process materials with a minimum amount of process-induced radiation damage,[2] such as that found in plasma etching or deposition, electron or ion beam lithography, and ion implantation.

One way to exert a high degree of spatial and temporal control over the energy input to any processing step is through the use of directed energy beams which are created with both high energy selectivity and spatial directionality. Ion beams, for example, have long been used to implant dopant species with repeatable doses and specific depth profiles. The recent use of ion beam lithography and direct write pattern transfer has also shown promise.[3,4] Photon and electron beams have also long been used for lithography and analytical measurements (optical and scanning electron microscopy, ellipsometry, etc.).

Herein we discuss recent progress in the application of photon beams (from laser sources) and electron beams (generated in an abnormal glow discharge) to semiconductor device fabrication. Specifically, we examine photon and electron beam assisted CVD of microelectronic films, electron beam induced annealing of ion implantation damage and associated dopant activation, electron beam induced alloying of silicide structures, self-developing electron beam proximity lithography of submicron features and traveling-melt photon and electron beam recrystallization of deposited polycrystalline silicon films on SiO_2.

Beam assisted process technology and the related equipment are in general still at an early development stage. However, this technology offers significant prospects for the future fabrication of microelectronic devices. The major advantages of the beam assisted technologies are based on:

- The high dimensional resolution capability,
- The ability to perform low temperature processes and to minimize the thermal exposure of the substrate,
- The ability to perform sequential process steps in situ without exposing the substrate to the atmosphere, thus minimizing the risk of potential surface contamination and oxidation.

The different beam assisted processes and their characteristics will be discussed in more detail in the following section.

BEAM ASSISTED CVD OF THIN FILMS

Conventional CVD Methods

The atmospheric or low pressure thermal CVD of microelectronic films occurs typically at 600° to 1000°C substrate temperature (corresponding to thermal energies below 100 meV). Thermal CVD relies on high substrate temperatures and catalytic reactions between donor molecules and substrate surfaces to achieve the dissociation of feedstock gases with binding energies in the 1 to 3 eV range. As a consequence thermal CVD deposition rates are low (100 to 1000 Å/min). One improvement over conventional thermal CVD techniques is plasma-enhanced chemical vapor deposition (PECVD). A plasma, while having a gas temperature of 10's of meV, contains energetic electrons with a distribution of energies which usually average several electron-volts. This plasma electron energy spectrum partially overlaps the electron-impact induced dissociation cross-section maxima for many polyatomic feedstock reactants. Thus, the burden of feedstock gas dissociation is carried by the plasma electrons and one can deposit films on substrates at a temperature below 400°C.[5] PECVD processes can lower substrate temperatures and raise deposition rates (>1000 Å/min) because the reactants are more directly and efficiently dissociated. However, PECVD may introduce radiation damage and electrode or vacuum chamber impurities to the substrate and the growing film. Ions in the plasma act as a possible source of physical damage to the substrate or deposited film if the plasma sheath surrounding the substrate accelerates ions to greater than several hundred eV. Radiation damage from plasma electrons and photons is not as severe as that from ions and can be annealed out with a relatively low temperature anneal.[2]

Electron Beam Assisted CVD

The plasma used to deposit films can be an abnormal glow discharge.[6] The combination of the abnormal glow discharge operation together with the use of high secondary electron emissivity cathode materials results in a glow discharge-created electron beam.[7] Conventionally, gamma (γ) is used for the secondary electron emission coefficient, where γ is the number of secondary electrons emitted from a cold cathode per incident ion. One can create an electron beam in an abnormal glow DC discharge by using, for example, a ceramic-metal compound as a cathode with $\gamma > 2$. The use of a thin oxide, nitride or oxynitride layer on a metallic cathode[7] can also result in a high γ. The cathodic voltage drop controls the electron beam energy. The cathode geometry determines the directionality of beam electrons because the plasma sheath conforms to the chosen cold cathode shape. Compared to conventional DC, RF, or microwave plasmas, the energy spectrum of the electrons created in these abnormal glow discharges has a much greater density of high energy electrons.[6,7] The presence of high energy electrons results in distinctive cracking patterns following electron-molecule collisions in electron beam created plasmas as compared to those found in conventional plasmas. By directing this electron beam into a mixture of feedstock gases, electron impact dissociation of gases occurs and film deposition results.

Laser Assisted CVD

We have used ultraviolet photons from an excimer laser source to photodissociate feedstock reactant molecules.[8] Photodissociation products then subsequently condense on a substrate as a solid film. The energy spectrum of the input laser beam is essentially monoenergetic as compared to electron-induced dissociation sources (conventional DC, RF, and microwave plasmas) which have statistical energy distributions of electron energies. Thus, laser driven photodissociation allows for fewer dissociation channels, better process repeatability and, in some cases, selective bond breakage. Laser-induced chemical vapor deposition (LCVD), unlike PECVD, also has the ability to localize the film deposition by use of simple optics to focus and direct the beam.

Experimental Apparati of Beam Assisted CVD

Laser-induced and electron beam-assisted CVD deposition reactors have many similar features. These apparati have been described previously[8,9] and are briefly reviewed here. The substrate upon which deposition is to occur is placed on a heater capable of temperatures from 50°C up to 450°C. Low substrate temperatures are desirable for submicron silicon processing as well as for processing materials which degrade at elevated temperatures (e.g., III-Vs, II-VIs and heavy metal-fluoride glasses). Once the desired substrate temperature is realized, feedstock gases containing the film precursors are introduced into the chamber. At the substrate temperatures employed negligible thermal CVD occurs. Feedstock gas flow is monitored using mass-flow control valves, while the deposition pressure is selected via a conventional downstream throttle valve/capacitance manometer control system. The electron or laser beam is then introduced either parallel or perpendicular to the substrate surface. A common feature of the parallel beam CVD technique is that the deposition-initiating particles, be they either electrons or photons, are directed near and above the substrate but do not impinge directly (except via scattering) upon the growing film. In this way, the film deposition occurs with reduced radiation damage as compared to perpendicular beam assisted CVD.

As shown in Figure 1, the dissociation volume in beam CVD is well-defined spatially, as compared to a typical PECVD method in which the plasma fills the entire interelectrode and CVD reactor volume. Using LCVD techniques, both insulating (SiO_2, Si_3N_4, Al_2O_3 and AlN) and conducting (Mo, Cr, W, Al) films have been deposited, while EBCVD has been used to deposit silicon dioxide and silicon nitride. The electrical, optical, chemical and physical properties of beam deposited films (LCVD and EBCVD) are contrasted and compared below to those obtained by conventional methods.

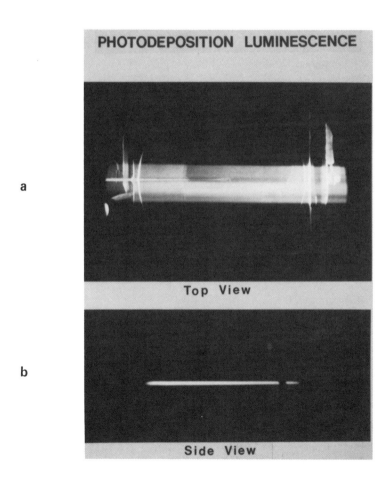

Figure 1: Photographs of laser CVD dissociation volume: (a) parallel to substrate, and (b) above substrate.

Comparison of Beam Deposited Film Properties

Laser-Deposited Dielectric Films. In Table 1, the typical deposition conditions for the laser assisted CVD of silicon-based insulating films are given. The laser, operating on a mixture of argon and fluorine, emits characteristic 193 nm photons. Tables 2 and 3 show comparisons of the properties of laser-deposited silicon dioxide and silicon nitride insulating films with these same materials deposited using conventional PECVD. The primary differences are that the LCVD films have increased incorporation of oxygen and carbon. This is attributed to the LCVD deposition system's poor vacuum base pressure (~1 mTorr) and inadequate backstreaming prevention in our initial vacuum system. Subsequent LCVD work by J.M. Jasinski at IBM has eliminated the C and O in Si_3N_4 films.[8]

Table 1: Deposition Conditions for LCVD of Silicon Dioxide and Silicon Nitride (ArF Laser Operating at 100 mJ/pulse, 100 Hz, 10 nsec/pulse).

Deposition Parameters	Silicon Dioxide	Silicon Nitride
Substrate temperature (°C)	380	380
Gas flow ratio	$N_2O/SiH_4/N_2$:80/1/40	$NH_3/SiH_4/N_2$:1/1/40
Total pressure (Torr)	6	2
Deposition rate (Å/min)	800	700

Table 2: Comparison of LCVD (Parallel Beam Configuration) and PECVD Silicon Dioxide Film Properties.

Film Property	Laser	RF Plasma
Impurities		
Nitrogen (at. %)	<4	3
Carbon (at. %)	<2	<0.1
Hydrogen bonding		
2270 cm^{-1} SiH (%)	2.3	2
3380 cm^{-1} H_2O (%)	~0.01	<0.001
3650 cm^{-1} OH (%)	0.6	0.002
Etch rate in 7:1		
Buffered HF (Å/s)		
as deposited	50-60	20
After 60 min 950°C		
N_2 anneal	~18	-
Dielectric strength (MV/cm)		
(1000 Å on Si)	6-8	-
(2000 Å on Si)	-	10
Resistivity at 5 MV/cm (Ω-cm)	10^{14}	10^{16}
Dielectric constant at 1 MHz		
Thermal oxide (3.9)	4	4.6
Pinholes (#/cm^2)		
(2000 Å on Si)	~1	~1
Refractive index (6328 Å)	~1.46	1.46

Table 3: Comparison of LCVD (Parallel Beam Configuration) and PECVD Silicon Nitride Film Properties.

Film Property	Laser Beam	RF Plasma
Impurities		
Oxygen (at. %)	5	<1
Carbon (at. %)	4	<1
Hydrogen bonding by FTIR		
Si-H (%)	12	12-16
N-H (%)	11	2-7
Etch rate: (Å/sec; 5:1 BOE)	20-50	1.7
Refractive index (6328 Å)	1.85	2
Density (g/cm^3)	2.4	2.8
Breakdown voltage (MV/cm)	2.5	4
Resistivity at 1 MV/cm (Ω-cm)	10^{14}-10^{15}	10^{15}-10^{16}
Dielectric constant at 1 MHz	7.1	7

Hydrogen incorporation, which is at similar levels for both methods, arises both from incomplete dissociation of feedstock gases as well as subsequent hydrogen-bearing free radical reactions. The high hydrogen content, which increases porosity of the film, can be remedied by subsequent thermal annealing as the etch rate comparison in Table 2 indicates. Alternatively, the impingement of laser radiation on the deposited film during deposition will also reduce the hydrogen content,[10] especially SiH and SiOH bondings. In the parallel beam technique, a high degree of film thickness uniformity can be achieved[11] because a wide area planar source is initiating the film deposition. This is easily appreciated if one considers the geometries of several possible deposition schemes where the source and substrate are separated by a distance R. A point source of species (e.g., evaporation) exhibits a $1/R^2$ (spherical) deposited film thickness uniformity dependence; whereas a line source provides a $1/R$ (radial) dependence on the uniformity of film deposition. With a uniform planar source of large area the deposited film uniformity shows no dependence on the source distance R since the total species flux from the beam dissociation volume averages to a conditionally fixed value. For this reason, superior vertical step coverage is obtained using the parallel beam assisted CVD technique.[11] In Figure 2, an example of conformal step coverage of LCVD SiO$_2$ is shown. In summary, silicon-based insulating films deposited using laser-induced CVD have properties which are comparable to those obtained in a plasma-enhanced deposition scheme.

Aluminum-based LCVD dielectric films are compared to sputtered alumina[12] and thermal CVD AlN[13,14] in Tables 4 and 5 respectively. The refractive index and electrical resistivity of LCVD alumina films are improved with increasing substrate temperature.[15] The LCVD alumina films deposited with substrate temperatures of 400°C compare well to RF sputter-deposited films with respect to the Al:O ratio, the dielectric constant, and the pinhole density. However, the LCVD films refractive index is slightly lower and the resistivities are an order of magnitude smaller.

Figure 2: Step coverage of LCVD silicon dioxide.

Table 4: Comparison of LCVD Alumina Films (Parallel Beam Configuration) With RF Sputtered Films.

Film Property	LCVD (400°C)	RF Sputtered[11] (150°C)
Al:O ratio	0.69*	0.68**
Impurities	C <1%*	Ar <5%**
Deposition rate (Å/min)	2000	200
Adhesion (dynes/cm^2)	>6.5 x 10^8	Strongly adherent
Pinhole defects	<1 in 5 cm^2 (thickness = 1500 Å)	31/cm^2 (thickness = 2500 Å)
Stress (dynes/cm^2)	<6 x 10^9 (tensile)	2.8 x 10^9 (compressive)
Refractive index	1.63	1.66
Etch rate in 10% HF (Å/min)	100	
Resistivity (Ω-cm)	10^{11} (@ 1 MV/cm)	10^{12} (@ 2.6 MV/cm)
Dielectric constant (@ 1 MHz)	10.2	9.96

*XPS
**RBS

The LCVD AlN films deposited at a substrate temperature of 200°C have substantial amounts of oxygen, hydrogen (bonded as N–H and O–H), and carbon. The hydrogen and oxygen content of the AlN films decreases with increasing substrate temperature. Table 5 summarizes properties for LCVD AlN films deposited at the highest substrate temperature used (450°C). The LCVD AlN film properties of refractive index, dielectric constant and electrical resistivity approach those values reported for thermal CVD films deposited at 800° to 1000°C[13,14] and RF sputter-deposited AlN films.[16,17,18] The LCVD films are lower in dielectric strength and dielectric constant[15] but contain up to 5% oxygen.

Table 5: Comparison of LCVD Aluminum Nitride Films (Parallel Beam Configuration) With TCVD (Thermal CVD) and RF Sputtered Films.

Film Property	LCVD (450°C)	TCVD[12,13] (800°-1200°C)	RF[15,16] Sputtering (<450°C)
Impurities			
Carbon	<5%*	<1%**	<1%*
Oxygen	<10%***	<5%**	<5%*
Refractive Index			
(@ 6328 Å)	1.95	2.05	2.0-2.1
Electrical resistivity			
(Ω-cm @ 1.5 MV/cm)	10^{15}	10^{15}	10^{15}
Dielectric strength			
(MV/cm)	3-5 (@ 1 μA/cm^2)	5-10	6-10
Dielectric constant			
(@ 1 MHz)	6-8	8.1-8.8	8.1-9.3

*ESCA
**Electron Microprobe
***RBS

Laser-Deposited Metallic Films. One can also deposit conducting films simply by changing the gas mixture in the chamber used for laser-deposited insulating films. The feedstock gases used for aluminum and refractory metal (Mo, Cr, W) deposition were trimethyl aluminum and the hexacarbonyls of the refractory metals [e.g., Mo(CO)$_6$]. In order that the photon wavelength would better match the donor gases' photon absorption cross section, a KrF laser operating at 248 nm was used for depositing the refractory metal films. Donor gases photodissociation cross sections versus wavelength are shown in Figure 3 (a),(b); the laser wavelengths used are also indicated.[19,20] Conducting films of molybdenum, tungsten, chromium and aluminum, deposited by photo-driven reactions in the volume, possess electrical resistivities within a factor of 2 to 8 of bulk values.[8] Properties of these LCVD metallic films appear in Table 6. Residual carbon and oxygen incorporation limits the electrical resistivity of the deposited films.

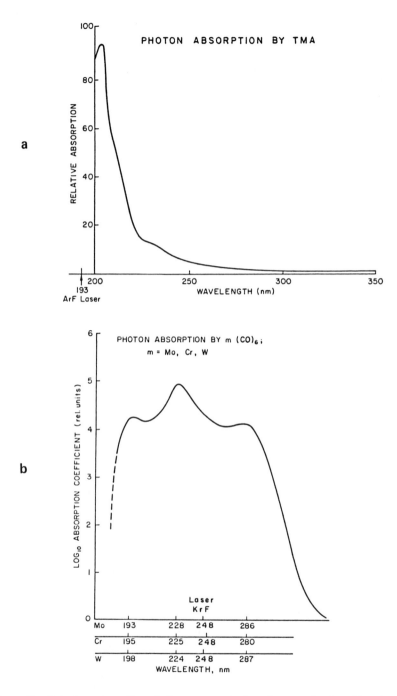

Figure 3: Relative photodissociation cross sections for conducting film donor gases versus photon wavelength. Laser wavelengths used are also indicated. (a) Trimethyl aluminum;[19] (b) Refractory metal hexacarbonyls.[20]

Table 6: The Properties of LCVD Metallic Films Deposited Using UV Photons From Excimer Laser.

Metal Film	Dep. Rate (Å/min)	Resistivity (Ω-cm) Bulk	Resistivity (Ω-cm) Film	% C in film (ESCA)	Adhesion to Quartz (dynes/cm^2)	Tensile Stress (dynes/cm^2)
Mo	2500	5.2	36	<0.9	>5.5 x 10^8	<3 x 10^9
W	1700	5.6	40	<0.7	>6.5 x 10^8	<2 x 10^9
Cr	2000	12.9	56	<0.8	>5.4 x 10^8	<7 x 10^9
Al	500	2.6	9	<7.0	>5.5 x 10^8	<1 x 10^9

Electron-Beam Deposited Dielectric Films. An abnormal glow discharge created electron beam can be injected into feedstock gases to dissociate them via electron-molecule collisions. The electron beam may be placed just above and parallel to the substrate or directed to impinge perpendicular to the substrate; for either case of electron beam assisted CVD (EBCVD), acceptable film properties result. Clearly, in the parallel beam technique electron radiation damage can be minimized as compared to the case of perpendicular arrangement. Physical damage by kilovolt electrons is less severe than that caused by ions of several hundred volts,[21] such as those present in conventional plasmas. Because the discharge voltages used (and hence, the maximum electron energies) are less than 3 kV, electron-induced damage can be annealed out with a low temperature heat treatment (<450°C) following deposition.[2,22]

Using both electron beam configurations (parallel and perpendicular), films of silicon dioxide and silicon nitride were deposited and the deposition conditions are summarized in Table 7. Note that while EBCVD creates films at a lower reactor pressure than PECVD the deposition rates are approximately the same or higher than those of PECVD. This is thought to occur because of the closer match under EBCVD conditions between the electron energy distribution and the electron impact dissociation cross section of the donor gases; this yields more efficient production of free radicals from the feedstock polyatomic gases following electron impact.

Table 7: Deposition Conditions for EBCVD of Silicon Dioxide and Silicon Nitride.

Process Variable	Silicon Dioxide	Silicon Nitride
Substrate temperature (°C)	350	400
Gas flow ratio	$N_2O/SiH_4/N_2$:75/1/75	$NH_3/SiH_4/N_2$:60/1/44
Total pressure (Torr)	0.25	0.35
Deposition rate (Å/min)	500	200
Discharge parameters	4.7 kV, 16 mA/cm^2	2.3 kV, 13 mA/cm^2

The properties of SiO_2 films deposited with the electron beam directed parallel to and above the substrate are given in Table 8 and are compared with conventional PECVD films. While the index of refraction and the resistivity are comparable, the EBCVD films are inferior with respect to wet chemical etch rate (up to 3 times higher) and pinhole density (10 times higher). These shortcomings are most likely due to film porosity (which is a manifestation of the high deposition rate) and low deposition temperature (which causes a low surface mobility of adsorbed species).

Table 8: Comparison of EBCVD Silicon Dioxide Films (Parallel Beam Configuration) and PECVD Deposited Films.

Film Property	Electron Beam	RF Plasma
Impurities		
Nitrogen (at. %)	<1	3
Carbon (at. %)	<2	0.1
Hydrogen bonding		
2270 cm^{-1} SiH (%)	<1	2
3380 cm^{-1} H_2O (%)	<0.001	<0.001
3650 cm^{-1} OH (%)	<0.01	0.002
Etch rate in 7:1		
Buffered HF (Å/sec)	30-60	20
Adhesion (10^8 dyne/cm^2)		
(1000 Å on Si)	>7	>7
Pinholes (#/cm^2)		
(2000 Å on Si)	10	~1
Refractive Index (6328 Å)	1.46	1.46
Stress (10^9 dyne/cm^2) on Si compressive	9.4	3.6
Dielectric strength (MV/cm)	2.6*	10**
Resistivity at 1 MV/cm (Ω-cm)	10^{14}-10^{16}	10^{16}
Dielectric constant at 1 MHz	3.5	4.6
[For comparison: Thermal oxide (3.9)]		

*1000 Å on Si
**2000 Å on Si

Films of silicon nitride (nitrogen rich oxynitride) were also deposited via the parallel electron beam-substrate geometry.[23] Silicon nitride films so deposited were also rich in hydrogen bonded to Si. When the EBCVD deposition geometry was altered, so that the beam electrons impinged directly upon the growing film, the Si–H bonding decreased well below (a factor of 20) that of the parallel EBCVD method. The hydrogen bonded to nitrogen level is nearly the same in PECVD, parallel EBCVD, and perpendicular irradiated EBCVD films. The shortcomings of the EBCVD silicon nitride films, as shown in Table 9, were the same as found in the EBCVD silicon dioxide (increased wet chemical etch rates and high pinhole densities). The high (13%) oxygen content of the silicon nitride films is judged to be due to two sources: primarily the poor base vacuum in the EBCVD reactor and secondly the much higher chemical reactivity of silane and associated free radicals (SiH_2, etc.) with background oxygen/water as compared to that with ammonia and its associated free radicals (NH_2, etc.)

Table 9: Comparison of EBCVD Silicon Nitride Films (Perpendicular Beam Configuration) With PECVD Deposited Films.

Film Property	Electron Beam	RF Plasma
Impurities		
Oxygen (at. %)	13	<1
Carbon (at. %)	<0.1	<1
Hydrogen bonding by FTIR		
Si–H (%)	<0.1	12-16
N–H (%)	8-10	2-7
Etch rate (Å/sec; 5:1 BOE)	3	1.7
Adhesion (10^8 dyne/cm^2)		
(1000 Å on Si)	>5.5	>6
Pinholes (#/cm^2) (1000 Å on Si)	5-100	2-3
Refractive index (6328 Å)	1.87	2
Dielectric strength (MV/cm)	5	4
Resistivity at 1 MV/cm (Ω-cm)	10^{15}	10^{15}-10^{16}
Dielectric constant at 1 MHz	7	7

A central concern of any deposition method is its ability to uniformly cover large area substrates ($\geq$15 cm dia.). The above EBCVD system deposits films with 4% uniformity over a 100 mm diameter substrate provided two conditions are met: (1) the cold cathode diameter is several centimeters larger than the substrate, and (2) the reactant gas flow was uniformly introduced into the reaction volume as shown in Figure 4. A plot of the spatial uniformity of the deposited film is shown in Figure 5.

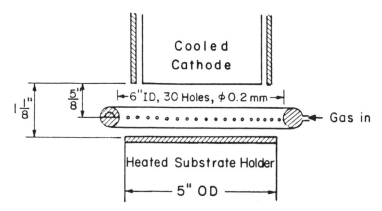

Figure 4: Technique of gas introduction in EBCVD to obtain high deposition uniformity.

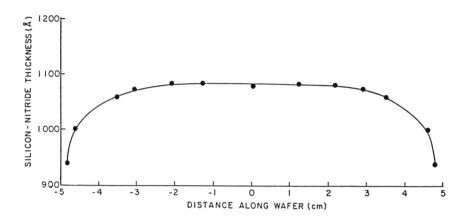

Figure 5: Deposition uniformity EBCVD silicon nitride on 100 mm wafer.

It is unlikely that EBCVD will displace conventional PECVD batch wafer processing, but it may have application as wafer dimensions increase to 20 centimeters in diameter and film fabrication becomes more oriented toward single-wafer (non-batch) in-line processing. Applications requiring a low Si–H content Si_3N_4 film would be well served by perpendicular EBCVD methods.

SUBMICRON PATTERN DELINEATION WITH LARGE AREA GLOW DISCHARGE PULSED ELECTRON-BEAMS

The fabrication limit of optical lithography is just under 0.75 μm, while x-ray lithography may be applied to feature sizes smaller than 0.5 μm.[24] Optical methods lack the sub-half micron replication capability of electron beam technology while x-ray lithography suffers from registration limits. Both these methods provide larger wafer throughput and better cost effectiveness than serial direct-write electron-beam lithography.

Direct write electron-beam lithography has two serious drawbacks: low wafer throughput and high equipment cost. Recently, we proposed a new method of implementing wide area electron-beam proximity lithography which eliminates the ultra-high vacuum constraint on the e-beam source.[25,26,27] The method proposed uses electrons produced in an abnormal glow discharge operating at 100 mTorr ambient[28] and preliminary experiments showed that adequate electron beam collimation to replicate feature sizes down to 5 μm is possible.[29] The glow discharge electron-beams not only exposed polymer resists such as PMMA but also self-developed these polymer resists. For those polymer resists which did not self-develop completely, we used subsequent wet or dry development to complete the process.

The glow discharge electron beams can be produced in either the CW or pulsed mode. High voltage pulsed beams in soft vacuum were found to be capable of fully self-developing submicron pattern delineation.[30] A typical experimental set-up is shown in Figure 6. A 25 kV pulsed electron-beam, produced by discharging as little as 2.5J of stored energy in 50 mTorr of helium or air, could replicate feature sizes 0.5 µm wide in 3.0 µm thick PMMA as shown in Figure 6.

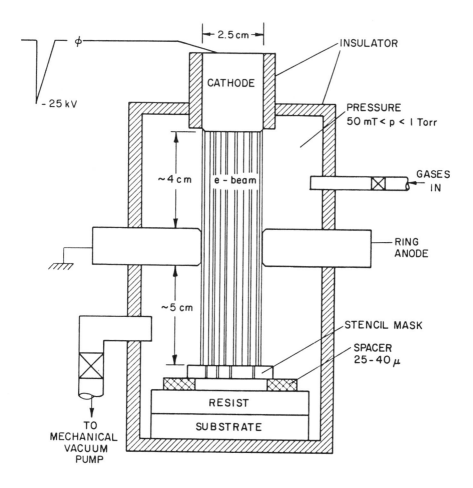

Figure 6: Electron beam proximity lithography system.

The chosen gaseous environment had a marked effect on the resist sidewall characteristics (see Figure 7). While resist exposed in helium showed vertical wall profiles (developed resist thickness is 3 µm), resist exposed in air (Figure 7b) showed sloped (~60°) walls. It is probable that by judicious mixing of these and other process gases, we can more fully control the wall profile.

332 Thin-Film Deposition Processes and Techniques

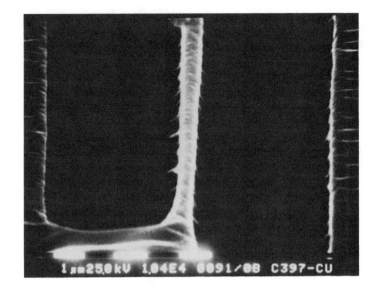

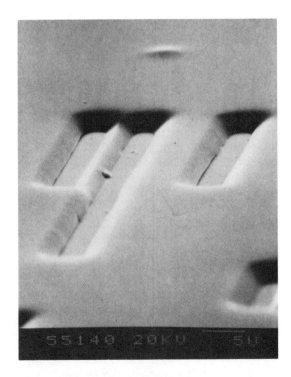

Figure 7: Submicron self-developed replication on PMMA: (a) in a helium ambient; (b) in an air ambient.

The outstanding features of this soft vacuum electron beam process may be summarized as: an electron beam operating in the soft vacuum pressure region (50 mTorr to 1 Torr); large area exposure (>50 cm^2) allowing for large field size, peak beam currents (I_{peak} > 100 A) allowing for self-development and in combination high wafer throughput. The short exposure time (1 μsec) and self-development capability makes it a contender to self-development methods employing energetic proton beams[31,32] or excimer laser beams.[33] Using proper resists optimized for the electron beam self-development process, these investigations could lead to a totally dry lithographic technique with submicron replication capable of operating around 1 Torr ambient pressure. The major drawback of wide area proximity electron beam lithography is the high cost and fragile nature of the required transmission masks.

BEAM INDUCED THERMAL PROCESSES

Overview

The application of heat to a semiconductor wafer can induce thermal processes such as annealing, alloying, and dopant diffusion. Conventional furnace-based thermal treatment is performed on 50 to 200 wafers per batch. Furnace processes, however, require initiation and termination times of one-half hour or more to ramp the wafers up to and down from the chosen process temperature. One can reduce the time-temperature cycling to which wafers are subjected and still achieve desired results,[34] provided one still reaches the threshold thermal energy input required to activate the physical processes. Independent of the heat source used, semiconductor annealing requires that the threshold temperature be induced only for milliseconds to seconds to achieve dopant activation and crystal lattice repair. One of these rapid thermal processing techniques is transient electron beam heating.[35,36,37]

To explore transient heat treating of four inch diameter wafers in a single exposure a large area (5 inch diameter) abnormal glow discharge electron beam was constructed. The wide area electron beam source, which is capable of radiating a power density of up to 120 watts/cm^2 at 4 to 8 keV continuously in an ambient of 0.1 to 1.0 Torr of helium, is shown schematically in Figure 8. It consists of a flat cold cathode made from a metal-ceramic of high secondary emission material (MgO–Mo). Electron emission occurs only from the front circular cathode surface because the remainder of the cathode structure is surrounded by a quartz shield. The MgO cathode component provides for high secondary electron emission following plasma ion bombardment while the molybdenum provides structural support and electrical conductivity. This wide area electron beam created in a glow discharge was used to anneal ion implant damage in single crystal silicon, and was also applied to the beam induced formation of silicide films.

In a later section, we will also compare laser and electron beam induced traveling-melt recrystallization of polysilicon deposited on four inch wafers. This purely thermal process consists of locally melting an area of the deposited polysilicon film followed by propagation of this melt front across the entire wafer surface. Progress in the application of both a scanned point source Ar$^+$

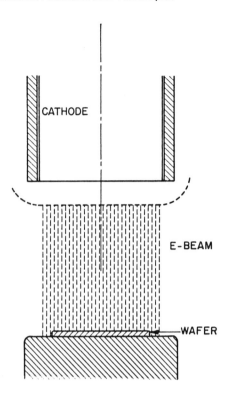

Figure 8: Large area electron beam annealing system.

laser beam and an extended line-source electron beam to this SOI (silicon on insulator) process are contrasted and compared.

Electron Beam Annealing of Ion-Implanted Silicon

To achieve the controllable introduction of dopant atoms into a semiconductor lattice at predetermined depths, dopant ions are accelerated and subsequently implanted in a thin region near the surface. The drawbacks of ion implantation are that: (1) the ions can displace lattice atoms (and in other ways disrupt lattice periodicity); and (2) that only a fraction of the implanted species come to rest on substitional (electrically active) lattice sites. Both lattice damage and interstitial dopants adversely affect carrier mobility and minority carrier lifetime in the crystal unless remedied by a post-implant oven anneal at temperatures up to 1000°C. These elevated temperatures, together with the required oven ramp cycles of one-half hour or more, cause dopants to diffuse away from the original as-implanted concentration profile during the oven anneal. This thermal redistribution becomes incompatible with the requirements of submicron VLSI processing when diffusion lengths become comparable to device dimensions. To reduce diffusion lengths, thermal anneals of shorter duration are required since the required threshold anneal temperature cannot be reduced.

The ability of a large-area abnormal glow discharge electron beam to rapidly anneal damage in crystalline silicon and activate dopants was ascertained in the following way. Boron, phosphorus, and arsenic ions were implanted into 8 to 21 ohm-cm <100> silicon wafers; in all cases, the substrate doping was opposite to that of the implanted species. Normalized sheet resistivity, ρ_A/ρ_B, was the parameter used to judge the extent of annealing. This ratio of sheet resistance before and after annealing minimizes geometry-dependent errors such as sample size and preanneal spatial doping variations. Samples with each implant species were annealed with increasing electron beam power densities until the ρ_A/ρ_B stabilized at a low value. Figure 9 shows the ρ_A/ρ_B versus beam power for various anneal times. Implant types and doses as well as varying implantation energy are also noted. The electron beam-induced process is capable of performing the anneal in a short time-high temperature manner.

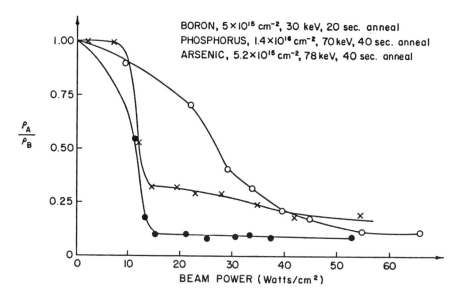

Figure 9: Normalized sheet resistivity versus beam power for boron, phosphorus, and arsenic ion implants in silicon. Implant and anneal conditions are noted.

The diffusion of the implanted species during transient beam annealing was examined using secondary ion mass spectrometry. In Figure 10, the electron beam-induced diffusion of boron implants in silicon is compared to the as-implanted profile and to that of a conventional furnace anneal. The ion implant and anneal conditions are noted in the figure. This shows that the electron beam can both repair the silicon lattice damage and activate dopants without appreciably altering the original as-implanted dopant profile.

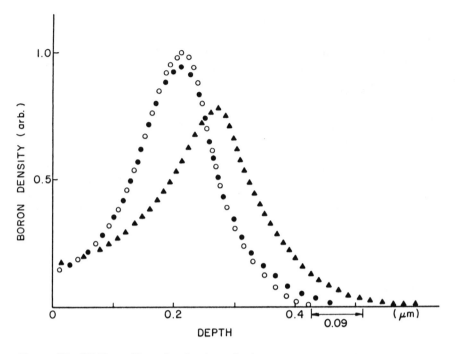

Figure 10: SIMS profiles of as-implanted, electron beam annealed and furnace annealed boron implanted silicon. The depth profile of B^+ implanted in silicon ("o": as implanted; "●": E-beam annealed at 3.5 keV, 25 MA/cm² for 40 seconds; "▲": furnace annealed at 960°C for 30 minutes).

Electron Beam Alloying of Silicides

Reactions of metals with silicon in the solid phase at elevated temperatures can form silicides, a set of materials whose properties are attractive for use as gates, contacts and interconnects in submicron VLSI fabrication. Silicides may be formed either by annealing of codeposited metal-silicon solid solutions or via interdiffusion of deposited metal films on silicon. In our work described below, tungsten and titanium silicides have been formed using transient exposure to an abnormal glow discharge electron beam as the process-driving heat input.

The starting materials were formed on <100> n-type silicon wafers by either: (1) sputter deposition of a 400 Å thick layer of Ti on bare Si or (2) PECVD of 1600 Å thick Ti-Si or W-Si films (of overall stoichiometry $TiSi_2$ or WSi_2) onto silicon wafers previously thermally oxidized to a thickness of 1000 Å. The wafers were then cleaved into smaller samples for ease of comparison of processing results. Prior to processing, Ti on Si samples had average sheet resistivities of 21 ohm/sq., the codeposited $TiSi_2$ had 10 ohm/sq., the codeposited WSi_2 had 40 ohm/sq. The ρ_A/ρ_B values versus beam power for these materials are shown in Figure 11.

For beam-induced interdiffusion of titanium and silicon layers (Figure 11), the ρ_A/ρ_B increases from 1 to 2.5 before reaching its ultimate value of 0.1 (which corresponds to a bulk resistivity of approximately 20 μ-ohm-cm). This effect was not observed for codeposited $TiSi_2$ which were electron beam alloyed. An explanation for this has been offered previously.[37,38] The spatial uniformity of the electron beam processing was examined by electron beam irradiating where codeposited $TiSi_2$. While the wafer and the cathode are both 7.5 cm in diameter, it was observed that only the central 5.5 cm of the wafer undergoes uniform heating due to the inhomogenous beam energy profile. Hence, the glow discharge electron beam source must be larger than the sample under process if wafer edge effects are to be avoided.

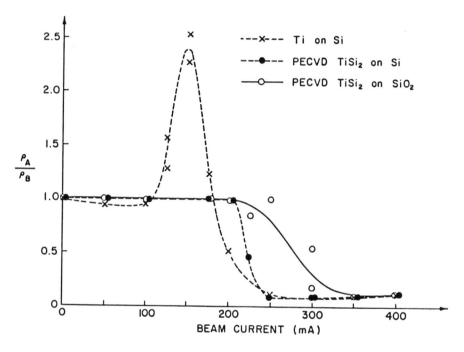

Figure 11: Normalized resistivity versus beam power for titanium silicide formation.

For electron beam induced alloying of WSi_2, the temporal nature of electron beam-induced silicide growth was examined. This was done by measuring the ρ_A/ρ_B as a function of beam exposure time for constant beam power densities of 75 and 100 W/cm^2. This is shown in Figure 12. At the higher beam power a low resistivity film is obtained after about 20 sec., while the lower beam power requires >100 sec. to reach a similar sheet resistivity. In either case, the ultimate film resistivity obtained is 125 μ-ohm-cm, about 2 to 3 times the bulk value; most likely due to oxygen incorporation from the SiO_2 underlayer via chemical reduction.[39]

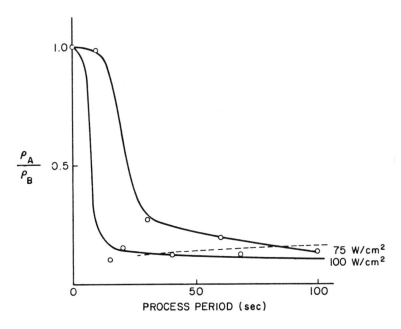

Figure 12: Temporal development of normalized resistivity during electron beam formation of tungsten silicide.

Laser and Electron Beam Recrystallization of Silicon on SiO_2

Processing microelectronic materials using either electron or photon beams can enable fabrication of structures which are either more costly or not possible to achieve using conventional processing methods. One such example is silicon-on-insulator (SOI) technology, in which one creates a thin crystalline silicon film which is dielectrically isolated from the underlying silicon substrate by a layer of SiO_2. The advantages of SOI over bulk silicon are that the devices and circuits in SOI layers have reduced parasitic capacitances (allowing faster circuit operation), increased radiation hardness (due to a thin active layer) and elimination of CMOS latch-up from undesired substrate currents. SOI may also make possible the vertical stacking of MOS devices (three-dimensional integration).

One of the most practical SOI fabrication techniques is traveling-melt recrystallization,[40] where a directed heat source locally melts a deposited polycrystalline film; an encapsulating film layer prevents beading of the silicon while it is molten. This molten zone of polysilicon is then propagated across the substrate. Upon contact of the molten silicon with the single crystal substrate, one is able to seed the resolidifying polysilicon film. This traveling-melt front technique is contingent on having a beam heat source capable of heating the silicon to its melting point (1415°C), a requirement which is met both by finely-focussed point source laser beams as well as wafer-size line source electron beams generated in abnormal glow discharges as described below.

An CW argon ion laser may focus to a spot approximately 40 μm in diameter.

The beam is then raster-scanned over an entire SOI wafer surface containing with periodic seed structures to melt and recrystallize the polysilicon film. The process is capable of creating 12 μm wide by 3 mm long SOI islands with carrier mobilities approaching bulk values.[41] A novel bilayer lateral CMOS structure was created as shown in Figure 13. This structure is inherently latch-up free. Laser induced traveling melt recrystallization most likely will not be incorporated into a manufacturing environment because to process one 100 mm diameter wafer with a scanned point source laser beam requires about 20 minutes. One practical alternative is to form a wafer size optical line source via lamps or lasers which can be swept across the upper polysilicon film. However, the total optical power required in such a laser source has not been achieved to date and lamp sources are limited to very slow propagation velocities (≤1 mm/sec).

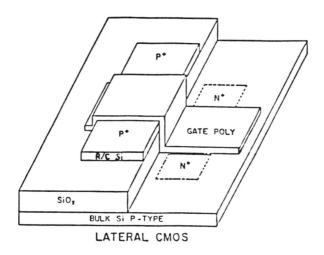

Figure 13: Bilayer lateral CMOS structure.

Toward this goal of line source recrystallization[42] we have developed a glow discharge created electron beam. Our 15 cm long line source cathode is capable of generating a multi-kilowatt line beam of ±1% intensity uniformity which extends across a 10 cm diameter wafer. This source alows for wafer scan velocities up to 10 cm/sec and still achieves melt zone formation during the short beam dwell time.[43] Using this method of line source recrystallization, single wafer SOI wafer process times are from 10 to 100 seconds. An optical microphotograph of line source recrystallized[43] silicon which has been decoratively etched is shown in Figure 14; a few grain boundaries can be seen. With substrate seeding the grain boundaries can be either spatially localized or eliminated.[42] Seeding to the underlying crystalline substrate provides both a preferred crystal orientation as well as a heat sink to create required thermal gradients between seed locations. Unseeded growth is desirable since wafer preparation is greatly simplified (because no lithographic definition is required), but this will require some other method of controlling grain boundary locations.[40]

340 Thin-Film Deposition Processes and Techniques

Figure 14: Optical micrograph of line-source recrystallized silicon.

SUMMARY AND CONCLUSIONS

The work described herein has demonstrated that photons from a laser are capable of photo-dissociating gaseous reactants whose photofragments can condense and/or react to form microelectronic films. Likewise, an abnormal glow discharge with an enhanced density of high energy electrons is capable of assisting chemical vapor deposition. In either case of beam assisted CVD, the films obtained are generally comparable to those deposited by conventional thermal and plasma-enhanced methods. Relative advantages and disadvantages of beam deposited films are outlined.

Beams of electrons and photons have also been used to induce some purely thermal processes, namely implant damage annealing, silicide alloying, and traveling-melt beam recrystallization. Ion implant damage in crystal silicon substrates can be annealed via transient electron beam exposure obtaining a high degree of electrical activation while minimally redistributing the original as-implanted dopant profiles. In a similar fashion, silicides of tungsten and titanium have been formed via electron beam initiated processes and silicide resistivities obtained approach bulk values. In either case, uniform treatments are possible if the beam diameter exceeds the wafer size. Electron beam proximity lithography is shown to self-develop submicron feature sizes in PMMA. Finally, laser and electron beams can be used to locally zone-melt a polysilicon film deposited on an insulating SiO_2 film which itself is grown on a crystalline silicon substrate. Via seed holes in the isolation SiO_2 down to the crystalline substrate the molten polysilicon contacts the solid substrate seed and forms a single crystal silicon island with little or no grain boundary defects.

In conclusion, photon and electron beam-induced processes have been shown to have applications to highly controlled spatial and temporal material fabrication. Whether or not these beam assisted techniques will better fulfill the needs of submicron VLSI circuit manufacture better than conventional or other alternative methods is not yet resolved.

Acknowledgements

This was supported by the Office of Naval Research, the National Science Foundation, NAVALEX, RADC, Applied Electron Corporation, and DARPA. Silicon wafers and processing were supplied by NCR-Microelectronics (P. Sullivan, D. Ellsworth), Tektronix Research Labs (H.K. Park), ASM America (D. Rosler), Hewlett-Packard Instrument Division (R. Toffness) and California Devices, Louisville, CO (M. Gullett).

REFERENCES

1. *VLSI Technology*, ed. by S.M. Sze, McGraw-Hill, New York (1983).
2. Fonash, S.J., *Solid State Technology*, p. 201, (April 1985), and Pang, S.W., *J. Electrochem. Soc.*, Vol. 133, p. 784 (1986).
3. Stengl, G., Loschner, H. and Muray, J., *Solid State Technology*, p. 119, (February 1986).
4. Bartelt, J., *Solid State Technology*, p. 215 (May 1986).
5. Dautremont-Smith, W.C., Gottscho, R.A. and Schutz, R.J., Chapter 5, *Semiconductor Materials and Process Technology*, ed. by G.E. McGuire, Noyes Publishing, Park Ridge, New Jersey (1986).
6. Chapman, B., *Glow Discharge Processes*, Wiley Interscience, New York (1980) or Von Engle, A., *Ionized Gases*, Oxford Univ. Press, Oxford (1965).
7. Shi, B., Yu, Z. and Collins, G.J., *IEEE J. Plasma Science*, (Aug. 1986). See also, Rocca, J.J., Meyer, J., Farrell, M. and Collins, G.J., *J. Appl. Phys.*, Vol. 56 (3), p. 790 (1984).
8. Solanki, R., Moore, C.A. and Collins, G.J., *Solid State Technology*, p. 220, (June 1985). See also, Jasinski, J.M., to be published, in *Appl. Phys. Letts.* (1986).
9. Thompson, L.R., Rocca, J.J., Emery, K., Boyer, P.K. and Collins, G.J., *Appl. Phys. Lett.*, Vol. 43 (8), p. 777 (1983).
10. Boyer, P.K., Emery, K.A., Zarnani, H. and Collins, G.J., *Appl. Phys. Lett.*, Vol. 45, p. 979 (1984).
11. Boyer, P.K., Ritchie, W.H. and Collins, G.J., *J. Electrochem. Soc.*, Vol. 129 (9), p. 2155 (1982). See also Thompson, L.. Gobis, L., Bishop, D., Emery, K. and Collins, G.J., *J. Electrochem. Soc.*, Vol. 131 (3), p. 462 (1984).

12. Nowicki, R.S., *J. Vac. Sci. Tech.*, Vol. 14 (1), p. 127 (1977).
13. Morita, M., Tsubouchi, K. and Mikoshiba, N., *Jpn. J. Appl. Phys.*, Vol. 21 (5), p. 728 (1982).
14. Irene, E.A., Silvestri, V.J. and Woolhouse, G.R., *J. Elect. Mater.*, Vol. 4 (3), p. 409 (1975).
15. Solanki, R., Ritchie, W.H. and Collins, G.J., *Appl. Phys. Lettr.*, Vol. 43 (5), p. 454 (1983). See also Minakata, M. and Furukawa, Y., *J. Electr. Mat.*, Vol. 15 (3), p. 159 (1986).
16. Fathimula, A. and Lakhani, A., *J. Appl. Phys.*, Vol. 54 (8), p. 4586 (1983).
17. Mirsch, S. and Reimer, H., *Phys. Stat. Sol.*, Vol. (A) 11, p. 631 (1972).
18. Gerova, E.V., Ivanov, A. and Kirov, K.I., *Thin Solid Films*, Vol. 81 (3), p. 201 (1981).
19. Ehrlich, D.J., Osgood, R.M. and Deutch, T.F., *IEEE J. Quant. Electron.*, Vol. QE-16 (11), p. 1233 (1980).
20. Hatzakis, M., *J. Electrochem. Soc.*, Vol. 116 (7), p. 1033 (1969).
21. Pang, S.W., *Solid State Technology*, p. 249 (April 1984).
22. Sah, C., Sun, J.Y., Tzou, J.J., *J. Appl. Phys.*, Vol. 54 (2), p. 944 and (8) p. 4378 (1983).
23. Bishop, D.C., Emery, K.A., Rocca, J.J., Thompson, L.R., Zarnani, H. and Collins, G.J., *Appl. Phys. Lett.*, Vol. 44 (6), p. 598 (1984).
24. Suzuki, K. and Matsui, J., *J. Vac. Sci. and Technol.*, Vol. B 4 (1), p. 221 (1986).
25. Scott, J.P., *J. Appl. Phys.*, Vol. 46 (2), p. 661 (1975).
26. Bohlen, H., Greschner, J., Keyser, J., Kulcke, W. and Nehimz, P., *IBM J. of Res. and Dev.*, Vol. 26 (5), p. 568 (1982).
27. Pfeiffer, H.C., *Proceedings in Microcircuit Engineering 83*, ed. H. Ahmed, J.R.A. Cleaver and G.A.C. Jones, p. 83, Academic Press, NY (1983).
28. Krishnaswamy, J. and Collins, G.J., Patent Disclosure Document, Colorado State University Research Foundation (August 1985) (unpublished).
29. Krishnaswamy, J., Li, L. and Collins, G.J., Colorado Microelectronics Conference, CMC-86, Colorado Springs, (March 24–25, 1986).
30. Krishnaswamy, J., Li, L., Collins, G.J. and Hiraoka, H., Ann. Mtg. of Mat. Res. Soc., Boston (1986).
31. Okada, I., Saitoh, Y., Itabashi, S. and Yoshihasa, H., *J. Vac. Sci. and Technol.*, Vol. B 4 (1), p. 243 (1986).
32. Rensch, D.B., Seliger, R.L., Csanky, G., Olney, R.D. and Stoves, H.L., *J. Vac. Sci. and Tech.*, Vol. 16 (6), p. 1897 (1979).
33. Lazare, S. and Srinivasan, R., *J. Phys. Chem.*, Vol. 90, p. 2124 (1986). See also, Srinivasan, V., Smrtic, M. and Babu, V., *J. Appl. Phys.*, Vol. 59 (11), p. 3861 (1986).
34. *Proceedings of the Materials Research Society on Transient Thermal Processing* Volumes 1, 4, 13, 23, 35, 51, 52, 74.
35. Moore, C.A., Rocca, J.J., Johnson, T., Collins, G.J. and Russell, P.E., *Appl. Phys.. Lett.*, Vol. 43 (3), p. 290 (1983).
36. Moore, C.A., Rocca, J.J., Collins, G.J., Russell, P.E. and Geller, J.D., *Proc. Mat. Res. Soc.*, Vol. 33, p. 803 (1984).
37. Moore, C.A., Rocca, J.J., Collins, G.J., Russell, P.E. and Geller, J.D., *Appl. Phys. Lett.*, Vol. 45 (2), p. 169 (1984).

38. Pramanik, D., Deal, M., Saxena, A.N. and Wu, K.T., *Semicon. Intern.*, p. 94, (May 1985).
39. Nicolet, M-A. and Lau, S.S., Chapter 6, *VLSI Electronics*, ed. by N.G. Einspruch and G.B. Larrabee, Academic Press, New York (1983).
40. Cullen, G.W., ed., *J. Crystal Growth*, Vol. 63 (3) (1983).
41. Sritharan, S., Solanki, R., Collins, G.J., Szluk, N., Fukumoto, J. and Ellsworth, D., Mat. Res. Soc. Ann. Mtg., Boston, MA (1985).
42. Knapp, J.A., *J. Appl. Phys*, Vol. 58 (7), p. 2584 (1985).
43. Moore, C.A., Meyer, J.D., Szluk, N.S. and Collins, G.J. (unpublished).

11
Ionized Cluster Beam Deposition

Isao Yamada, Toshinori Takagi, and Peter Younger

INTRODUCTION

As a result of the unique properties of small clusters which have physical and chemical characteristics unlike those of the bulk and liquid states, numerous new applications in plasma physics, atomic and molecular physics, surface science, and thin film formation have become available. In particular, a cluster is an aggregate of only a few hundred to a few thousand atoms, and most of the atoms are situated on the cluster surface. Therefore, the overall structure of the cluster is dominated by the surface atoms, and we should consequently expect that the physical and chemical properties of the cluster are much different from those of bulk and liquid.[1] Extensive theoretical and experimental investigations of these cluster properties are being conducted in various fields.[2] Film formation using cluster beams is one of the most practical and promising applications in semiconductor microelectronics.[3]

Research of Ionized Cluster Beam (ICB) technology was started early in 1972 at Kyoto University.[4] The purpose of this research was to overcome problems in the film deposition process technology. The major problem to be initially addressed was the production of high-quality thin films for functional devices at high deposition rates by utilizing the effects of ion bombardment as in ion plating.

Clusters can be created by condensation of supersaturated vapor atoms produced by adiabatic expansion through a small nozzle into a high vacuum region. In ICB technology, the source of atoms which form the clusters can be simply a solid heated to sufficient temperature to cause the pressure differential for expansion through the nozzle. Once formed, the cluster beam can be partially ionized to positive single charge state by electron impact. Then energy can be added to ionized clusters by use of acceleration potentials. Unique capabilities of ICB are attributed to the properties of the cluster state. Enhancement of the adatom migration effect is one of the most significant properties of cluster beam deposition. Properties such as extremely low charge to mass ratios to prevent space charge problems and equivalently high current and low energy ion beam transport in a range from above thermal energy to a few hundred eV are important characteristics of ICB for forming films by this technology.

In ion beam deposition, various effects of ion bombardment can be expected in different energy ranges. For example, in the energy range of 1 to 5 keV, the sputtering effect, which is widely used in semiconductor fabrication processes, is dominant. In the energy range of 5 to 20 KeV, the so-called predeposition process, in which ions are implanted to shallow depths, occurs and the implanted low energy ions then become an impurity source for thermal diffusion. When beam energies of a few tens to hundreds of keV are used, the standard ion implantation process predominates.

Kinetic energies corresponding to thermal energies given by kT are approximately 0.01 to 0.1 eV, or much lower than binding energies in the temperature range below 1000°C. A strong effect can be expected, however, as the result of bombarding by accelerated ion beams, even at energies of only a few eV which correspond to binding energies.

These considerations show that an ion beam with an energy range from near thermal to a few hundred eV would be extremely useful for many industrial purposes. Practical low energy ion beams, however, have been difficult to produce. Although the plasma technique, in which ions of a few eV move randomly, is one of the popular methods of surface treatment, the low energy ions associated with this technique cannot be extracted into a vacuum as a beam. This is due to space charge effects.

In the cluster beam, the cluster is an aggregate of about 100 to 2,000 atoms. Therefore, energies of the constituent atoms are in a range from a few to a few tens of eV, which is comparable to the binding energy of the atoms in the solid, even when the clusters themselves are accelerated to kilovolts. Therefore, the ICB technique has the feature of transferring low-energy and equivalently high-current beams. This feature allows high quality deposition and epitaxy of materials at low temperature onto a wide variety of substrates and even permits the formation of thin film materials not previously possible.

FORMATION OF CLUSTERS AND PROPERTIES OF THE CLUSTER

The cluster source used is a cylindrical nozzle source. The dimensions are shown in Figure 1, where D is the nozzle diameter, λ is the mean free path of the atoms in the crucible, P_o is the inner pressure of the crucible, P is the vapor pressure outside the crucible and L is the thickness of the nozzle. Typical operating conditions are P_o = 10 Torr, D = 0.5 - 2 mm and L/D = 1.[5]

In nucleation theory of cluster formation, the change in the Gibbs free energy ΔG is given by:

$$(\Delta G)_{max} = \Delta G^* = \frac{16 \pi \sigma^3}{3 \left\{ \frac{kT}{V_c} \ln S \right\}^2} \quad (1)$$

where σ is the surface tension, V_c is the volume/molecule, and S the saturation ratio P/P_∞, P_∞ being the equilibrium vapor pressure. Assuming thermally stable equilibrium, the nucleation rate J is expressed as:

$$(2) \qquad J = K \exp\left(-\frac{\Delta G^*}{kT}\right)$$

where K is a factor which varies much slower with P and T than the exponential term. It is given by Frenkel[6] as:

$$(3) \qquad K = (P/kT)^2 \, V_c \, (2\sigma/\pi m)^{1/2}$$

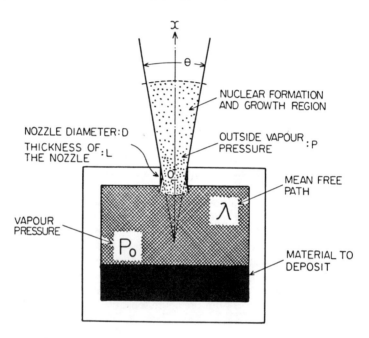

Figure 1: Dimension and parameters of ICB source.

Figure 2 shows the calculated change of ΔG for several materials. Obviously, there is no drastic difference in the nucleation barrier for metals as compared to gases.[7] But Hg is an exception.

The processes of metal cluster formation during supersonic expansion through the nozzle were analyzed by a computer simulation. A cluster is assumed to start to grow from the critical radius with maximum value of ΔG. The nuclei grow through a process in which impinging vapor atoms are balanced against reevaporation at a rate given by:

$$(4) \qquad \frac{dr}{dt} = \left(\frac{\xi}{\rho_c}\right)(P/\sqrt{2\pi RT} - P_c/\sqrt{2\pi RT_c})$$

where ξ is the sticking coefficient, ρ the density of the cluster and T_c the temperature of the cluster. Here P_c is the saturated vapor pressure at the surface of the cluster and is given by Thomson[8] as:

(5) $$P_c = P_\infty \exp(2\sigma / \rho_c RT_c r)$$

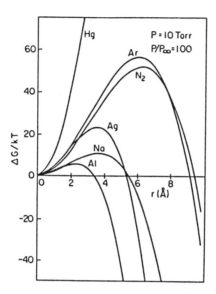

Figure 2: $\Delta G/kT$ as a function of cluster radius for several materials. Pressure and supersaturation ratio are fixed to $P = 10$ Torr and $P/P_\infty = 100$, respectively.

The one dimensional flow equations for the conservation of mass, momentum and energy are:

(6) $$\frac{d}{dx}\left(\frac{\rho A u}{1-\mu}\right) = 0$$

(7) $$u\frac{du}{dx} + \frac{1-\mu}{\rho}\frac{dP}{dx} = 0$$

(8) $$u\frac{du}{dx} + (1-\mu) C_p \frac{dT}{dx} = h_{fg}\frac{du}{dx}$$

where ρ is the vapor density, A the cross-sectional area of the flow, u the velocity, μ the ratio by mass of the condensed and vapor phase, C_p the specific heat and h_{fg} the latent heat. The source condition was set to a saturated vapor and the flat plane surface tension σ_∞ was used in the calculation. The area of the nozzle as a function of the x axis was taken as:

(9) $$A = (\pi D^2 / 4) [1 + (2x \tan(\tfrac{\theta}{2}) / D)^2]$$

where D is the throat diameter and θ is the nozzle opening angle.

Figure 3 shows the variation of relative flow temperature T/T_0 and relative flow pressure P/P_0 for Al. An increase of P_0 over several Torr causes cluster formation, as can be seen by the increase in the relative flow temperature caused by the release of the latent heat of condensation. The calculation shows that the condensation of Al vapor starts upstream of the throat. Figure 4 shows the cluster size distribution for Al 6 mm from the throat for different P_0. The cluster size increased with increasing P_0.[7]

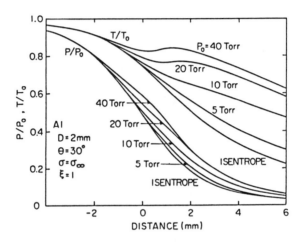

Figure 3: Variations of relative flow temperature T/T_0 and pressure P/P_0 of Al nozzle flow along the axis.

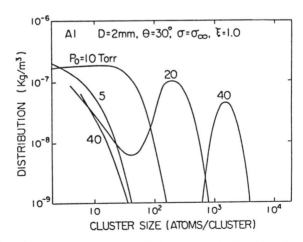

Figure 4: Cluster size distribution of Al clusters obtained by simulation under different source pressures P_0.

The most common method for experimentally evaluating cluster sizes is energy analysis either by the retarding field method or by the electrostatic energy analyzer. Since the kinetic energy of a cluster is proportional to its mass and square velocity, it is possible to estimate the cluster size distribution by measuring the retarding field method. Figure 5 shows Ag cluster size distribution measured by the retarding field method. The size ranges broadly from several hundreds to two thousand atoms/cluster. Beam intensity and cluster size were increased by increasing the crucible temperature.

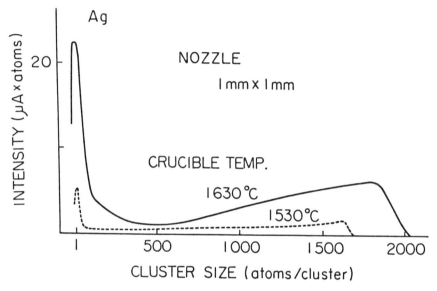

Figure 5: Size distribution of Ag clusters measured by the retarding field method.

IONIZED CLUSTER BEAM DEPOSITION EQUIPMENT

The basic construction of an ICB ion source for film formation is shown in Figure 6. The source is divided into three parts: crucible heating, ionization and acceleration. The crucible can be heated up to 2000°C. The ratio of ionized to total cluster flow can be adjusted by changing the ionization electron current. For example, the degree of ionization is 5 to 7% at ionization current Ie = 100 mA, 7 to 15% at Ie = 150 mA and 30 to 35% at Ie = 300 mA.[9]

A view of an ICB experimental deposition system now commercially available is shown in Figure 7.[10] This system has four crucibles which can be moved sequentially into the ion source by remote control. Deposition conditions such as crucible temperature, deposition rate, acceleration voltage, electron ionization current, substrate temperature, and film thickness can be controlled automatically by the system computer. A triple cluster beam system having three crucibles has also been developed. A single crucible system with multiple nozzles which forms a curtain beam with high uniformity is shown in Figure 8.

350 Thin-Film Deposition Processes and Techniques

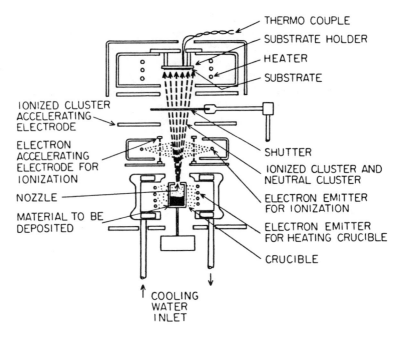

Figure 6: Schematic diagram of the ICB deposition apparatus.

Figure 7: Laboratory ICB system (EATON ICB-100).

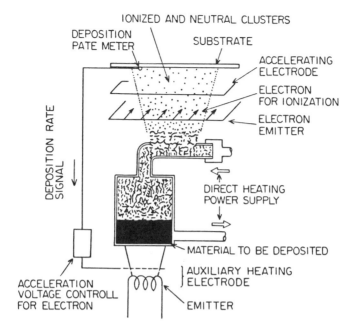

Figure 8: ICB source with single crucible and multiple nozzles.

FILM FORMATION KINETICS

Generally, the influence of the deposition parameters on film growth can be explained in terms of the effects on the sticking coefficient, the nucleation density, and the surface mobility of adatoms. Some of the related fundamental effects in ICB deposition are:

- Enhancement of adatom migration
- Creation of activated centers for nucleation
- Desorption of surface impurities
- Change of sticking probability
- Surface heating
- Enhancement of chemical reactions
- Ion implantation

These are due to the kinetic energy of the accelerated ionized clusters, electric charge of the cluster and cluster state.

In film formation by ICB, the following fundamental processes are to be considered: when the clusters bombard the substrate surface, both ionized and neutral clusters are broken up into atoms which are then scattered over the surface with high surface diffusion energy. A scattered atom is physically attracted to the substrate surface but it may move over the surface (adatom migration) because of high kinetic energy parallel to the surface obtained from the incident kinetic energy. The adatom interacts with other adatoms and/or substrate atoms to form a stable nucleus and, finally, to cease its movement and become a chemically adsorbed atom. In Figure 9, a conceptual illustration of the film deposition process is shown.

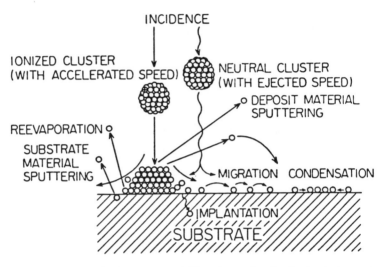

Figure 9: Conceptional illustration of film formation by ionized cluster beam deposition.

Kinetic energy and electric charge of the ICB influence these dynamic processes. Migration of atoms on the substrate surface was observed by electron microscopy of the films.[11] The results showed that migration can be enhanced by increasing the acceleration voltage.

Other effects have been confirmed. The deposited mass M on the substrate as a function of substrate temperature is given by:

$$(10) \qquad M = \dot{M}t - \frac{\dot{M} N_o}{I^*} \exp(-U/kT)$$

where $U = \phi_{ad} - \phi_d$, ϕ_{ad} is the activation energy for desorption and ϕ_d in the activation energy for surface diffusion, $\dot{M}$ is the mass impingement rate on the substrate, I^* is the rate of formation of critical nuclei, and N_o is the density of adsorption sites on the substrate surface. Since ϕ_{ad} is generally larger than ϕ_d, the curve of the deposited mass versus the reciprocal of the substrate temperature will have a positive slope. Figure 10 shows experimental results of deposited mass versus the reciprocal of the substrate temperature. Dashed lines indicate the different crystalline states of the films deposited at different acceleration voltages (V_a). For the nonionized clusters, the mass deposited increases with decreasing substrate temperature, corresponding to conventional deposition methods.

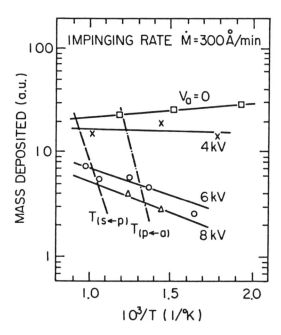

Figure 10: Relation between deposited mass and reciprocal substrate temperature, and the change of transition temperature for different acceleration voltages.

For the ionized cluster beams, the mass deposited at a given substrate temperature decreases with increasing acceleration voltage because of sputtering and reevaporation, and the slope changes with acceleration voltage. The change of the slope from the positive to the negative sign with increasing acceleration voltage is considered to be due to the change in the values of the parameters N_o, I^*, and U. This is one of the important features of the ICB deposition process which differentiates it from conventional deposition processes using atomic, neutral or ionized particles, where the energy U does not change.

To understand the effect of migration enhancement by increasing acceleration voltage, the dependence of the mass impingement rate $\dot{M}$ on the substrate temperature was measured as shown in Figure 11. The figure also shows the polycrystalline to single-crystalline transition temperature $T_{(s \leftarrow p)}$ for silicon films prepared at $V_a = 0$ kV (no acceleration and no ionization) and $V_a = 8$ kV, respectively. The slopes of both lines in the figure yield the activation energy for surface diffusion (ϕ_d) from the following relation:

(11) $$\dot{M} \leq C \exp(-\phi_d / kT_{(s \leftarrow p)})$$

where C is a constant and k is Boltzmann's constant. The result shows that ϕ_d decreases with increasing acceleration voltage. This is explained by relating that the surface diffusion energy can be enhanced by increasing the acceleration voltage.

354 Thin-Film Deposition Processes and Techniques

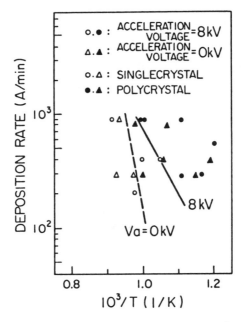

Figure 11: Characteristics of impinging rate versus reciprocal substrate temperature, and the change of transition temperature of the crystalline state at different acceleration voltages.

The density of nuclei can also be controlled by changing the energy of the clusters as increasing the kinetic energy of the ionized clusters yields a higher density of nuclei. Nucleation density is considered to be determined mainly by such factors as the deposition rate, the presence of electric charges, structural defects, ion bombardment, and statistical fluctuations in the supersaturation conditions.

The results above show that ICB bombardment influences the dynamic processes in the film formation. These dynamic processes include the breaking of clusters into atoms upon bombarding the substrate surface, formation of activation centers for nucleation, adatom migration, and shallow implantation. In ICB deposition, these processes can be controlled by changing the acceleration of ionized clusters and the content of ionized clusters in the total flux, and consequently, the film properties can be controlled.

FILM PROPERTIES

Many kinds of films have been deposited on various kinds of substrates by ICB and by reactive ICB (RICB). The films deposited by these methods and their features are summarized in Table 1. In this section, only novel and important results which elucidate the ICB characteristics are described. These results show that ICB can be used to control the deposition processes, consequently, the structure of the films and their electrical, optical and mechanical properties.

Table 1: Film Properties and Their Features Prepared by ICB.

July 1, 1986

Films/Substrate	Accel. Volt. (kV)*	Subs. Temp. (°C)	Advantages (Applications)
METALS			
Au,Cu/glass, Kapton	1-10	RT	Strong adhesion, high packing density, good electrical conduction in a very thin film(high resolution flexible circuits, optical coating)
Pb/glass	5	RT	Controllable crystal structure, strong adhesion, smooth surface, improved stability on thermal cycling(superconductive device)
Ag/Si(n-type)	5	RT	Ohmic contact without alloying
Ag/Si(p-type)		400***	Ohmic contact by single layer at low temperature (semiconductor metallization)
Ag-Sb/GaP	2	400	Ohmic contact, strong adhesion(semiconductor metallization)
Al/SiO$_2$,Si	0**-5	RT-200	Controllable crystal structure strong adhesion, ohmic contact at low temperature, electro-migration resistant(semiconductor metallization)
INTERMETALLIC COMPOUNDS			
CuAl/glass	0**-5	RT	Controllable crystal structure, smooth surface (shape memory device)
CuNi/glass	0**-5	RT	Controllable crystal structure, high strain gage (strain gage device)
C-axis preferentially oriented MnBi/glass	0**	300	Spatially uniform magnetic domain, high density optical memory(magneto-optical memory)
GdFe/glass	0**	200	Thermally stable and uniform, amorphous (magneto-optical memory)
C-axis preferentially oriented PbTe/glass	0**-3	200	High Seebeck coefficient, low thermal conductivity (high efficiency thermo-electrical converter)
Preferentially oriented ZnSb/glass	1	140	Thermally stable amorphous, high Seebeck coefficient (high efficiency thermo-electric converter)
ζ-FeSi$_2$/glass	0**-5	150	Thermally stable, amorphous, p- or n-type controllable, high Seebeck coefficient (high efficient thermo-electric converter)
SEMICONDUCTORS			
Si/(111)Si Si/(100)Si Si/(1$\bar{1}$02)sapphire	6-8	400-620	Low temperature epitaxy in a high vacuum(10^{-7}-10^{-6} Torr) shallow and sharp p-n junction(semiconductor devices)
Amorphous Si/glass	0**-6	200	Thermally stable(solar cell, thin film transistor)
Ge/(100)Si Ge/(111)Si	0.5-1	300-500	Low temperature epitaxy in a high vacuum(10^{-7}-10^{-6} Torr) shallow and sharp p-n junction (semiconductor devices)
GaAs/GaAs, Cr-doped GaAs	1-6	550-600	Epitaxy in a high vacuum(10^{-7}-10^{-6} Torr) (semiconductor devices)
GaP/GaP	4	550	Low temperature epitaxy in a high vacuum (10^{-7}-10^{-6} Torr) (low cost LED)
GaP/Si		450	
ZnS/NaCl	1	200	Single crystal formation(optical coating, ac-dc electroluminescent cells of low impedance)
ZnS:Mn/glass	1	200	

(continued)

Table 1: (continued)

Films/Substrate	Accel. Volt. (kV)[*]	Subs. Temp. (°C)	Advantages (Applications)
SEMICONDUCTORS (Continued)			
InSb/sapphire	3	250	Controllable crystal structure (magnetic sensor)
CdTe/glass	2	250	Improved monocrystalline domains (infrared detector)
CdTe-PbTe/Si, InSb	5	250	Very thin and multilayered films, low temperature growth (opto-electronic devices, superlattice)
$Cd_{1-x}Mn_xTe$/glass	0[**]–5	200–300	Controllable optical band gap and Faraday rotation angle (magneto-optical devices, optical wave guide)
OXIDES, NITRIDES, CARBIDES, AND FLUORIDES			
CaF_2/Si(111), Si(100)	0[**]–5	400–700	Low temperature growth, smooth surface (three dimensional device)
MgF_2/glass	0[**]–5	100–300	Low scattering, strong adhesion, moistureproof (optical coating)
FeOx/Si, glass	3	250	Controllable optical band gap (photovoltaic cell)
Li-doped ZnO/($1\bar{1}02$)sapphire	0.5–1	230	Single crystal formation (optical wave guide)
C-axis preferentially oriented ZnO/glass	0[**]	150	Controllable crystal transmission (optical wave guide, SAW)
BeO/(0001)sapphire	0[**]	400	Single crystal formation, transparent film (semiconductor device, heat sink, electrical insulator)
C-axis preferentially oriented BeO/glass	0[**]	400	High electrical resistivity and high thermal conductivity coating, low temperature growth (semiconductor device, heat sink, electrically insulatng and thermal conduction)
PbO/glass	3	RT	Low temperature growth, smooth surface, accurately controllable thickness, strong adhesion (superconductor)
SnO_2/glass	0[**]	400	Transparent low resistivity film, strong adhesion (NESA glass)
SiO_2/Si	0[**]–2	200	Extremely low temperature growth (Si devices)
TiO_2/glass	0[**]–7	300–350	Controllable crystall structure, high refractive index (multilayered optical films)
GaN/ZnO/glass	0[**]	450	Low temperature growth of GaN film on amorphous substrate (low cost LED, photocathodic electrode)
SiC/Si, glass	0[**]–8	600	Controllable crystal state (energy converter, surface protective coating)
ORGANIC MATERIALS			
Anthracene/glass	0[**]–2	–10	Controllable crystal structure (detectors)
Cu-Phtalocyanine/glass	0[**]–2	200	Controllable crystal structure (photovoltaic devices, FET)
Polyethylene/glass	0[**]–1	0–110	Controllable crystal structure (high insulating film formation, FET)

[*] Since a cluster consists of 100 ~ 2000 atoms, one atom has an incident energy corresponding to $\frac{1}{100} \sim \frac{1}{2000}$ of the acceleration voltage. For example, in a case of 5kV acceleration voltage, one constituent atom has an energy of 2.5 ~ 50eV, which is coincident to the optimum energy for film formation.

[**] Ejection velocity only

[***] Annealing temperature

Metals

One of the necessary conditions for obtaining epitaxial films is generally known to be that the lattice misfit should be less than a few percent. For Al deposition on the Si(111) substrate, the lattice misfit is -25% at least in one direction. However, ICB permits epitaxial growth of Al on both clean Si(111) and Si(100) substrates at room temperature.[12]

Single crystal Al films 400 nm thick were grown on Si(111) and Si(100) substrates at room temperature. Figure 12 shows a typical 75 keV Reflection High Energy Electron Diffraction (RHEED) pattern from Al films on Si(111) (a) and Si(100) (b) substrates. The crystalline orientation of the Al films on the Si(111) substrate was determined as Al(111)//Si(111), Al[$\bar{1}$10]//Si[$\bar{1}$10]. Two orthogonal orientations, Al(110)//Si(100), Al[001]//Si[011] denoted as Al(110) and Al(110)//Si(100). Al[$\bar{1}$10]//Si[011] denoted as Al(110)R were observed in the film on the Si(100) substrate. Crystalline quality and epitaxial relation were determined by Rutherford Backscattering Spectroscopy (RBS) and Reflection High Energy Electron Diffraction (RHEED). After the films were annealed for 30 minutes at temperatures in the 450° to 550°C range, there were no hillocks and valleys as normally observed in Al films prepared by conventional vacuum deposition. Figure 13 shows the SEM images of the surface and the interface, revealed by phosphoric acid etching after annealing at 450°C, for ICB and conventional vacuum deposited films. Auger Electron Spectroscopy (AES) of the ICB films also showed that the interface was very abrupt, and RBS measurements showed that the crystalline quality remained stable after annealing.[13,14] The electromigration effect has been tested with films of 10 micron width, 1,000 micron length and 400 nm thickness. When they were passed by current at a density of 10^6 A/cm^2 at 250°C, there was no change in resistance after 400 hours of operation, as opposed to sputtered Al films which normally fail at 10^6 A/cm^2 after 20 hours.[15]

The thermal stability of the electrical characteristics was evaluated by fabricating Schottky barriers. The barrier height and the n value are 0.75 eV and 1.17, respectively and the values showed little change after annealing to temperatures up to 550°C.[14] This behavior is in stark contrast to that of the films made by sputtering or evaporation which show changes of more than 0.1 eV in barrier height and 0.1 in n value. The electrical resistivity of 2.7 $\mu\Omega$-cm for the 400 nm thick film is comparable to the bulk value.[16]

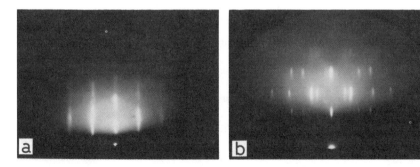

Figure 12: 75 keV RHEED patterns for epitaxial Al films deposited by ICB on Si(111) substrate (a), and Si(100) substrate (b).

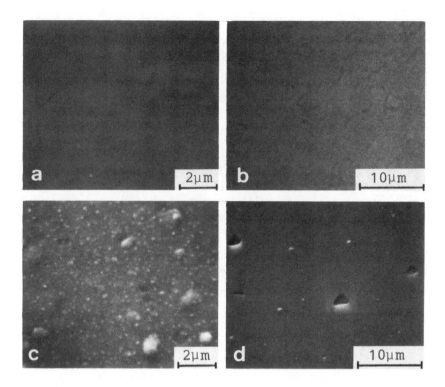

Figure 13: SEM images of surface and interface of Al films on Si(111) formed by ICB deposition and by conventional vacuum deposition after annealing at 500°C for 30 minutes.

On amorphous substrates, for example on SiO_2, preferentially oriented Al(111) is formed. Figure 14 shows the X-ray diffraction intensity ratio, I_{111}/I_{200} as a function of the acceleration voltage. The ratio increased as the acceleration voltage is increased.

Metal-Insulator-Semiconductor Structures

An epitaxial metal-insulator-semiconductor structure was fabricated for the first time by ICB deposition exclusively using CaF_2 as an insulator material. This demonstrates the possibility of constructing a completely epitaxial metal/insulator/semiconductor system, which brings about high performance solid state devices. A CaF_2 film was deposited under a vacuum of 1×10^{-4} Pa produced by an oil diffusion pump. RBS and RHEED observations showed that the CaF_2 grows with an orientation $CaF_2(111)//Si(111)$, $CaF_2[1\bar{1}0]//Si[\bar{1}10]$ (so-called type B orientation). χ_{min} of the grown film was 2.0% on a Si(111) substrate. No cracks or fractures could be seen in the film deposited at 5 kV. The surface roughness of the CaF_2 films on Si(111) and Si(100) has been observed as a func-

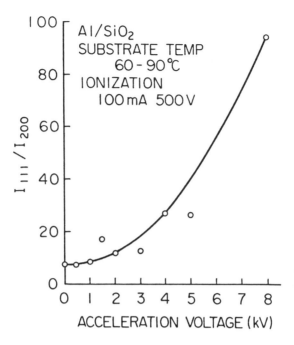

Figure 14: Ratio of X-ray <111> diffraction intensity I_{111} and <200> diffraction intensity I_{200} as a function of acceleration voltage.

tion of acceleration voltage during the deposition. Both films were deposited to a thickness of approximately 300 nm. Usually a rough, columnar structure is grown on Si(100) substrate, whereas films grown on Si(111) are smooth because the (111) CaF_2 surface has the lowest free energy of all orientations. It was shown that the surface roughness of the ICB deposited CaF_2 could be changed by changing the acceleration voltage. Nomarski micrographs of the films have shown that a smooth surface could be obtained with deposition at higher acceleration voltages. The smooth surface has also been verified by the streaky RHEED pattern from Si(111) epitaxial films.[17]

Al films were deposited on the CaF_2 films grown on Si by ICB. RHEED patterns indicated that a good Al crystal grew with (111) parallel to the substrate surface. By taking different RHEED patterns from Si substrate and grown CaF_2 and Al films at several azimuth angles, the epitaxial relation of the stacked film could be determined as: Al(111)//CaF_2(111)//Si(111), Al[$\bar{1}$10]//Si[$\bar{1}$10].

Changes in the morphology of the surface of the Al on CaF_2/Si(111) were examined before and after annealing. Figure 15 shows the SEM images of the film surface. In the film deposited by conventional vacuum deposition and annealed at 450°C, there are annealing hillocks and voids as often observed, while nothing can be seen on the surface of the Al film deposited by ICB and annealed identically.

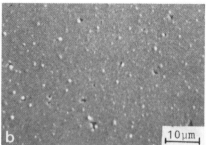

Figure 15: SEM images of the surface of Al films on CaF$_2$/Si(111), postannealed at 450°C for 30 minutes in vacuum. (a) Surface of an Al film deposited by ICB (V_a = 5.0 kV, I_e = 100 mA, and T_s = room temperature) and (b) surface of an Al film deposited by conventional vacuum deposition.

Semiconductors

Single crystalline gallium arsenide has been recently grown epitaxially on GaAs(100) LEC substrates at temperatures below 400°C under residual gas pressures of only 10^{-4} Pa.[18,19] Films ranging from 100 nm to 1 micron thickness were grown over the substrate area of 1 cm x 1 cm at a deposition rate of 15 nm/min. This result shows that ICB could be more suitable for industrial production because of much higher allowable deposition rates for crystal growth, and possibility of using lower substrate temperatures and more versatile doping methods. In ICB deposition of GaAs, it was shown that the sticking coefficient of ICB deposited As is enhanced. The experiment showed that the ionization and acceleration of the Ga beam was necessary to obtain smooth and stoichiometric GaAs films. The RHEED pattern of the film deposited at an acceleration voltage of 1 kV showed a streaked pattern and Kikuchi lines, which indicated the surface is extremely smooth in addition to having excellent crystallinity. RBS channeling spectra showin in Figure 16 demonstrate that there is no measurable difference in the spectrum of the film compared to that of the substrate.

In ICB epitaxy of compound materials, the crystal composition can be controlled by the acceleration voltage during the deposition. Demonstration has been made by $Cd_{1-x}Mn_xTe$ epitaxy on a sapphire (0001) substrate.[20] By using a dual source ICB system, ionized CdTe and neutral MnTe closter beams were produced. When the acceleration voltage of CdTe ICB was changed from 0 to 7 kV, the film composition x varied from 0.58 to 0.86. When MnTe beam was ionized and accelerated, the crystalline quality could be improved but the film composition x did not change. Films deposited by ICB showed sharp dispersion characteristic for Faraday rotation, and the dispersion became much sharper and larger when the crystalline state of the film was improved.

CdTe and PbTe are well known as materials of poor crystallographic quality. To see the effect of ICB, a CdTe/PbTe multi-layer structure was formed on a InSb(111) substrate by using a dual ICB deposition system where CdTe and PbTe beams were used. Deposition was made at a substrate temperature of 250°C with a background pressure of 2×10^{-4} Pa. Twenty periods of CdTe-PbTe bilayers with 15 nm for each layer were deposited on a 150 nm CdTe buffer layer.

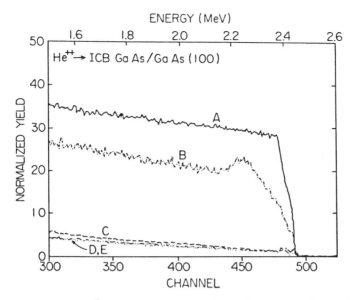

Figure 16: 3.0 MeV He^{++} backscattering and channeling spectra for GaAs films deposited by ICB. (A) Random spectrum, (B) V_a = 500 V, T_s = 500°C (Ga:Neutral), (C) V_a = 1 kV, T_s = 500°C, (D) V_a = 1 kV, T_s = 600°C, and (E) substrate.

Application of high acceleration contributes to form high quality crystalline films with smooth and uniform surfaces in each film layer. When 60 periods of CdTe/PbTe bilayers of thickness of 8 nm and 6.5 nm, respectively, were deposited on Si(111) substrates, optical absorption spectra corresponding to the n = 1 energy level could be observed.[21]

Oxides, Nitrides and Others

Other applications exist where there is a crucial need for dense, adherent films with well-controlled microstructure. One example is the optical coatings industry where multilayer laser mirrors and antireflection coatings frequently display variations in scattering, sensitivity to moisture, and (for laser mirrors) uncontrolled damage threshold. Oxide films for sensors is also an important application field where tight, dense films of special types of crystalline orientation or structure are necessary. Aiming at these applications, ICB techniques for the deposition of TiO_2, SiO_2 and Al_2O_3 have been developed.[3] In this reactive ICB (RICB) methods, reactive gas of oxygen or nitrogen is introduced to a pressure of 10^{-5} to 10^{-4} Torr. At these pressures, no plasma is produced in the deposition chamber.

A set of TiO_2 films was prepared at different acceleration voltages in a range of 1 to 6.6 KV keeping I_e = 300 mA, T_s = 300° to 350°C, P_o = 2 x 10^{-4} Torr constant.[22] Figure 17 shows the change of the crystal structure examined by the x-ray diffractometer. A phase transition from the anatase structure to the rutile structure can be seen by increasing the acceleration voltage to about 3 kV. A change of the structure back to the anatase type occurred as a result of a further increase of the acceleration voltage.

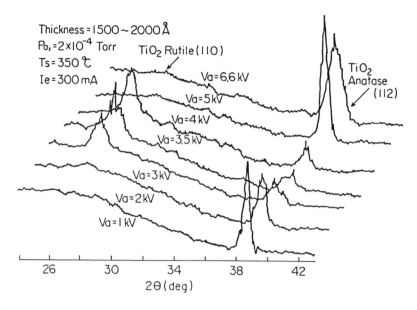

Figure 17: X-ray diffraction traces of TiO_2 films deposited at various acceleration voltages.

Just as the formation of highly oriented inorganic films has been made possible by ICB, the fundamental effects associated with ICB could be exploited to produce high quality organic film, too.[23] Recent researches on anthracene, polyethylene and Cu-phthalocyanine film depositions have proven these possibilities for organic film formation.

CONCLUSIONS

The ICB technology offers low temperature process capabilities especially for the semiconductor industry. Low temperature semiconductor film formation will also require complementary low temperature oxide, nitride, and carbide film formation abilities to complete an overall low temperature approach. Some of the results presented here show the possibility of doing this. In optical coating applications, a low temperature process also offers depositions on a wide variety of substrates. Recent results of organic film formation have shown that even in an organic film it is possible to control the crystal structure. This opens the possibility of producing very strong, chemically stable, light structural materials.

REFERENCES

1. Borel, J.P. and Buffet, J., ed., *Surface Sci.*, Vol. 106 (1981).
2. Frank Trägger and Gisbert zu Putlitz, ed., *Metal Clusters* (Springer-Verlag, Heidelberg, 1986).

3. *Proc. Int. Workshop on Ionized Cluster Beam Technique*, ed., T. Takagi, Research Group of Ion Eng., Kyoto Univ. (1986).
4. Takagi, T., Yamada, I., Kunori, M. and Kobiyama, S., *Proc. 2nd Int. Conf. Ion Sources*, p. 790 (1972) (Oesterreichische Studiengesellschaft für Atomenergie, Vienna).
5. Yamada, I., Takaoka, H., Inokawa, H., Usui, H., Cheng, S.C. and Takagi, T., *Thin Solid Films*, Vol. 92, p. 137 (1982).
6. Frenkel, J., *Kinetic Theory of Liquids*, New York, Dover Publication (1955).
7. Yamada, I., Usui, H. and Takagi, T., *Proc. Int. Symp. Metal Clusters*, Heidelberg, p. 50 (1986) (Ruprecht-Karls-Universität), to be published in *Atoms, Molecules and Clusters* (Springer-Verlag).
8. Thomson, W., *Proc. R. Soc. Edinburgh*, Vol. 7, p. 63 (1870).
9. Takagi, T., Yamada, I. and Sasaki, A., *Inst. Phys. Conf. Ser.*, Vol. 38, p. 142 (1978).
10. Technical Data, Eaton Corporation, 33 Cherry Hill Drive, Danvers, Massachusetts 01923, USA.
11. Takagi, T., Yamada, I. and Sasaki, A., *Thin Solid Films*, Vol. 39, p. 207 (1976).
12. Yamada, I., Inokawa, H. and Takagi, T., *J. Appl. Phys.*, Vol. 56, p. 2746 (1984).
13. Yamada, I., Inokawa, H. and Takagi, T., *Thin Solid Films*, Vol. 124, (1985).
14. Yamada, I., Palmstrm, Kennedy, E., Mayer, J.W., Inokawa, H. and Takagi, T., *Mat. Res. Soc. Symp. Proc.*, Vol. 37, p. 402 (1985).
15. Chen, G., (EATON) Private communication.
16. Casey, D., (GTE) Private communication.
17. Yamada, I., Takaoka, H., Usui, H. and Takagi, T., *J. Vac. Sci. Technol.*, Vol. A4 (3), p. 722 (1986).
18. Younger, P., *J. Vac. Sci. Technol.*, Vol. A3, p. 588 (1985).
19. Yamada, I., Takagi, T., Younger, P.R. and Blake, J., *SPIE*, Vol. 530, p. 75 (1985).
20. Koyanagi, T., Matsubara, K., Takaoka, H. and Takagi, T., *Proceedings of the 9th Symposium on Ion Source and Ion Assisted Technology 1985*, p. 397, Tokyo (IEEJ, Tokyo 1985).
21. Takaoka, H., Kuriyama, Y., Matsubara, K. and Takagi, T., Extended Abstract of the 16th (1984 International) Conf. on Solid State Devices and Materials, p. 423, Kobe (Jpn. Appl. Phys. Tokyo 1984).
22. Fukushima, K., Yamada, I. and Takagi, T., *J. Appl. Phys.*, Vol. 58, p. 4146 (1985).
23. Usui, H., Yamada, I. and Takagi, T., *J. Vac. Sci. Technol.*, Vol. A4 (1), p. 52 (1986).

12

Ion Beam Deposition

John R. McNeil, James J. McNally and Paul D. Reader

INTRODUCTION

Ion bombardment of surfaces has been used in connection with thin film deposition for decades. The amount and type of ion bombardment greatly influences film structure and composition; this in turn determines optical, mechanical and electrical properties of the film. The bombardment can be achieved using a number of techniques, some of which are very simple. However, the highest degree of control of critical parameters (e.g., ion flux, energy, species and angle of incidence) is provided by broad-beam Kaufman ion sources. This is due to the gas discharge of the ion source being separate and removed from the rest of the deposition system. This is not the case with some other ion bombardment techniques used in film deposition. The discussion here is restricted to applications of the broad-beam Kaufman ion source. For discussion of other ion sources, see References 1-3.

The following material is consistent with the general nature of this volume, in that application, or "how-to", aspects of ion beams for thin film deposition are emphasized. References are given to more in-depth technical treatments of several topics. For example, Harper et al[3] give an excellent review of modification of film properties provided by ion bombardment. The first two sections which follow are fairly general, while the section discussing applications is somewhat specific. This is necessary, first because of the backgrounds of two of the authors in optical materials. Second, the approach was taken that it is more beneficial to discuss a few applications in detail rather than many applications in a cursory manner.

OVERVIEW OF ION BEAM APPLICATIONS

The general technology of broad-beam ion sources has been discussed very adequately in the literature (see, for example, Reference 1). The emphasis of this section is a brief description of the construction of ion sources and to cite specific examples of operational characteristics that impact the design of processing tasks. The typical ion source consists of an enclosure containing electrodes, magnetic fields, ion accelerating grids and emitters suitably arranged to sustain electron-bombardment ionization of a working gas. The arrangement of these components has been widely varied by many experimenters and analytically investigated by even more. The results invariably support the local mythology of what a "good" ion source should be. The many configurations and concepts incorporated in both commercially available and custom built ion sources tend to overwhelm users upon initial exposure to the technology.

Categories of Kaufman Ion Sources

The following comments do not deal with the full range of ion source possibilities. They merely attempt to describe some of the design trends in an overview, using details only to illustrate extremes of the trends. Ion sources are like most apparatus, in that when they are highly optimized for a given process, they tend to sacrifice performance in other areas. This tendency leads to grouping of types of sources that conform to the process parameters of interest.

Sputtering sources, those used to remove the maximum amount of material from a target in the minimum time, are one example. In general, beam uniformity is not of great importance. The source is usually operated at 1 to 2 kV. It is important that the beam strike the target and that the source not produce contaminants. The lower the gas flow rate, the better. Frequently, operation with reactive gases is desirable – or even necessary – to control stoichiometry. These parameters usually dictate small to moderate diameter (2 to 15 cm) sources operating at close to maximum conditions. When large numbers of substrates need to be coated, larger diameter sources are used, and the source-to-target and target-to-substrate distances are increased. This geometric scaling results in inverse square losses, with a corresponding reduction in coating rate. Linear sources have been applied to this task to partially eliminate the losses associated with increasing distance. The linear sources permit reduced geometries. While the number of workpieces coated is also decreased, there can be a net gain in throughput in a significantly smaller working volume.

Another example is etching sources, which have significantly different characteristics. They generally operate at voltages below 1 kV, with some even optimized for voltages between 50 and 200 eV. Beam uniformity is normally critical, with a typical specification of ±5% across a significant fraction of the beam. Etching sources are characterized by highly developed grid systems which seek to present a high current-density beam of acceptable flatness to the workpiece. The workpiece is frequently mounted on a fixture which produces a complex, non-integer motion to compensate for the remaining non-uniformities of the beam. Etching sources are frequently required to operate with gases that react with the workpiece to enhance etch rate or selectivity.

Peculiar ion source shapes occasionally appear among etching sources because of the need to be consistent with the geometry of the workpiece or that swept by the fixturing. Beams and sources shaped to segments of circles or annular segments are available. Linear sources are used to sweep, in a pushbroom fashion, around cylindrical fixtures or along continuous rolls of material (e.g., webs).

Similar to the etching sources are sources used for cleaning workpieces prior to deposition or used to simultaneously bombard coatings during the deposition process (enhancement sources). These sources usually have less uniformity requirements. For example, if approximately 50 A of material is removed to clean the workpiece, then a large percentage variation in this amount can be tolerated. Relatively small sources can be used for the cleaning task, but they are generally operated at maximum performance to minimize cleaning time. The voltage and current density used for cleaning are sometimes determined by the material to be cleaned. As an example, a web of plastic materials passing under a high performance linear source moves at a rate such that the web is in the beam for about one second. If the web motion is interrupted for 3 seconds, the web is melted. Further discussion of ion beam cleaning is given in a later section.

Enhancement sources provide ion bombardment to a surface being coated with material produced by magnetron, thermal, electron-gun, or ion sputter techniques. Enhancement sources are typically operated at low beam energies and current densities. This energy input into the growing film modifies the physical structure and growth characteristics of the film. It is possible to alter the crystalline phase of some films. Enhancement sources are frequently required to operate with reactive gas to control the stoichiometry of the film. Some arrangements permit constant exposure of the substrate to the beam, while other arrangements allow the substrate to pass through the beam occasionally due to planetary rotation. The effects of the ion bombardment might not be the same for the two situations.

The point of this discussion is to make the reader aware of the wide variety of potential tasks that can be accomplished using broad-beam ion sources, and that the ion sources can be optimized for various tasks. Frequently, the optimization from one specialized task to another can be accomplished by something as simple as a grid change. Other optimization can require extensive hardware changes and include significant variations in control systems and operating procedure.

A good example of major change is the method of producing the ionizing electrons for ion source operation. The most popular method of producing ionizing electrons is by thermionic emission. One method to achieve this is the use of a bare refractory wire as an emitter. Another method makes use of a low work-function material dispersed through a tungsten slug that is heated inside an inert-gas-buffered tube. The latter device is known as a hollow-cathode. The simple thoriated tungsten wire has given good service as an emitter for many investigators. However, the wire is sputtered by the ions in the discharge, and this limits the cathode lifetime. The wire emitter also presents a problem when operated with reactive gases. In addition, some users find problems with both the heat and photon flux on sensitive workpieces, as well as contamination due to evaporated and sputtered filament material. The hollow-cathode emitter can be used for both the source discharge chamber emitter and the neutralizer emitter to solve these lifetime problems. Use of the hollow cathode does impose heat-up and cool-down constraints prior to venting the work chamber, but for long runs these factors become less critical. The hollow-cathode requires a supply of inert gas, usually argon or xenon, to buffer the low work-function insert. Any gas may be used in the source discharge chamber, giving wide flexibility to the use of the hollow-cathode ion source.

Radio frequency discharges can be used to produce the ions that are accelerated to produce the beam; this has been under investigation since the early 1960's. Recent advances in efficient RF circuit components have allowed the control and impedance matching problems to be reduced such that reasonably reliable operation can be achieved. While the RF sources eliminate the hot emitter from the chamber, many electronic design, control and stability problems remain to be investigated.

Operational Considerations

Ion sources that produce beams using electrostatic acceleration present certain potential problems. The beam is extracted and accelerated by significant DC potentials (200 to 2,000 V typically). The work chamber contains a moderate current ($\geqslant$1A) beam and is filled with a dilute, conducting plasma. Care must be taken to electrostatically shield the high potential leads to the source. Electrical breakdown is possible between closely spaced leads, and this is enhanced by the dilute plasma. Some ion sources are designed in which gas flow lines are isolated from ground potential near the ion source. Other sources have a conductive gas line at high voltage that passes through an insulator in the vacuum system wall and is electrically isolated outside the chamber. This line is often overlooked as a high voltage electrode.

The pressure inside the gas line goes from a few psi above atmospheric pressure at the gas bottle, to approximately 10^{-3} Torr in the ion source plasma region. Screens and metal wool pads are frequently used to provide large area surfaces on which ions can recombine. This prevents electrical breakdown in the gas feed tubes.

Oxidation can occur if a hot ion source is vented at atmosphere too soon after operation. The oxide films on the various electrodes can become sufficiently insulating to prevent source operation on a subsequent pump down. Similarly, sputtering or etching of materials in a reactive atmosphere can produce insulating coatings. To avoid problems due to this, regular cleaning of the ion source is important. This can be accomplished mechanically, for example, using bead blasting, or chemically using acids to remove thin film material.

ION BEAM PROBING

The applications of ion beams to coating discussed below require knowledge of the ion flux at the target. In some applications such as ion assisted deposition (IAD) in which the target is the substrates being coated, this is very important; knowing just the total beam current is not sufficient. In this section, probe construction and operation are first discussed, and then use of the probe as related to film deposition is discussed. Only applications of the probe to measure particle flux are discussed here. Probes can also be used to determine charged particle energy. Typically ions from a Kaufman ion source are nearly monenergetic; they have an energy spread of approximately 10 eV centered about the anode potential (i.e., beam energy) for the dual grid extraction arrangement.[3] Ion energy spread in the beam from a source with a single grid arrangement is somewhat larger.[3] However, because beam energy characteristics are fixed by discharge conditions and are rarely measured by the user, this probe technique will not be discussed in further detail.

The ions and neutralizing electrons in a beam can be considered a weakly ionized plasma, or gas discharge. Plasma probe technology can be applied to monitor the beam conditions. For details, many references on plasma probes are available.[4] Only the relevant technology will be discussed here. A simple arrangement which is widely used is the planar Faraday probe illustrated in Figure 1a. The probe is connected to an electrical vacuum feedthrough with a shielded conductor. The shield is connected to the part of the probe which encloses the probe element. Note that the back of the probe element is also shielded to avoid ion current contribution from this region. The probe element diameter is typically 5 to 10 mm, and it is recessed into the shield enclosure by approximately the same amount. This maintains approximately the same sheath dimensions and effective probe area with minor variations in probe bias voltage that will be used. Insulators of alumina can be used to separate the probe element and the shield enclosure. This separation should be as small as convenient, with approximately 0.5 mm being typical. A simpler, though perhaps less accurate version of probe which can be easily constructed is shown in Figure 1b. In this case, the probe element is mounted on a sheet of insulating material such as mica, and this is fastened to a piece of metal (e.g., Al) plate. Again, the back of the probe and the electrical conductor are shielded to avoid extraneous sources of ion current. The entire probe arrangement, whichever form is used, should be constructed such that it is easily disassembled for cleaning.

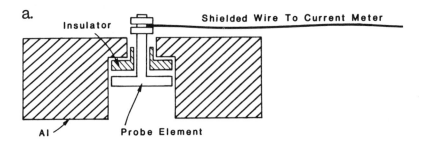

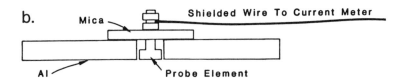

Figure 1: Two arrangements of a Faraday probe to monitor ion beam current; (a) more accurate version and (b) simple version. Note, in both arrangements the back side of the probe should be shielded (e.g., with Al foil) to prevent ion current contribution from this region.

Outside the vacuum system, the probe can be biased, and probe current can be measured with a meter connected in series with the arrangement. If the probe is to be used to measure ion current density, and therefore to calculate ion flux, the bias should be approximately -30 V. This potential repels electrons from the beam, and the measured current is primarily due to the ions. If the probe is perpendicular to the beam, then the ion current density is approximately equal to the measured probe current divided by the probe element area. The ion flux (Γ) is calculated by dividing the current density by the ion charge q. For example, if the probe current is 25 μA and the probe element is 1 cm in diameter, the current density and ion flux are given by:

$$J = I_p/A_p$$
$$= 25/\pi(0.5)^2 \; \mu A/cm^2$$
$$= 31.8 \; \mu A/cm^2$$

and

$$\Gamma = J/q$$
$$= 31.8 \times 10^{-6}/1.6 \times 10^{-19}$$
$$= 19.9 \times 10^{13} \; ions/cm^2\text{-sec}$$

The latter could, for example, be used for comparison with the flux of thin film molecules at the substrate.

As Kaufman discusses,[3] there are two sources of error in the probe current measurements described above. First, ions striking the probe surface can cause secondary electrons to be emitted from the probe. Because the probe is biased negatively, the electrons are repelled, and this constitutes a current which is additive with that of incident ions. Because the probability of secondary electron emission from the metal surfaces involved (e.g., Al, stainless steel, etc.) at the ion energies employed is of the order of 0.1, there is little error introduced by this effect. The second source of error involves secondary (charge exchange) ions produced by the high energy ions of the beam. The probe will collect both types of ions, and therefore give an erroneously high current reading. Unless the background pressure in the vacuum chamber is abnormally high, this effect is generally negligible except at the outer edges of the ion beam. At this point, the density of the secondary ions can be relatively high compared to that of the beam ions, and one might think the extent of the beam is greater than it actually is.

When this arrangement is used while depositing conducting film materials, no special precautions are necessary. The probe element should remain insulated from the rest of the probe body, and the electrical characteristics of the probe element should remain constant. The probe might be moved into the ion beam just prior to substrate cleaning, sputtering or whatever the application, in order to check the beam condition. The probe would then be moved out of the beam.

When insulating materials are being deposited, care should be exercised to prevent the probe element from becoming coated. This would change the probe characteristics to provide incorrect indications of ion beam current density. During coating the probe might be moved to an area where it is shielded from coating material, or the probe might have a shield attached. If the beam is to be monitored during coating, some deposition on the probe element is inevitable, and the probe should be later cleaned.

SUBSTRATE CLEANING WITH ION BEAMS

Substrate condition prior to coating is extremely important for proper film adhesion; in addition, it has an influence on subsequent film growth characteristics. The general topic of substrate cleaning is very diverse and will not be treated here. What will be discussed is the application of ion beam apparatus as the final step of cleaning a surface immediately before film deposition. Results of ion beam cleaning and the improved adhesion of films which this can provide, such as Au on BK-7 glass, are discussed in Reference 5. For a general treatment of substrate cleaning techniques, see for example, Reference 6.

Bombardment of a surface with ions and electrons to clean a surface has been commonly used for many years primarily by employing a glow discharge in the deposition system. However, with this so-called glow discharge technique, one has very little idea of what species are actually bombarding the substrate, nor the energy and flux of the bombarding species. Use of ion beam apparatus to provide the particle bombardment affords more control of the important system parameters, such as ion species (reactive or inert), flux and energy. The same parameters are not only known, they are in large part independently controllable. This, combined with proper beam probing techniques discussed previously, make the process very useful for cleaning and bombardment during early stages of film growth. With a neutralized beam, conducting and insulating substrates can be treated. Ion beam cleaning obviously represents no added equipment to the deposition system if ion assisted deposition (IAD) is employed. See a later section.

The processes which occur on the substrate due to ion and electron bombardment include a number of things. This includes desorption of adsorbed water vapor, hydrocarbons and other gas atoms. Adsorbed gases are very efficiently desorbed at low energies (~300 eV). Chemisorbed species and possible occluded gases are sputtered. If the bombarding ion is an oxygen species, chemical reactions with organic species on the surface can result in compounds which are more volatile and hence more easily removed.[6] For the levels of ion flux and energy employed, and for the small time required for cleaning, the substrate temperature rise is usually negligible, and this has little effect on substrate outgassing. Ion bombardment causes substrate defect production which in general is beneficial for subsequent film nucleation and adhesion.

Typically, the level of ion flux at the substrates used during precleaning is several 10's of $\mu A/cm^2$. In this application, uniformity of current density is not as important as it is for other applications such as IAD. Because of this, small ion sources are useful provided they have sufficient total beam current. Relatively low energy (~300 eV) ions are preferable to higher energy ions to minimize substrate sputtering, while still being sufficiently energetic to produce the effects mentioned above. A mixture of Ar and O_2 can be used as the gas species in the ion sources. As an example, consider a 30 µA, 300 eV beam from a 2.5 cm diameter ion source which is directed at substrates approximately 30 cm away. At the substrates, depending somewhat upon the vacuum chamber pressure, the peak current density would be approximately 160 µA/cm^2; the width of the beam (FWHM beam current density) would be approximately 17 cm. If the deposition chamber dimensions permit, the ion source should be placed farther away to provide a larger, more uniform beam at the substrates. If there is only single-axis rotation of the substrates, the ion source might be placed directly below the center of the substrates; if it is not directly below the substrates, then the beam pattern is simply spread over a larger area. In either case, it might be advantageous to direct the beam off-center of the substrate fixturing if the fixturing is much larger than 17 cm in diameter. If the fixturing involves planetary rotation, then the beam should be directed to a region that will ultimately provide bombardment of all substrates. In this case, the ion beam is obviously not on a particular substrate continuously; it is "time-shared" between all substrates. This, however, does not present the potential problems that are discussed in connection with IAD of substrates in planetary arrangements.

To estimate the time required for cleaning substrates, one should consider the amount of material which is sputtered. This is determined in part by the sputter yield of the substrate material and that of the unknown species on the substrate. Sputter yields for many materials are well characterized for various incident ions and ion energies; typical values range from 0.1 to 3 atoms/ion for inert gas ions of approximately 0.5 to 2 keV energy.[7] If a sputter yield (S) of 0.1 atoms/ion is assumed, the thickness (L) of the layer of material removed in 5 minutes (t) due to a beam of 50 µA/cm^2 current density (J) is given by:

$$L = SJVt/q$$

$$= (0.1)(50 \times 10^{-6})(27)(300)(10^{-16})/1.6 \times 10^{-19}$$

$$= 25 \text{ Å}$$

Here V is an assumed volume of the sputtered atoms, taken to be 27 Å^3, and q is the charge of the ion in Coulombs. The factor of 10^{-16} in the equation is included to express the layer thickness in Angstroms. This calculation illustrates how several atomic layers of material can be quickly removed from the substrate to leave it atomically clean. This should actually be considered a maximum estimate of substrate material removal. The adsorbed species on the surface are, in general, much easier to remove and have a large effective sputter yield. If the beam is "time-shared" as described above in connection with planetary fixturing, then one must estimate the fraction of time F the substrates are in the ion beam; in the expression above, t would become t/F.

There are four considerations one should be aware of in applying ion bombardment, regardless of the technique, to clean substrates. The first has to do with the influence of ion bombardment on the substrate microroughness. If the substrate is polycrystalline and composed of relatively large crystallites, the microroughness of the substrate might increase significantly if the substrate is ion bombarded for a very long time. This is due to the crystals in the substrate sputtering at different rates and therefore different amounts being removed, depending upon their orientation to the incident ions. See Figure 2. For example, Cu substrates have demonstrated an increase in rms roughness from approximately 25 Å to nearly 50 Å when 1200 Å of the substrate was removed using ion beam sputtering.[5] This amount of material removal is somewhat severe in terms of substrate cleaning, but it illustrates the point. On the other hand, amorphous or very small-grained polycrystalline substrates as well as single crystalline materials do not present a problem. For example, glass substrates (e.g., fused silica, zerodur, and BK-7) have been ion beam milled several microns deep with no noticeable increase in roughness, starting with substrates as smooth as 7 Å rms roughness.[8] Electroless Ni films have been examined in this same manner, and they do not display an increase in roughness with ion milling.

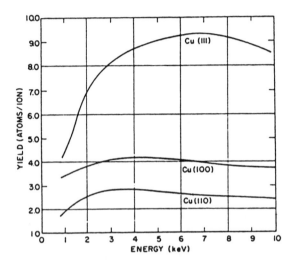

Figure 2: Sputter yield S of three planes of Cu bombarded by normal-incidence Ar^+ ions.[7]

The second consideration when applying ion precleaning is that the stoichiometry of the top few atomic layers of substrates composed of molecular species can be altered. This is due to the ions preferentially sputtering one atomic species from the top molecules, leaving the surface rich in the other atomic species. The extent of this altered layer production is naturally minimum with minimum ion energy (for energies less than approximately 2 keV) and flux, while the impact of it depends upon the application of the coated substrate. In the case of fused silica and other glass substrates, the effect is not noticeable in optics used in routine as well as severe applications, such as high energy laser optics.[9] Similarly, fused silica samples which were ion milled several microns and examined with an ellipsometer displayed no change in optical constants. In this case, the effects of the altered layer were either negligible or exposure to atmosphere made them so. The situation in the case of other substrate materials might not be as favorable as that for glass materials, and one should be aware of potential problems.

A third consideration when applying ion beam precleaning involves the potential damage to a crystalline substrate surface, such as semiconductor materials. It is very difficult to obtain a clean semiconductor surface and maintain the bulk properties at the surface; atoms in the crystal lattice structure are dislocated due to the ion bombardment. A general guideline is to reduce the ion energy to the lowest level possible if substrate damage is a concern. The dislocation energy of an atom is typically a few 10's of eV. The sputter yield of adsorbed molecules should be sufficient at the reduced ion energy to provide the desired cleaning with minimal substrate damage.

A fourth consideration in applying ion precleaning involves contaminating the surface of the substrate being cleaned with the material from the fixturing holding the substrate. Slight contamination, especially around the periphery of the substrate is unavoidable. When possible, the fixturing material might be constructed of material which is compatible with the coating. For example, in deposition of oxides for optical applications in which a high background pressure of O_2 is present, the fixturing might be constructed of Al; in this case, the contaminant from the fixturing could be present in the form of Al_2O_3 which is harmless in most cases.

APPLICATIONS

Two general arrangements are most popular for application of the ion source to thin film coating. Ions from the source can be directed at a target which is sputtered, and the sputtered material is deposited as a thin film. This is termed ion beam sputter deposition (IBS). Second, ions from the source can be directed to the substrate which is being coated with material generated by some independent technique. This is termed ion assisted deposition (IAD). In both cases, the intent is to provide the thin film adatoms on the substrate with increased energy and therefore more mobility prior to nucleation. This leads to the modifications in film properties mentioned previously. Two or more ion sources can be used simultaneously to provide both IBS and IAD. These two processes are discussed in detail in this section.

Ion Beam Sputtering

Aspects of ion beam sputtering which will be described here first include a brief review of sputtering parameters and results. This is necessary, for example, to describe some of the advantages and disadvantages of ion beam sputtering compared to other forms of sputtering. Detailed treatments of sputtering are given in References 1, 7 and 10. Next, several aspects of ion beam apparatus and other experimental details are described. Last, a summary of film properties provided by ion beam sputtering is given.

Aspects of Sputtering. When an energetic ion or neutral atom strikes a surface, many processes can occur. One of these is the ejection of atoms of the surface target material, and this is known as sputtering. Of primary concern in sputtering is the sputter yield (S), in atoms/ion, and the relation of S to parameters involved such as ion and target species, ion energy, ion incident angle and target condition. In addition, it is very important to have an idea of the average ejection energy of sputtered atoms, as this has a significant influence on resulting thin film properties. Similarly, the angular distribution of the sputtered atoms is of importance in determining film deposition rate and uniformity. The following is a summary of relevant aspects of sputter yield and sputtered particle properties taken from References 1, 7 and 10.

Sputter Yield.

(a) S has a threshold level of approximately 10 to 30 eV for most ion/target combinations. It increases exponentially with ion energy for energies up to 200 or 300 eV (see Figure 3) and increases approximately as E^x, where $0.5 \leqslant x \leqslant 1$ for ion energies up to several keV.[1] At higher ion energies, S decreases most likely due to implantation.

(b) S increases for many target elements with increasing mass of rare gas ions. For example, for 500 eV ions incident on Ta, S is 0.01, 0.28, 0.57, 0.87 and 0.88, respectively, for He, Ne, Ar, Kr, and Xe.[11]

(c) S increases with ion angle of incidence to the target, up to angles of approximately 45 to 60 degrees. The increase can be as great as a factor of two or more; at larger angles of incidence, S decreases. See Figure 4.[1,10]

(d) S can be different for different crystal orientations of the same material.

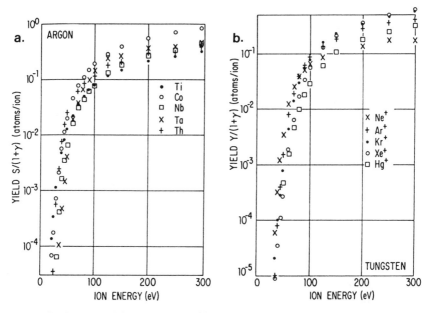

Figure 3: Sputter yield S versus ion energy; (a) shown for several materials with Ar^+ bombardment, and (b) for W bombarded by different ion species.[10] Reproduced with permission from Maissel and Glang, *Handbook of Thin Film Technology*, McGraw-Hill (1970).

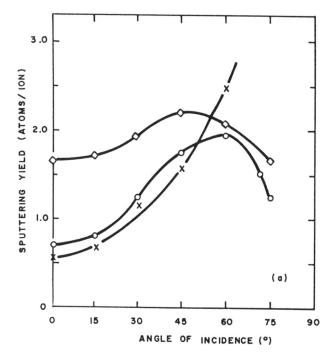

Figure 4: Effect of angle of incidence on sputter yield for ◊ Gold, ○ aluminum, and x photoresist.[30] Reproduced with permission from Vossen and Kern, *Thin Film Processes*, Academic Press (1978).

Sputtered Particle Energy.

(a) Average ejection energies increase with incident ion energy up to roughly 0.5 to 1.0 keV, above which the energy of the sputtered particle increases little.

(b) For the same incident ion species, materials with lower sputter yield tend to have higher ejection energies.

(c) For the same target material, average ejection energies tend to increase for heavier bombarding ions.

(d) Average ejection energy tends to increase with ion angle of incidence.

(e) There is not a significant dependence of the sputtered atom ejection energy upon crystal orientation of the target.

Angular Distribution of Sputtered Atoms. In general, the angular distribution of sputtered particles is peaked in a forward direction, at an angle roughly comparable to that of the incident ions. A cosine distribution has been used to describe this, although several factors have an influence. In particular, the target condition is very important. For example, the target can be single crystal or polycrystalline, and there will be an effect on the distribution pattern. In addition, the distribution changes in the case of a new target that is being used for the first few hours. A target of solid material will provide a distribution pattern different from that of one made of pressed powder. In addition, if the target is arranged at a different angle with respect to the ion beam, the distribution pattern would be different as well. Energy of the incident ions has been observed to have an effect on the sputtered particle distribution.

From what has been described, it is apparent that the best approach to characterize the distribution pattern in applications in which it is critical is to directly measure it. For example, sample substrates might be located at various positions relative to the target and then coated. Film deposition rate and uniformity could be determined from coating thickness measurements using a spectrophotometer or other types of interferometry.

Advantages/Disadvantages of Ion Beams for Sputtering. In sputter applications, ion beams provide control of ion energy, flux, species and angle of incidence. These parameters are not only controllable over a wide range, they can be controlled nearly independently of each other. This represents significant advantages compared to other forms of sputtering. In addition, the gas discharge of the ion source is contained and separate from the rest of the deposition system. From the discussion above, it can be seen that having an ion energy of 1 to 2 keV is desirable for higher sputter yield and higher sputtered atom energy. In the case of magnetron sputtering, for example, the incident ion energy is at most equal to the energy provided by the cathode fall voltage (typically 600 V), and it could be less due to charge transfer collisions in the region above the cathode. In addition, the ion energy can not be controlled over a wide range, and it is closely coupled to the magnetron discharge current. Being able to arrange the sputter target at an angle to the incident ion beam is advantageous both in terms of sputter yield and sputtered atom energy, as well as allowing flexibility in the experimental arrangement. In the case of magnetron sputtering or other processes in which the sputter target is in contact with the gas discharge, there is no control of the ion incident angle. Ion beam sputtering allows greater flexibility in target material and composition than in other forms of sputtering. This is discussed below in connection with targets. In addition, the species in the ion beam can be easily and accurately controlled by adjusting the flow of one or more types of gas through the source; reactive (e.g., O_2, N_2, etc.) or inert ion species can be present in practically any ratio desired. This is not the case with other forms of sputtering. The low background pressure present during ion beam sputtering might lead to less gas inclusion compared to other processes. However, the deposition geometry should be such that the substrates are not bombarded with energetic neutralized ions which are reflected from the target. In this case, the incorporation could increase significantly due to a larger sticking coefficient of the energetic atom.

An obvious disadvantage of the application of ion beams in general is the added expense and complexity of the coating process. However, neither of these is severe. Another disadvantage is that it is more difficult to scale the ion beam sputter process compared to magnetron or some other sputter process. Geometries of the ion beam sputter arrangement must sometimes be small in order to provide reasonable deposition rates, and this might be a problem.

Aspects of Ion Beam Sputter Apparatus. Figure 5 illustrates the apparatus used in ion beam sputtering. Two ion sources are shown in the figure; one is used

Ion Beam Deposition 377

for sputtering and the other is used for bombarding the substrate prior to coating or during deposition (i.e., IAD). Also shown is a probe to characterize the beam in the vicinity of the target, as well as the other shutters typically found in a deposition system.

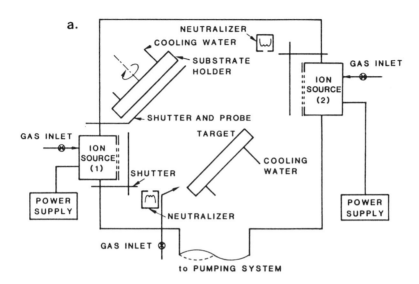

Figure 5: Ion beam sputter apparatus. Although the vacuum system shown here is designed specifically for IBS, the same ion beam apparatus could be added to an existing vacuum system.

Vacuum Chamber. Chamber size requirements are influenced little by the addition of a small ion source; more elaborate arrangements such as illustrated in Figure 5 require a specially designed system. Typical size of the arrangement shown in the figure is roughly 50 cm in diameter. Similarly, vacuum pumping requirements are not influenced greatly with the addition of a small ion source; typical added gas load is 2 to 4 cm^3/min for a 2.5 cm diameter ion beam source. A 6-inch diffusion-pumped or cryogenically-pumped system will typically operate in the 10^{-5} to low 10^{-4} Torr region with this gas load. Incorporating large ion sources into a system may require additional pump capability.

The arrangement of the sputter target and the substrates is often as illustrated in Figure 5. A typical separation between the ion source and the target is 30 cm. The target is arranged at an angle of approximately 45 degrees to the ion beam to optimize the sputter process as discussed above. The energetic neutralized ions which are reflected from the target can, in some situations, be detrimental to the film if they are allowed to bombard the substrate. This is the case, for example, when high quality oxide films are being deposited using Ar ions to sputter a target and with an O_2 partial background pressure.[12] To minimize this effect, the substrates are arranged as shown in Figure 5. Although the neutralized ions do not all reflect in a specular sense, it is believed they are predominantly directed in this direction.

Ion Sources. For ion beam sputtering it is usually desirable to deliver as much ion current to the sputter target as possible, and this dictates the use of large ion sources (e.g., 10 to 15 cm diameter). Wire filament cathodes and neutralizer can be used which are made of W or Ta typically 0.010 to 0.015 inches in diameter. Both have lifetime limits of several (one to over ten) hours. Lifetime of the neutralizer is increased if it is placed near the edge of the ion beam and not through the center. This can be tedious to achieve, as the beam profile and therefore the position of the neutralizer are somewhat dependent on operating conditions (pressure, beam energy, etc.). Both the cathode and neutralizer lifetime are greatly reduced if a reactive gas is used in the ion source. In addition, the W or Ta material can contaminate the target and be deposited in the film at a low level. In place of the wire cathode, a hollow cathode discharge device can be used to supply electrons to the main discharge of the ion source. Although this operates very well with inert gas, operation with O_2 has not been satisfactory. Similarly, the wire neutralizer can be replaced with an auxiliary discharge device, a plasma bridge neutralizer, to supply neutralizing electrons to the beam. This must remain outside of the beam to avoid being sputtered and contaminating the target material. Another approach is to construct the housing of the plasma bridge neutralizer of a material which is compatible with the target and film.

Targets. Target size should be sufficient to insure that none of the beam passes the target and sputters other material. In optical coating arrangements in which large ion sources are used, target size is typically 20 cm wide and 30 cm long. Placement of the target and the influence of this on the distribution of sputtered atoms has been discussed previously.

In the case of optical coatings, elemental targets of Si, Ti, Ta, etc., are often preferred because of the higher sputter yields and therefore large deposition rates provided, compared to targets composed of the oxide materials. In this particular situation, Ar ions are often used to sputter the target; a second ion source, possibly with a single-grid arrangement to provide low energy ions, is often used with O_2 to improve film stoichiometry by IAD. Deposition rates of 1 to 2 Å/sec can be achieved.[12] Similar techniques can be used in the case of nitride film deposition. This arrangement is illustrated in Figure 5.

In general, there is no restriction placed on a target used in ion beam sputtering; the target can be an insulator since the ion beam can be neutralized. Sputter targets are often water-cooled due to the incident power (e.g., 0.25 A at 1,500 V in vacuum).

A technique which has been used to ion beam sputter compounds of materials involves placing pieces of one target material on a second target material.[1,13] The film composition depends on the relative sputter yields, the relative areas of the targets which are sputtered and the relative sticking coefficients of the atoms at the substrate.

Properties of Ion Beam Sputtered Films. Harper[1] has described the properties of thin films deposited by ion beam sputtering. These include superior adhesion and denser film structure, both attributed to the higher energy of the sputtered particles. The improved adhesion was observed for films produced with and without ion precleaning. The superconducting transition temperature, T_c, of ion beam deposited transition materials has been found to increase over bulk values for some metals and decrease for others. The effects were correlated with the size of the sputtering ion used and attributed to the influence on lattice expansion.

More recently, ion beam sputtered oxides have been found to exhibit optical properties which are superior to those of films of the same material deposited using e-beam evaporation. This is illustrated in Figure 6[12] which shows the refractive indices of several film materials deposited using ion beam sputtering as described above. The total losses (absorptance, scatter and transmittance) from high reflectance mirrors constructed from these films is 100 parts per million or less, as determined by performance in a ringdown laser arrangement. The films are stable with temperature cycling to 300°C in atmosphere for several hours.

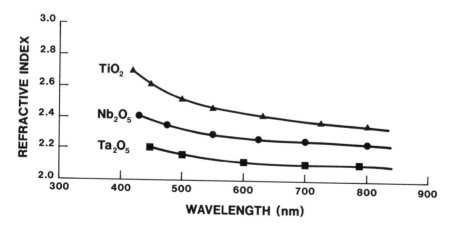

Figure 6: Values of refractive index obtained from examining thin film materials which were ion beam sputtered.[12]

380   Thin-Film Deposition Processes and Techniques

The internal stress of ion beam sputtered films is compressive, and it is generally greater than that of evaporated films of the same material. In the case of some oxide materials, the compressive stress is sufficient to cause film delamination from the substrate.[12] The amount of film stress changes with the location of the substrates relative to the sputter target. This suggests the dependence might be related to substrate bombardment during film deposition.

Ion Assisted Deposition

Ion assisted deposition (IAD) employs an ion source to direct a beam of ions at the substrates during deposition. The source of film material could be an evaporation source or a sputter arrangement, such as the dual ion beam deposition arrangement mentioned in the previous section. An obvious advantage to utilizing IAD is that the process can be easily incorporated into an existing vacuum deposition system.

In this section, we describe the equipment used to deposit thin films using IAD, the advantages inherent in using IAD and the limitations in applying IAD to deposit thin films. This technique has been employed in the fabrication of multilayer optical coatings, in the deposition of compound thin films at reduced substrate temperature and to modify film properties during deposition. We review results of employing IAD and illustrate how control of the ion energy and flux can be utilized to tailor the properties of thin films.

Equipment. A schematic of a vacuum deposition system configured for IAD is shown in Figure 7.[9] The operation of the source of thin film material is totally independent from the ion source. The Kaufman ion source provides a monoenergetic, neutralized ion beam which is directed at the substrates being coated. This configuration has a number of advantages over conventional plasma configurations used to provide ion bombardment of the substrate. Three advantages provided by ion beams include the independence of the bombardment process from the deposition process, the inherent control provided by the ion beam and the ability to accurately measure all critical deposition parameters. An additional feature concerning IAD is that substrate cleaning can be readily accomplished prior to deposition by controlling the appropriate gas species through the ion source.

As illustrated in Figure 7, the gas is introduced through the ion source into the deposition chamber. IAD is inherently a low pressure process (10^{-5} to 10^{-4} Torr) that permits a great deal of flexibility in deposition configuration. Any number of gas species can be introduced, depending upon the complexity of the gas manifold. Neutral background gas can also be introduced into the vacuum chamber through another inlet if desired.

As described in a previous section, it is essential that the ion beam is properly probed prior to deposition. For some materials, film properties are sensitive to bombarding ion flux levels; examples are given later in this section. As illustrated in Figure 7, an ion current probe is located just below the substrate. The probe platform is mounted on an aluminum rod that is attached to a rotary vacuum feedthrough. The probe is rotated into position immediately below the substrates to be cleaned or coated. After the ion flux value is measured, the probe is rotated away from the substrate area to minimize coating of the probe. Not shown in the Figure 7 is a shield attached to the probe platform which protects the probe from film material during deposition. This shield prolongs the lifetime of the probe when depositing insulating film materials.

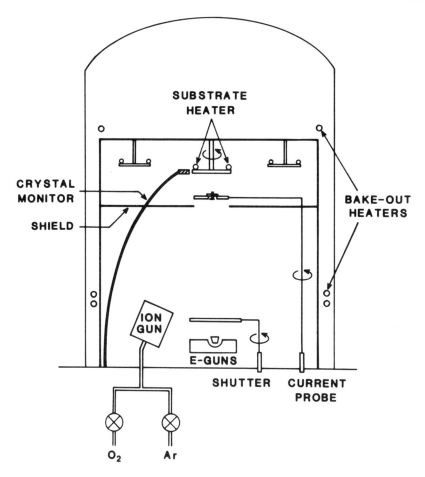

Figure 7: Schematic of an IAD deposition system.

An important parameter in depositing thin films using IAD is the ratio of the ion-to-molecule arrival rates. Proper beam probing allows accurate determination of the ion flux. In the case where substrates are continuously exposed to the ion beam, modification of film properties occurs on a continual basis. One method to determine the flux of arriving thin film molecules is measurement of the film deposition rate using a crystal monitor. In the data presented below, film properties are plotted versus ion beam current density and for a constant deposition rate. Thus the ratio of ion-to-molecule arrival rates is a variable. As illustrated in Figure 7, the monitor can be located very close to the substrates. The monitor must be calibrated to determine an accurate tooling factor for different film materials and for various ion source conditions.

If the substrate fixturing involves planetary rotation, the monitor cannot be located immediately adjacent to the substrates. In this arrangement, determining the proper flux of ions and film molecules is more difficult. Usually the substrates are exposed to the ion beam for only a fraction of time, F. The effect of the ion bombardment during the fraction of time F may not be the same as if the substrate were bombarded continuously. This might be the case even if the average levels of ion flux are kept constant, for example, by increasing the ion current density by a factor of 1/F. Therefore, when using IAD with a planetary rotation geometry, care must be exercised in selecting the ion beam parameters which result in the desired thin film properties. This is best achieved by first coating sample substrates with single layer films (if multiple layered structures are ultimately desired). For each film material, optimal ion beam current density and energy can be determined by examining samples deposited under different conditions.

Another important consideration when applying IAD is the respective arrival directions of the bombarding ions and the film molecules. As illustrated in Figure 7, when employing a fixed-position substrate geometry with an ion source located off-axis, it is important that substrate rotation be employed. The rotation of the substrates is necessary to minimize anisotropic properties in the films due to the different arrival directions of the ions and the molecular species.

Proper selection of the discharge voltage is important. The value should be high enough to sustain an appropriate discharge in the source, but not too high a value which will cause an increased production of doubly ionized ions. The doubly ionized species will accelerate to twice the beam energy, and this reduces the level of control the operator has over the ion beam parameters. In addition, high values of discharge voltage decrease the lifetime of the cathode filament in the source. For a 10 cm source, cathode filament lifetime for typical IAD conditions (500 eV, 30 $\mu A/cm^2$) operating in oxygen is approximately 4 hours for a 0.015 inch diameter tungsten wire. The lifetime for a 0.020 inch diameter tantalum wire is only 1 to 1.5 hours under similar conditions.

As mentioned in a previous section, it is necessary to provide an equal number of electrons to neutralize the ion beam, especially when bombarding an insulating target. This is often accomplished using a thermionic neutralizer filament which emits the appropriate electron current into the beam. The value of this neutralizer current is made approximately equal to that of the beam current. Note that the ions are not each neutralized, but rather the net charge flux to the target is zero. When using this simple arrangement for beam neutralization, consideration must be given to problems arising due to neutralizer filament sputtering. Another technique for beam neutralization is the use of a plasma bridge neutralizer arrangement (see a previous section).

Procedures. The thin film material can be preheated prior to coating and, as illustrated in Figure 7, a shutter can be employed to shield the chamber from the vapor material. Immediately upon completion of the cleaning process, film deposition can commence. The time between the two processes is limited to the delay, if any, associated with changing the gas species flowing through the ion source.

It is critical that the ion flux is stable and accurately measured before deposition begins. The gas discharge of the ion source should be established and the desired beam parameters (ion energy, flux and neutralizer current) set at the ion source power supply. When the ion flux is measured by the probe, an accurate record of its value and all ion source parameters should be made. When the probe is rotated out of position, deposition can commence. During deposition, the beam energy and current must be monitored on the power supply to ensure stable ion source operation. At the completion of deposition, the ion flux should be checked by rotating the probe back to the original measurement position.

Examples of Applications of IAD to Optical Coatings. Early work utilizing a separate ion source to bombard evaporated SiO with 5 keV oxygen ions during deposition was examined by Dudonis and Prannevicius.[14] They investigated the effects of ion flux on film stoichiometry and found that for increasing flux values, the O/Si ratio approached two. They also simultaneously bombarded with oxygen ions during the evaporation of aluminum and measured changes in film resistivity as function of ion flux; resistivity changes for 10^{17} were obtained using the ion bombardment. Changes in resistivity are an extremely sensitive measure of film stoichiometry. Cuomo, et al,[15] obtained superconducting NbN films by bombarding evaporated Nb with N_2^+ during deposition. They also reported a change in stress from tensile to compressive for evaporated Nb films bombarded with 100 to 800 eV Ar^+ during deposition. Hoffman and Gaerttner[16] bombarded evaporated Cr films during deposition with 11.5 keV Ar^+. They reported changes in stress from tensile to compressive and attributed the results to modifications in film microstructure. In these studies, an important parameter affecting film properties was the ion-to-atom arrival rate ratio. This ratio can be used to normalize results from various experiments and is discussed in more detail later in this section.

Recently, IAD has been applied in the production of optical coatings. Netterfield and Martin have studied the effects on the properties of ZrO_2 and CeO_2 films bombarded with O_2^+ and Ar^+ during deposition.[17-19] They reported increased packing densities, reduced permeability to water and changes in film crystal structure. Bombardment with Ar^+ produced changes in film stoichiometry. Allen applied IAD to obtain increased values of refractive index in TiO_2 films.[20] The increase was larger at 300 eV bombardment than at 500 eV. McNeil et al[21,22] examined the effects of 30 and 500 eV oxygen ion bombardment on the properties of TiO_2 and SiO_2 films deposited using TiO and SiO starting materials. They reported improvements in film properties were obtained with 30 eV bombardment in the case of ambient substrate temperatures. In particular, they obtained films with improved stoichiometry. Al-Jumaily, McNally and McNeil[23-25] examined the effects of ion energy and flux on optical scatter and crystalline phase in Cu, TiO_2 and SiO_2 films. They reported reduction in optical scatter and crystal phase transitions for bombarded films. The modifications in Cu films were obtained with 500 eV Ar^+ bombardment. McNally et al [26,27] reported on the effects of oxygen ion bombardment on the properties of Ta_2O_5, Al_2O_3, TiO_2 and SiO_2 films. Increases in the values of refractive index were reported for 200, 300, 500 and 1,000 eV bombardment. The increase was found to be directly dependent on the ion flux, which is consistent with the comments in a previous section regarding the importance of proper beam probing. Changes in film stoichiometry and crystal phase were also reported.

As a final example of recent IAD work, the deposition of protective coatings on sensitive substrates at low temperatures has been demonstrated.[28] Films of MgF_2 were successfully deposited on heavy metal fluoride glass substrates at 100°C temperature. The films were bombarded with 300 eV Ar^+ during deposition. Conventionally evaporated MgF_2 films are soft when deposited at this substrate temperature, while the IAD MgF_2 films were hard and robust. This represents a significant advantage in using IAD to deposit films.

IAD Results. Some sample results from applying IAD to produce coatings are given in this section. McNally et al[9] produced Ta_2O_5 coatings which were electron-beam evaporated at a rate of 0.30 nm sec^{-1} with oxygen backfill pressure of 1.0×10^{-4} Torr. The coatings were deposited onto heated substrates (275°C), and were bombarded with oxygen ions during deposition. The transmittance specta for two Ta_2O_5 coatings are given in Figure 8. The curve labeled J=0 corresponds to a coating deposited with no ion bombardment; the curve labeled J=5 is for a coating bombarded during deposition with 500 eV O_2^+ at a current density of 5 μA cm^{-2}. The curve for the ion assisted coating contains larger differences in transmittance extrema than the curve for the conventional e-beam evaporated coating. These larger differences indicate a larger value of refractive index (n) for the ion assisted coating. Good film stoichiometry was obtained for the ion bombardment conditions employed, as indicated by an absence of measurable absorption at wavelengths for which the film is multiple half-wave in optical thickness down to 340 nm. Examination of the reflectance spectra for these coatings indicated an absence of any refractive index inhomogeneity.

The values of n (at λ = 400 nm) for Ta_2O_5 coatings bombarded with 200, 300 and 500 eV oxygen ions are plotted in Figure 9 as a function of O_2^+ current density. The error bars indicate the uncertainty in the index measurements. The values increase from 2.16 for the coating deposited without bombardment to maximum values of 2.25, 2.28 and 2.19 for films bombarded with 500, 300 and 200 eV O_2^+, respectively.

The increase in the values of n with increasing O_2^+ current density indicates that ion bombardment during deposition modifies the growth of film columnar microstructure resulting in film densification. The results indicate that the coatings bombarded with 300 eV O_2^+ had larger values of n than those bombarded with 500 eV O_2^+. The results in Figure 9 indicate that only minor densification occurred for coatings bombarded with 200 eV oxygen ions during deposition. A similar dependence of refractive index on bombarding ion energy has been reported for ion assisted CeO_2 films.[19]

The current density value (for a fixed ion energy) at which the maximum n occurs is termed the critical value. The results in Figure 9 illustrates that film index values decrease for ion current densities greater than the critical values. The decrease in index may be explained as a result of degradation in film stoichiometry, creation of closed isolated voids, or oxygen incorporation into the films. The decrease was largest for bombardment with 500 eV ions and least for 200 eV ions in the case of Ta_2O_5. This energy dependent decrease is consistent with the dependence on ion energy of the average ion penetration depth and preferential sputtering yield. Similar results for which the values of refractive index decrease for current densities greater than the critical values have been reported for ion assisted ZrO_2 films[18] and CeO_2 films.[19]

The coatings bombarded during deposition at oxygen ion current densities up to approximately the critical values exhibit good optical characteristics. For higher levels of bombardment, the optical absorption of the coatings increase. In Figure 10, values of extinction coefficient (k) for Ta_2O_5 coatings (300 nm thick) bombarded with 500, 300 and 200 eV oxygen ions are plotted as a function of O_2^+ current density. The values of k were calculated at λ = 400 nm. The error bars indicate the uncertainty in the measurements. The dashed line across the bottom at $k = 2.0 \times 10^{-4}$ indicates the level below which the values of k were too small to be regarded as reliable due to the minimum sensitivity of the measurement technique.

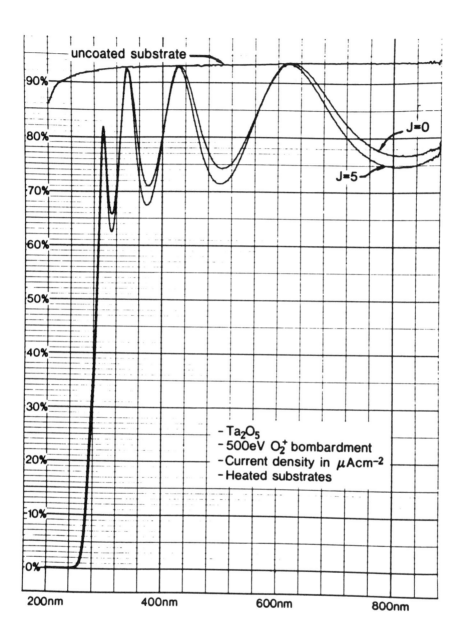

Figure 8: Transmittance spectra of two Ta_2O_5 coatings, one deposited without ion bombardment (J = 0) and the other deposited with ion bombardment (J = 5 $\mu A/cm^2$).

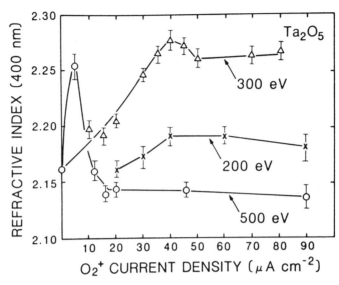

Figure 9: Values of refractive index for Ta_2O_5 coatings deposited under different levels of O_2^+ current density and energy.

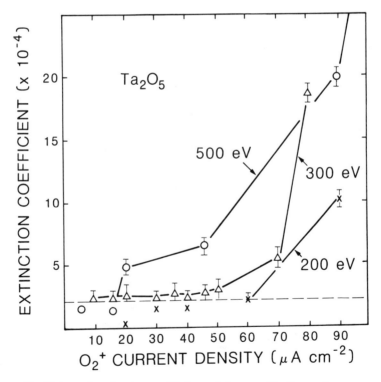

Figure 10: Values of extinction coefficient of the same Ta_2O_5 coatings represented by Figure 9.

As illustrated in Figure 10, film optical absorption increased with higher levels of oxygen bombardment. The most probable mechanism for this is the preferential sputtering of oxygen in the Ta_2O_5 molecule. Preferential sputtering would result in oxygen-deficient layers continuously integrated into the coatings as deposition occurs. Values of k for coatings bombarded at a fixed current density were the lowest for 200 eV O_2^+, higher for 300 eV and highest for 500 eV O_2^+. This result is consistent with increasing preferential sputtering yields for higher energy ions. This damage mechanism has been observed in other IAD films.[18] This highlights the requirement for proper beam probing to avoid compositional change by preferential sputtering.

The environmental stability of optical coatings is in large part limited by the porosity of the film microstructure. Figures 11 and 12 are the transmittance curves for two Ta_2O_5 coatings exposed to humidity testing. The spectra in Figure 11 are for a coating deposited onto a heated silica substrate with no ion bombardment (J=0). The spectra in Figure 12 are for a coating bombarded during deposition with 300 eV O_2^+ at a current density of 20 μA cm^{-2} The curve labeled "AIR" is the spectrum measured after the coating has been removed from the vacuum chamber and exposed to the ambient atmosphere. The curve labeled "POST HUMIDITY" is the spectrum for the same coating remeasured after exposure to 97% relative humidity at 35°C for 6 hours.

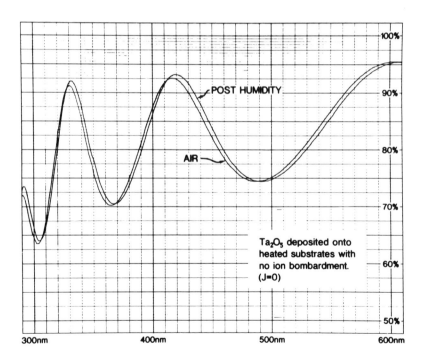

Figure 11: Transmittance spectra of a Ta_2O_5 coating deposited without ion bombardment subjected to humidity test.

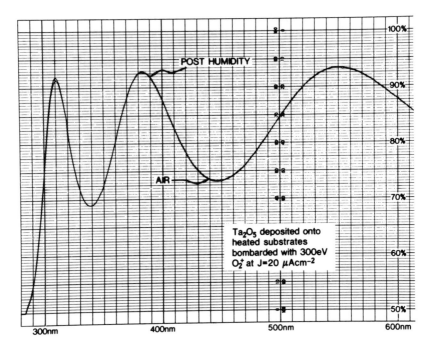

Figure 12: Transmittance spectra of a Ta_2O_5 coating bombarded during deposition with 300 eV O_2^+ at $J = 20$ μA cm^{-2} subjected to humidity test.

The curves in Figure 11 illustrate a spectral shift to longer wavelengths of 1% for the coating deposited without bombardment. The shift is most likely due to water adsorption into the microvoids in the film microstructure. This increases the effective index value of the coating and, in turn, increases the optical thickness (nt). Figure 12 is typical of the results obtained for IAD coatings. No spectral shifts within the measurement precision of the spectrophotometer are observed. Similar results have been reported for IAD ZrO_2 and TiO_2 coatings.[17,18]

IAD can affect intrinsic stress in films. Changes in stress for Ta_2O_5 coatings have been reported.[9] The stress was measured interferometrically. The stress was computed from measured substrate bending, and because film thickness was much less (factor of 1000) than the substrate thickness, the elastic constants of the coating were not required.

Figure 13 illustrates the values of film stress plotted versus oxygen ion current density. The triangles and circles represent stress values for coatings bombarded during deposition with 300 and 500 eV O_2^+, respectively. The results indicate film stress was compressive and increased for increasing levels of ion bombardment.

There have been a number of studies in which the effects of ion bombardment on stress in metal films were examined. Cuomo et al[15] reported a change in stress from tensile to compressive for evaporated Nb films bombarded with 100 to 800 ev Ar^+ during deposition. They attributed the changes in stress to modifications in film microstructure and incorporation of argon. Hirsh and Varga[29] measured stress relief in evaporated Ge films bombarded with 100 to 300 eV Ar$^+$ during deposition. The stress changed toward compressive values. Hoffman and Gaerttner[16] bombarded evaporated Cr films during deposition with 11.5 keV Ar$^+$. They reported changes in stress from tensile to compressive and attributed the results to modification in film microstructure. Thus, the results illustrated in Figure 13 are consistent with the trend toward increasing compressive stress for increasing ion bombardment.

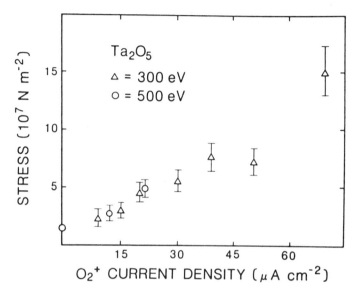

Figure 13: Values of stress of Ta_2O_5 coatings for films deposited with different levels of O_2^+ current density and energy.

IAD has also been used to produce Al_2O_3 optical coatings.[9] The Al_2O_3 coatings were electron-beam evaporated at a rate of 0.40 nm sec^{-1} with oxygen backfill pressure of 1.0 x 10^{-4} Torr. The coatings were deposited onto heated substrates (275°C), and were bombarded with oxygen ions during deposition.

The values of n (at λ = 350 nm) for Al_2O_3 coatings bombarded with 300, 500 and 1,000 eV oxygen ions are plotted in Figure 14 as a function of O_2^+ current density. The values increase from 1.64 for coatings deposited without bombardment to maximum values of 1.70, 1.68 and 1.68 for films bombarded with 100, 500 and 300 eV O_2^+, respectively.

The increase in the values of n for increasing levels of O_2^+ current density is similar to Ta_2O_5 results (Figure 9) and indicates that ion bombardment during deposition results in film densification. The results indicate that the Al_2O_3 coatings bombarded with 1,000 eV O_2^+ had larger values of n than those bombarded with 500 and 300 eV O_2^+. These results illustrate that the effects of ion bombardment on the values of n are material dependent. The ion energy at which the largest value of n occurred for Ta_2O_5 was 300 eV, yet for Al_2O_3 it was 1,000 eV.

The results in Figure 14 illustrate that values of n decreased for ion current densities larger than the critical values. The decrease was largest for bombardment with 1,000 eV ions and least for 300 eV ions. This energy dependent decrease is consistent with the results for IAD Ta_2O_5 coatings (Figure 9).

Attempts to measure the values of extinction coefficient for the Al_2O_3 coatings (400 nm thick) were limited due to the minimum sensitivity of the measurement equipment. All computed values of k were less than 2.0×10^{-4}. The low values of k calculated for all the Al_2O_3 coatings indicate that for the conditions examined, preferential oxygen sputtering is not a dominant mechanism for Al_2O_3 as was the case for Ta_2O_5 (Figure 10). This result again illustrates that the effects of ion bombardment on film properties are material dependent.

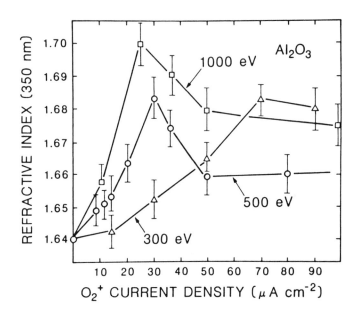

Figure 14: Values of refractive index of Al_2O_3 coatings deposited under different levels of O_2^+ current density and energy.

Application Summary

The two configurations for using ion beams for thin film deposition (Figures 5 & 7) each contain their distinct advantages and disadvantages. Direct comparison of the two configurations to determine which provides the "best" coatings is difficult. The decision as to which configuration to employ must be made with consideration of the end application, the existing deposition system (if any), total costs and the issues discussed in previous sections. In this section, a discussion summarizing the advantages and disadvantages of IBS and IAD is given with the emphasis on listing information to aid the users in deciding which configuration would better serve their needs.

The improvements in the properties of thin films produced using IBS are detailed in a previous section. Similar improvements in the properties of IAD films are given in a previous section. It appears that each technique results in thin films of similar quality. Thus, a decision based on film quality would need to be made on a case-by-case basis.

A direct advantage of IAD is that the technique is readily implemented with the addition of an ion source apparatus to an existing vacuum deposition system. This offers significant savings in both time and money. On the other hand, IBS may require more modifications to an existing deposition system, or the purchase of a totally dedicated one. Either choice requires a major investment. Another advantage of IAD is the ease in which the technique can be scaled to large geometries. Scaling the IBS process is more limited than the IAD process by the size of ion sources available. A number of trade-offs among the deposition parameters (deposition rate, uniformity, etc.) exist, and there must be considered before choosing which technique would best achieve the desired results.

CONCLUDING COMMENTS

In this chapter, a general description and two specific applications of ion beams for thin film coating have been described. A common advantage of ion beam techniques is the degree of flexibility and control provided compared to other techniques that incorporate a gas discharge. Ion beam sources will continue to be incorporated into deposition arrangements and applied to other materials to provide films with improved properties.

Acknowledgements

The authors would like to thank S.J. Holmes for a critical reading of the manuscript and for helpful comments. In addition, two of the authors (J.R.M. and J.J.M.) gratefully acknowledge the contribution of our coworkers, K.C. Jungling, S.R. Wilson, A.C. Barron, G.A. Al-Jumaily and F.L. Williams.

REFERENCES

1. Harper, J.M.E., *Thin Film Processes*, J.L. Vossen and W. Kern, eds., pp. 175-206, Academic Press, New York (1978).
2. Wilson, R.G. and Brewer, G.R., *Ion Beams with Applications to Ion Implantation*, New York:Wiley (1973).
3. Harper, J.M.E., Cuomo, J.J., Gambrino, R.I. and Kaufman, H.R., in: *Ion Bombardment of Surfaces: Fundamentals and Applications*, O. Auciello and R. Kelly, eds., pp. 127-162, Elsevier, Amsterdam (1984); see also, Kaufman, H.R. and Robinson, R.S., *AIAA Journal*, Vol. 20, p. 745 (1982).
4. Chen, F.F., *Plasma Diagnostic Techniques*, R.H. Huddlestone and S.L. Leonard, eds., pp. 113-200, Academic Press, New York (1965).
5. Herrmann, W.C., Jr. and McNeil, J.R., *Proc. of SPIE*, Vol. 325, p. 101 (1982).
6. Pulker, H.K., *Coatings on Glass*, pp. 52-62, Elsevier, Amsterdam (1984).
7. Vossen, J.L. and Cuomo, J.J., in: *Thin Film Processes*, J.L. Vossen and W. Kern, eds., pp. 11-73, Academic Press, New York (1978); see also, Chapman, B., *Glow Discharge Processes*, pp. 177-284, Wiley, New York (1980).

8. McNeil, J.R. and Herrmann, W.C., Jr., *J. Vac. Sci. Technol.*, Vol. 20, pp. 324-326 (1982), see also, McNeil, J.R., Wilson, S.R. and Barron, A.C., "Ion Beam Milling for Rapid Optical Fabrication," to be presented at the Workshop in Optical Fabrication and Testing of the Optical Society of America, Seattle, WA (October 21-25, 1986).
9. McNally, J.J., Jungling, K.C., Williams, F.L. and McNeil, J.R., *J. Vac. Sci. Technol.*, Vol. A5, p. 2145 (1987).
10. Wehner, G.K. and Anderson, G.S., *Handbook of Thin Film Technology*, L.I. Maissel and R. Glang, eds., pp. 3.1-3.38, McGraw-Hill, New York (1970).
11. Wehner, G.K., Report No. 2309, General Mills, Minneapolis, MN (1962).
12. Holmes, S.J., Personal communication.
13. Masaki, S. and Morisaki, H., *Proc. 10th Symp. on ISIAT '86*, pp. 427-431, Tokyo (1986).
14. Dudonis, J. and Prannevicius, L., *Thin Solid Films*, Vol. 36, p. 117 (1976).
15. Cuomo, J.J., Harper, J.M.E., Guarnieri, C.R., Yee, D.S., Attanasio, L.J., Angilello, J., Wu, C.T. and Hammond, R.H., *J. Vac. Sci. Technol.*, Vol. 20, p. 349 (1982).
16. Hoffman, D.W. and Gaerttner, M.R., *J. Vac. Sci. Technol.*, Vol. 17, p. 425 (1980).
17. Martin, P.J., Macleod, H.A., Netterfield, R.P., Pacey, C.G. and Sainty, W.G., *Appl. Opt.*, Vol. 22, p. 178 (1983).
18. Martin, P.J., Netterfield, R.P. and Sainty, W.G., *J. Appl. Phys.*, Vol. 55, p. 235 (1984).
19. Netterfield, R.P., Sainty, W.G., Martin, P.J. and Sie, S.H., *Appl. Opt.*, Vol. 24, p. 2267 (1985).
20. Allen, T., *Proc. Soc. Photo-Opt. Instrum, Eng.*, Vol. 325, p. 93 (1982).
21. McNeil, J.R., Al-Jumaily, G.A., Jungling, K.C. and Barron, A.C., *Appl. Opt.*, Vol. 24, p. 486 (1985).
22. McNeil, J.R., Barron, A.C., Wilson, S.R. and Herrmann, W.C., Jr., *Appl. Opt.*, Vol. 23, p. 552 (1984).
23. Al-Jumaily, G.A., McNally, J.J., McNeil, J.R. and Herrmann, W.C., Jr., *J. Vac. Sci. Technol.*, Vol. A3, p. 651 (1985).
24. Al-Jumaily, G.A., Wilson, S.R., McNally, J.J., Jungling, K.C. and McNeil, J.R., *J. Vac. Sci. Technol.*, Vol. A4, p. 439 (1986).
25. Al-Jumaily, G.A., Wilson, S.R., McNally, J.J., McNeil, J.R., Bennett, J.M. and Hurt, H.H., "Influence of Metal Films on the Optical Scatter and Related Microstructure of Coated Surfaces," *Appl. Opt.*, Vol. 25, p. 3631 (1986).
26. McNally, J.J., Al-Jumaily, G.A. and McNeil, J.R., *J. Vac. Sci. Technol.*, Vol. A4, p. 437 (1986).
27. McNally, J.J., Al-Jumaily, G.A., Wilson, S.R. and McNeil, J.R., *Proc. Soc. Photo-Opt. Instrum. Eng.*, Vol. 540, p. 479 (1985).
28. McNally, J.J., Al-Jumaily, G.A., McNeil, J.R. and Bendow, B., *Appl. Opt.*, Vol. 25, p. 1973 (1986).
29. Hirsh, E.H. and Varga, I.K., *Thin Solid Films*, Vol. 69, p. 99 (1980).
30. Melliar-Smith, C.M. and Mogab, C.J., *Thin Film Processes*, J.L. Vossen and W. Kern, eds., p. 524, Academic Press, New York (1978).

13

Plasma and Elevated Pressure Oxidation in Very Large Scale Integration and Ultra Large Scale Integration*

Arnold Reisman

INTRODUCTION

According to a recent article,[1] by the end of the 1980's one can expect to see 1 μm lithographic dimensions in full scale manufacturing. At present, every major company has under advanced development, and in some instances even in manufacturing, 1.25 μm lithographic processes for MOSFETs in NMOS and CMOS implementations. In general, however, the state of manufacturing art in advanced manufacturing employs 1.5 to 2.0 μm lithography. Such design-rules enable the fabrication of dynamic RAM, DRAM, at densities of 250,000 bits per chip in an area about 1 cm on an edge. This is equivalent to 500,000 devices/chip, since each MOSFET consists of a transfer device and a storage capacitor. According to Reference 1, Very Large Scale Integration, VLSI, covers the regime 2^{16-21} devices/chip, roughly 64K-2M devices/chip, while the Ultra Large Scale Integration, ULSI, regime covers the range from 2^{21-26} devices/chip. We see therefore that the state of manufacturing art is already well within the VLSI regime. Oxide thicknesses at 2 μm design-rules are of the order of 35 to 40 nm for the gate insulator, and some 400 to 450 nm for the field oxides.

At 1 μm design-rules, we may expect approximatey 1 M DRAM bits in a chip some 8 to 10 mm on an edge, with gate insulators some 22.5 nm thick, and field oxides <350 nm. At 0.5 μm dimensions, the same area should accommodate 4 M DRAM bits (8 M devices), and utilize 10 nm gate insulators, and field insulators about 300 to 350 nm thick.

Junction depths will shrink from approximately 500 nm at 2 μm design-rules to as little as 200 nm, or less at 0.5 μm design-rules. While this scaling is in itself

---

* Supported in part by SRC Contract 83-01-040, and Ceramics and Electronic Materials Program NSF Contract # DMR-8502388

a formidable task, an even greater challenge is the equivalent scaling of tolerances associated with these dimensions. Thus, a + or -50 nm tolerance from a nominal value at 2 μm will have to be held to a + or - 12.5 nm tolerance at 0.5 μm design-rules to maintain performance scaling in circuits. Similar considerations apply to logic chips where device densities are a factor of perhaps 6 to 8 smaller.

In a MOSFET fabrication process, two general steps impede achievement of the required scaled tolerances in junction depths. The first of these is ion implantation impurity activation, and the concomitant removal of ion implantation damage. The second is that involving oxidation, particularly of the field (isolation) oxide. A third process, namely post metal annealing poses a problem of a somewhat different nature. With the increased use of processes involving ionizing radiation, e.g., X-ray and electron beam lithography, the gate insulator of a MOSFET is subjected to as much as 10^8 rads SiO_2.[2]

Such exposure generates volts of threshold shift due to the formation of charged defects. In addition, equivalent volts of potential threshold shift, due to the generation of neutral centers which can trap electrons during device use create a potential reliability problem. Not all of these generated defects can be removed readily using conventional hydrogen containing ambient atmosphere annealing at elevated temperatures. As a consequence, threshold time zero, and long term threshold tolerance control pose potential problems as device dimensions are decreased.

The fundamental relationship which impacts the ability to scale dimensions and tolerances using constant processing temperatures is given by equation 1:

(1) $$x = (Dt)^{1/2}$$

In equation 1, x is the average distance an atom with diffusion constant D would move in a time t.

If we include junction drive-in plus all additional high temperature thermal treatments, and assume a value of $D = 10^{-12}$ cm^2/sec, and solve for Δt with $\Delta x = 12.5$ nm (1.25×10^{-6} cm), we find that the allowable excess time at temperature (approx. at 1000°C) is only about 50 sec. This means that this is the time tolerance for the cumulative set of processes involved which subject the device wafers to a temperature of approximately 1000°C. The actual time tolerance will, of course, depend on the magnitude of D, and could be an order of magnitude greater or smaller. In any event, this is why there is an increasing preoccupation with lower temperature processing of semiconductor wafers. In addition, during the formation of the field oxide using conventional thermal oxidation procedures, the resulting structure exhibits a "so-called" bird's beak topographical feature. This occurs because of the lateral growth of oxide under the field oxidation mask resulting in a cross section reminiscent of a bird's head and beak.[3] This encroachment process results in loss of valuable "real estate" to an extent as much as 30%, and has led to the development of esoteric fabrication approaches to minimize the problem.[4]

In order to ameliorate the time parameter constraint on tolerance control, it is evident that the simplest approach conceptually is to decrease processing temperatures when possible. This concept cannot, however, be employed to remove ion implantation damage, and rapid thermal annealing techniques, RTA, are being explored to economize on the "thermal budget" by minimizing ramp-up and ramp-down steps. This approach is an alternative to lowering processing temperatures based on reducing processing time-at-temperature. Where applicable, "assisted processes" involving the use of a plasma which are based on changing the chemistry of the process are being explored to enable reduction of processing temperatures. Plasma enhanced CVD, PECVD, for example, is used extensively in semiconductor manufacturing for silicon nitride deposition at temperatures <300°C.[5]

In oxidation, RTA may be of value in gate insulator formation, but is probably not useful for the growth of field oxides. "Assisted-process" alternatives to conventional oxidation approaches, involve the use of either elevated pressure or a plasma. The first of these is based on the argument that the oxidation process at the silicon dioxide-silicon interface is governed by an equation of the type depicted in equation 2:

(2) $$R = kp^n$$

where R is the rate of oxidation, k is the specific rate constant, p is the partial pressure of oxygen, and n is the order of reaction.

If p is increased, the rate of this oxidation process increases. The second method, generally in the form of a plasma anodization, attempts to counteract diffusion limitations with the assistance of field enhanced motion of ions through the growing oxide.

While both processes have received attention in semiconductor applications for perhaps two decades, neither has achieved widespread acceptance in manufacturing. Safety and equipment design considerations have limited the use of elevated temperature oxidation processes to 20 atmospheres in western nations, and only eight atmospheres in Japan. However, there are commercial systems available for elevated pressure semiconductor processing which is a big plus. As far as plasma oxidation processes are concerned, however, there does not exist a commercially manufactured tool for use in semiconductor technology, although anodization processes have been shown to have broad potential for metal oxidation applications.[6-10]

Removal of radiation induced defects using elevated pressure annealing in nonexplosive hydrogen containing ambient atmospheres has been shown to be a commercially viable approach,[11] but here again, there does not exist a manufacturing tool. It may be argued on theoretical grounds that radiation damage poses less of a problem as insulator thicknesses decrease. However, it should be remembered that tolerances must be scaled in terms of allowable device threshold shifts. In any event, the path is clear, and should the need arise, a manufacturing tool could be developed quickly.

The following will delve into the history and current status of elevated pressure and plasma assisted processes, focusing on the oxidation of silicon. There has in recent times been considerable work on plasma assisted nitridation of silicon. A review of this area is not possible here because of space limitations. However, when there is some unusual aspect of a nitridation "assisted-process," such as an interesting equipment variation, such work will be discussed. The emphasis in the following will be on experimental techniques, and the resulting structural and physical properties of the resulting oxides.

PLASMA ASSISTED OXIDATION PROCESSES

It is evident from an examination of the literature that detailed understanding of the mechanism(s) of plasma assisted oxidation is still in a primitive state, and that the reported results are noteworthy, more for the conflicting opinions and data presented than for any consistency.[12] This is not surprising, since even with nonassisted oxidation processes the kinetics are not resolved, particularly for the early stages of oxidation.

Equipment for conducting plasma oxidations is quite varied. Gaseous discharges may be created by ionizing the gas using either dc, or ac (rf and microwave frequencies) excitation. In dc approaches, the electrodes used for creating the discharge must be in contact with the oxidizing gas. In ac approaches, the system may be "electrodeless," employing capacitive, and/or inductive coupling of the high frequency energy to the gas through the walls of the system. Alternating current approaches may also employ electrodes contacting the oxidizing gas, as in the dc case. With internal electrodes, however, the likelihood of contamination due to electrode evaporation and sputtering is greater than in an electrodeless system. Electrodeless system designs may also lead to contamination, due to sputtering from the chamber walls by the energetic ions generated. Such a case has been described in detail,[13] where it was shown that at pressures below 10 mTorr, SiO_2 can be sputtered from the wall of a cylindrical fused silica reactor, and deposit on the silicon substrate.

DC techniques may employ "cold," hollow or hot cathode approaches. The latter uses a cathode heated by a filament behind it, and requires very low voltages (<50 V) during operation.[14] These low voltages do not generate energetic ions, so wall sputtering is not a problem. Electrode material contamination is, however, potentially severe.

In addition to the several forms of gas excitation, oxidations may be conducted using biasing potentials between the sample which contacts an anode, and a noncontiguous cathode, or without intentional electric field assistance, using electrically isolated or "floating" samples. The first of these approaches is termed plasma anodization, and is by far the most commonly employed method. Biasing is employed, because without it, it has been claimed[15] that oxidation appears to be self limiting in samples immersed in oxygen plasmas. The biasing electrodes, of course, increase the possibility of contamination. In all reported instances, except one,[16-18] independent of the method for gas excitation, and independent of how intentional biasing is achieved, or even when biasing is not intentionally applied, oxidation is observed to occur on the sample surface facing the plasma. This feature is undesirable since the growing oxide is subject to intense radiation damage, causing high levels (ca. $10^{13}/cm^2$) of charged and neutral defects, as well as like amounts of interface states, all of which require annealing at 1000°C or higher for removal which defeats to a degree, depending on the time-at temperature, the use of "assisted" processing.[19-21]

Finally, the other major viariation in plasma oxidation equipment is how many wafers can be processed at a given time. For the majority of cases, where anodization techniques are employed, only a single sample can be oxidized at a time, because each sample requires its own electrode. Several systems have been described, however, which enable plasma oxidation (or nitridation) of a number of silicon wafers at one time, either in biased or electrically isolated mode.[18,22-26]

Figure 1 depicts the general design of a microwave discharge system, using internal dc biased electrodes for anodization, based on the general features described in References 15, 27 and 28. The discharge may be excited in oxygen at a pressure between 0.2 to 1.5 Torr using 2.45 GHz microwaves, providing between 300 and 1,000 W CW (nominal value 300 W) for acceptable growth rates, e.g., 600 nm in 2 hours at <300°C with >50 V dc bias). The biasing electrodes have voltages impressed between them, ranging between 30 to 90 V which is below the excitation voltage of the gas when the electrodes are 20 to 25 cm apart.

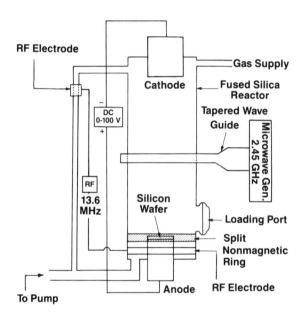

Figure 1: Microwave DC biased anodization system.

An interesting feature of the system described in Reference 15 is the use of a 27 MHz rf sputtering capability in conjunction with the microwave discharge to enable simultaneous partial sputtering of the growing oxide. This feature is also shown in Figure 1.

It enables sputtering over a wide range during the growth process. An even more interesting feature described in Reference 27 is the use of a split nonmagnetic band of metal around the silica tube at the height of the silicon wafer. This band is shown in place in Figure 1 also. With this band, equivalent growth rates are obtained with and without dc biasing of the sample as seen from Figure 2.

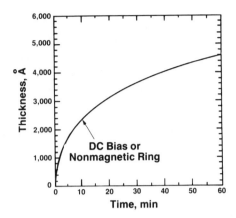

Figure 2: Thickness-time data using DC bias or nonmagnetic ring.

Without the band, it was found that oxide growth in self limiting at about 400 to 500 nm.[15,27] In Reference 15, anodization was found to follow a parabolic thickness-time dependency (equation 1), while in References 27 and 28 it was not. This deviation was assumed to be due to concomitant unintentional sputtering of the oxide during growth. In addition, Reference 15 reported no microwave power dependency, but did see a temperature dependency. Films were nonuniform in all instances, even when 1 cm diameter wafers were employed, and were found in Reference 28 to be highly "pinhole" defected. In addition, Reference 28 reported that the introduction of water vapor into the system reduced growth rates dramatically to less than half, and References 27 and 29 concluded that negative ions cannot be responsible for plasma oxidation, because in the absence of biasing voltages, when the sample would be self-biased negative to the plasma, growth was still found to be rapid.

Another common approach for generating plasmas involves the use of dc discharges. As noted, these may employ cold cathodes,[30] hot cathodes[14] and hollow cathodes.[22,31] Cold and hollow cathode systems are similar, in that a high voltage dc supply (several hundred to 1,000 V) is used to strike the discharge, while a second supply is used to bias the sample. In semiconductor applications, contamination is a serious problem when employing dc discharge techniques, so the method is not of great use. Figure 3 shows the hollow cathode design, as applied to a multiwafer reactor design.[22]

Such a design might be useful using other methods for striking the discharge. For example, as shown in Figure 4, if the bell jar is configured as shown, an rf coil could be used to generate the plasma.[32-34]

The third general method used for exciting the discharge in a plasma oxidation system involves the use of radio frequencies.[16-18,23,32-34] Frequencies employed ranged from about 0.4 MHz to 3 MHz. In general, the reported methods have utilized anodization approaches,[32-34] and oxidation temperatures have ranged between 300°-600°C. With anodization, 1 μm films could be grown in

Plasma and Elevated Pressure Oxidation in VLSI and ULSI 399

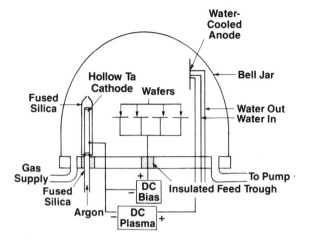

Figure 3: Multiwafer DC hollow cathode anodization system.

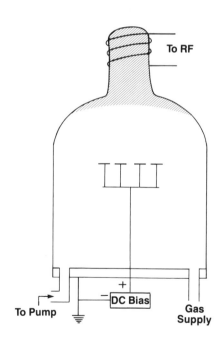

Figure 4: Multiwafer RF internal DC biased electrode anodization system.

two hours, but high fields (1.5 MV) were required across the growing oxide, as were high current densities (30 to <200 ma/cm^2.[35] Without anodization, growth rates were considerably less, 3 μm in 5 hours.[16] It is to be expected that, if the split-ring approach described in Reference 27 is utilized, growth in the absence of biasing electrodes would approximate those obtained with biasing. Anodized samples were found to exhibit high interface state and fixed charge levels, as mentioned. It has been claimed that annealing in CCl_4 at 900°C for 20 minutes reduced defect levels to the 10^{10} level for both fixed charge and interface state levels.[36] Samples oxidized using electrically floating techniques,[16,17,23] by the unique method described in these references, provided defect levels in the 2 to 6 x 10^{10} level without annealing, and in the 1 x 10^{10} range following brief annealing in oxygen or argon (20 minutes at 1000°C).

It is to be noted, that excepting for the work described in References 16, 17 and 23, oxidation, however the discharge was initiated, and whether or not samples were biased, resulted in the grown oxides being on the surface of the wafer facing the plasma. In References 16, 17 and 23, oxidation of the electrically floating samples always occurred on the surface of the wafer facing away from the plasma. Similar behavior was observed by these workers in plasma assisted nitridation of silicon in NH_3 or N_2 to H_2 mixtures.[18] Figure 5 shows schematically the plasma system configuration employed in References 16 to 18 and 23.

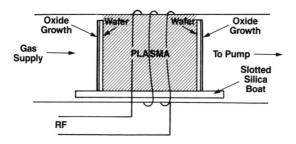

Figure 5: Multiwafer electrodeless floating wafer system.

Each pair of silicon wafers is configured so that it sits on either side of a few turns of an induction coil which surrounds the reactor. Under proper conditions (gas pressure between 10 and 200 mTorr), the discharge is ignited between the wafers, and the rf energy does not couple to wafers. The wafers are positioned so that their polished surfaces face away from the plasma (into the dark region), and oxidation occurs on these polished surfaces. Uniform oxides are grown, whose properties are essentially indistinguishable from high quality thermal oxides. In device fabrication,[16,17] it was found that the grown oxides did not exhibit bird's beak topography, oxidation enhanced diffusion, oxidation induced stacking faults, or orientation and impurity dependent oxidation rates.

The final approach of interest in plasma assisted oxidation and nitridation is that employed for multiwafer systems. Very little has been published in this area,[18,22-26] and it is not clear that of the four approaches reported, any has manufacturing potential in its present state. For example, in References 18 and 23, the concept depicted in Figure 5 is expanded by a series connection of a number of multiturn induction coils, each surrounded by a pair of wafers. Reisman and Berkenblit[37] have strung together eight such coils, allowing for simultaneous oxidation of 16 wafers, and in Reference 23, it was reported that scaling of wafer size did not seem to pose a problem, up to 81 mm at least. Scaling of the technique to accommodate 50 or 100 wafers appears awkward, using the general design shown. However, alternate scaling approaches are being examined which may make the method of greater practical importance. In any event, as noted, the advantages of this approach are that the system is electrodeless, the quality of the oxides is excellent, wafer size scaling appears feasible, oxide growth rates parallel those of dry oxidation at 1100°C, and the use of the method in device fabrication results in bird's beak free structures.

The approach described in Reference 22 is not believed to have too much manufacturing potential because of the use of exposed electrodes in the system, and because the potential number of wafers that can be oxidized simultaneously is small, especially when wafer sizes are increased. However, because growth rates are high, wafer throughput may be adequate if oxides of decent quality can be obtained. As mentioned, if the configuration employed in Figure 3, coupled with a rf discharge approach[34] is employed, as depicted in Figure 4, a "bell jar" system may be useful.

The multiwafer system design reported in Reference 25 is depicted schematically in Figure 6. It consists of two graphite rods, each contained in a slotted fused silica tube. Pairs of wafers are positioned close together so that their unpolished sides face each other. Each pair of wafers contact one electrode only.

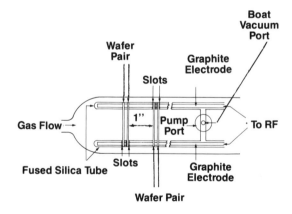

Figure 6: Multiwafer RF internal floating electrode oxidation system.

The first pair of wafers contact one electrode, the second pair of wafers contact the second electrode, the third pair of wafers contact the first electrode, etc. A 13.56 MHz power supply is connected across the graphite rods. Opposing sets of wafers are driven to excite the plasma using up to 130 V rf peak voltage in a 2 Torr oxygen plasma. At 600°C, a 25 nm thick oxide film was grown in 2 hours.

This is a much slower growth rate than reported in Reference 16, and very much slower than obtainable in a biased sample system. The authors assumed that the wafers were self-biased relative to the plasma, because when rf peak voltages were increased above 100 V, growth rate was observed to increase rapidly in nitridation studies. This method is interesting because it enables a wafer packing density of 2/inch, and because of the electrode configuration, it enables the reactor to be placed within a furnace. However, the growth rates are unacceptably low for field, or source-drain oxidations. It appears that separate biasing of the electrodes might enhance growth rates to acceptable values. Of course, the presence of electrodes exposed to the plasma is undesirable. However, the authors addressed this problem by constructing an independent vacuum outlet for the slotted boat. Thus, oxygen flowing into the system is presumably pulled through the slotted tubes, minimizing contamination of the plasma.

A final multiwafer system design is described in References 24 and 26. The system is not unlike that described in Reference 25, but was used only for nitridation. It employed exposed SiC electrodes connected to a 13.56 MHz power supply, and an additional dc biasing supply (up to 250 V). Silicon wafers straddled the electrodes as in Reference 25, and were allowed to contact only the anode electrode, being insulated from the other electrode by a fused silica spacer. Unlike the wafer configuration employed in Reference 25, single silicon wafers were alternated with SiC "wafer" discs along the electrodes. It would appear that a combination of the design features reported in References 24, 25 and 26 might offer potential for a manufacturing system design if the quality of the resultant oxides, both physical and electrical, is of device caliber, and if the method can be scaled to large wafer sizes, and acceptable growth rates.

Thus we see that while plasma assisted oxidation processes offer promise for conserving thermal budget in small dimension device fabrication, the promise has yet to be realized in practical terms. It is not crucial that the plasma mechanisms be understood fundamentally to achieve a practical implementation, although such understanding is highly desirable. Many processes employed today are deficient in understanding. What is missing in the interim is a demonstrated empirical scaling approach. A number of workers have and are exploring the mechanisms of plasma oxidation involving biased and "floating wafers,"[7,9,14,16,27,29,32,34,35,38-51] and it is not unlikely that, if such understanding is achieved, manufacturing systems will follow. There is still uncertainty as to whether negative ions are responsible,[15,41] neutral oxygen molecules and electrons are responsible,[12,46] or whether oxygen alone, or oxygen and silicon are both mobile species,[34] or whether multiple mechanisms are involved. Understanding of the kinetics is in no better shape.

ELEVATED PRESSURE OXIDATION

Compared to the literature on plasma assisted oxidation, that concerned with the use of elevated pressures is small. The early work,[52-55] and almost all of the later reports focused on steam oxidation which is to be expected, since by analogy with atmospheric steam oxidation, higher rates are to be expected than with dry oxygen. It was found that the oxidation kinetics at 650°C with pressures up to 90 atm. in heated, gold lined, sealed Inconel X bombs exhibited linear thickness increase with time. This is indicative of a surface controlled reaction, and the rate was found to be proportional to steam pressure.[55] The constant oxidation rates for three different silicon crystal orientations were (110) > (311) > (111), and the activation energies for the three orientations were (110) < (311) < (111).[54] The mobile species was assumed to be interstitial water.[53] In a study conducted in similar fashion more than a decade later,[56] it was demonstrated that 110 to 120 nm thick CVD Si_3N_4 could be used effectively as an oxidation mask for oxidations at 675°C, and 700°C in steam pressures between 60 and 90 atm. At 675°C and 60 atm., it was found that (111) silicon oxidized at a constant rate of 5.4 nm/min. At 90 atm., and the same temperature, the rate increased to 8 nm/min. (100) silicon was found to oxidize at about 2.7 nm/min at 675°C and 60 atm. Thus, the (111)/(100) oxidation rate ratio is about 2/1. It was found in Reference 56, surprisingly, that constant growth rates were obtained up to 6 μm, after which growth appeared to become diffusion limited. One interesting aside of the work reported in Reference 56 is that the authors predicted, based on calculated values for the diffusion and surface rate constants and the activation energy for the surface reaction, that a crossover in the rate limiting step will occur at above 675°C, and that as a consequence, at 675°C, oxidation at 1 atm. will be surface limited. Such an occurrence is indeed possible, and is depicted schematically in Figure 7. Such behavior has been reported in solution etching kinetics,[57] but has yet to be verified in the silicon oxidation case. The point is mentioned, since it points the way to better understanding of atmospheric oxidation of silicon, about which there is still considerable confusion.

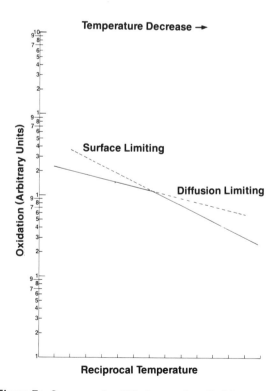

Figure 7: Crossover in diffusion-surface limiting steps.

Reports on the design and use of a multiwafer, elevated pressure pyrogenic oxidation system first appeared in the time frame from 1973 to 1977.[58-60] It consisted of an inner fused silica reactor surrounded by a furnace, this entire assembly being contained in a pressurized metal chamber. Steam was formed within the reactor either by the reaction of H_2 (carried in N_2) with oxygen,[59,60] or via the use of a waterboiler.[58] The resultant steam pressure within the reactor was about 6.4 atm., and was maintained by flow rates of between 2 and 3.54 SLM of O_2 and H_2. Oxidations were conducted between 800° and 1000°C. On log-log plots, the thickness-time behavior was found to be linear, the slopes of the curves decreasing with increasing temperature, and with (111) rates > (100) rates. Compared to atmospheric pressure oxidation, the authors found the elevated pressure oxidation at 1000°C to be about a factor of 5 greater. Slopes calculated from their log-log plots by the present author showed that thickness varied as time$^{0.6}$ at 1000°C to time$^{1.0}$ at 800°C for the (100) orientation which is consistent with the lower temperature results described above,[57] and indicates that the crossover between surface and nonlinear growth occurs above 800°C. It was shown the following year,[61] that as reported in the earlier "bomb" experiments, Si_3N_4 was an effective oxidation mask at 6.4 atm., up to 1100°C.

The first commercial elevated pressure pyrogenic system, whose basic features are somewhat similar to those reported in References 58 to 60, was reported in 1977.[62] It is shown schematically in Figure 8. It was intended for use at about 900° to 1000°C, up to perhaps 25 atm., can accommodate up to 200 wafers/load, exhibits a claimed uniformity of better than 5% across a wafer, and gases are metered by the use of sonic orifices which nullify the effects of back pressure. An HCl supply is provided for cleaning up the system periodically, and to the author's knowledge, the system is now available for up to 200 nm wafers, in both pyrogenic and water pumped versions.[63] Detailed studies of pyrogenic steam oxidation using this system have been reported by two groups[64,65] which with minor variations support earlier published data with noncommercial systems. As might have been anticipated, it has been found that using 700°C, 20 atm. steam oxidation conditions, oxidation induced stacking faults are suppressed on both (100) and (111) wafers of both n and p type silicon.[66,67] It has also been reported that similar to atmospheric steam oxidation,[68,69] heavy n doping enhances 10 atm. steam oxide growth rate, but less than observed for atmospheric conditions. Boron doping was not found to enhance the oxidation rate very much during either 10 or 1 atm. steam oxidation.[70] Electrical and physical characterization of 10 atm. steam oxides grown in the temperature range 650° to 800°C[71] indicated that fixed oxide charge was in the low 10^{11} range, and that interface states were of the same order of magnitude.[72] In addition, interface state densities increased with oxidation temperature, breakdown strengths were in the high MV range, and the index of refraction of the grown oxides was a function of temperature, increasing with decreasing temperature. The final data to be discussed are those concerned with bird's beak formation, and the effects of elevated pressure steam oxidation on boron outdiffusion effects. Detailed studies[73,74] have shown that there is little advantage to be gained in minimizing bird's beak at elevated pressure, but a decided advantage in terms of both lateral boron motion along the beak, and outdiffusion into the oxide. According to Reference 73, bird's beak may even be a little poorer than in atmospheric steam oxidation.

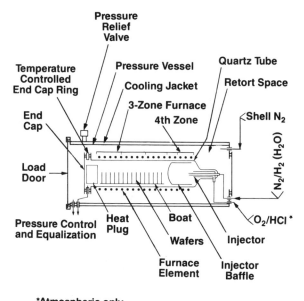

Figure 8: Schematic of Gasonics high pressure system.

There has been only one report to date on the elevated pressure oxidation of silicon in dry oxygen in a multiwafer system.[75] The studies were conducted on (100) and (111) wafers, between 800° and 1000°C in the pressure range 1 to 20 atm. The kinetics appear to be similar to those in steam elevated pressure oxidation, but it was found that neutral trap content measured by avalanche injection techniques was greater than obtained with 1 atm. oxidations, by about a factor of 5. Fixed charge levels were comparable at all pressures. These authors found the kinetics to be consistent with the Deal-Grove Model.[76]

CONCLUSIONS

It is evident that the status of assisted oxidation processes is such that elevated pressure techniques are considerably further along than their plasma counterparts. Multiwafer commercial systems exist, and there is good agreement between results obtained in such systems, and "home-built" units in several laboratories. There appears to be significant advantage to be gained for conserving thermal budget in small dimension device fabrication. Technologically, it appears to the present author, that the greatest attribute of elevated pressure approaches might be the linear kinetics obtainable at lower temperatures, while preserving throughput. Not enough appears to have been made of this attribute. Based on the studies reported in Reference 73, gate, source/drain and field oxidations can all be effected at elevated pressures equally well, with the potential for extremely tight tolerance control on thickness. There exists a little uncertainty about the electron trapping equivalency of elevated pressure oxides to atmospheric grown ones, but it seems that this is an easily resolvable issue. One deficiency is the shortcoming in suppressing bird's beak topography, but at least the problem does not appear to be aggravated. Plasma processes continue, after two decades, to offer potential for even lower temperature capabilities which could find use in conventional MOSFET and bipolar fabrication processes. An equally important application could be in thin film silicon transistor applications for large area displays where fabrication temperature minimization is imperative. In at least one approach to plasma oxidation, it has been shown that bird's beak could be eliminated completely, and this would be a major plus. However, no clear path has been demonstrated to cope with the throughput problem using this technique, and therefore, for the present, plasma approaches must be viewed as still possessing untapped potential.

REFERENCES

1. Reisman, A., *Proc. IEEE*, Vol. 71, p. 550 (1983).
2. Reisman, A., Merz, C.J., Maldonado, J.R. and Molzen, W.W., Jr., *J. Electrochem. Soc.*, Vol. 131, p. 1404 (1984).
3. Bassous, E., Yu, H.N. and Maniscalco, V., ibid., Vol. 123, p. 1729 (1976).
4. Chiu, K.Y., Moll, J.L. and Manoliu, J., *IEEE Trans. Elect. Dev.*, Vol. ED-29, p. 536 (1982).
5. Zhou, N.S., Fujita, S. and Sasaki, A., *J. Electronic Mats.*, Vol. 14, p. 55 (1985).
6. Miles, J.L., U.S. Patent No. 3,394,066 (July 1968).
7. Miles, J.L. and Smith, P.H., *J. Electrochem. Soc.*, Vol. 110, p. 1240 (1963).
8. Tibol, G.L. and Hull, R.H., ibid., Vol. 111, p. 1368 (1964).
9. Jennings, T.A. and McNeill, W., ibid., Vol. 114, p. 1134 (1967).
10. O'Hanlon, J.P., *Oxides and Oxides Films*, Vol. 5, p. 105, A.K. Vijh, ed., Marcel Dekker, N.Y. (1977).
11. Reisman, A. and Merz, C.J., *J. Electrochem. Soc.*, Vol. 130, p. 1384 (1983).
12. Friedel, P. and Gourrier, S., *J. Phys. Chem. Solids*, Vol. 44, p. 353 (1983), see section 3.1, p. 357 in particular.
13. Ray, A.K. and Reisman, A., *J. Electrochem. Soc.*, Vol. 128, p. 2460 (1981).
14. Gourrier, S., Mircea, A. and Bacal, M., *Thin Solid Films*, Vol. 65, p. 315 (1980).
15. Ligenza, J.R., *J. Appl. Phys.*, Vol. 36, p. 2703 (1965).
16. Ray, A.K. and Reisman, A., *J. Electrochem. Soc.*, Vol. 128, p. 2466 (1981).
17. Ray, A.K. and Reisman, A., ibid., Vol. 128, p. 2424 (1981).
18. Reisman, A., Berkenblit, M., Ray, A.K. and Merz, C.J., *J. Electron. Mater.*, Vol. 13, p. 505 (1984).
19. McCaughan, D.V. and Heilig, J.A., Jr., *Int. J. Electron.*, Vol. 34, p. 737 (1973).
20. Sugano, T. and Ho, V.Q., *Annual Report of the Engineering Research Institute Faculty of Engineering*, University of Tokyo, Vol. 41, p. 117 (1982).
21. Kassabov, J., Atanassova, E.D. and Goranova, E., *Solid-State Electronics*, Vol. 27, p. 13 (1984).

22. Orcutt, W.B., Thesis, M.S., Air Force University, Faculty of the School of Engineering, Air Force Institute of Technology (1972).
23. Ray, A.K., Reisman, A. and Berkenblit, M., Abs. No. 89, Electrochem. Soc. Meeting (May 8–13, 1983).
24. Hirayama, M., Matsukawa, T., Arima, H., Ohno, Y., Tsubouchi, N. and Nakata, H., *J. Electrochem. Soc.*, Vol. 131, p. 664 (1984).
25. Wong, S.S. and Oldham, W.G., *IEEE Electron. Dev. Letters*, Vol. EDL-5, p. 175 (1984).
26. Hirayama, M., Matsukawa, T., Arima, H., Ohno, Y. and Nakata, H., *J. Electrochem. Soc.*, Vol. 132, p. 2494 (1985).
27. Kraitchman, J., *J. Appl. Phys.*, Vol. 38, p. 4323 (1967).
28. Skelt, E.R. and Howells, G.M., *Surface Science*, Vol. 7, p. 490 (1967).
29. Olive, G., Pulfrey, D.L. and Young, L., *Thin Solid Films*, Vol. 12, p. 427 (1972).
30. O'Hanlon, J.F., *J. Vac. Sci. and Technol.*, Vol. 7, p. 330 (1970).
31. Ligenza, J.R. and Kuhn, M., *Solid State Technology*, p. 33 (Dec. 1970).
32. Pulfrey, D.L., Hathorn, F.G.M. and Young, L., *J. Electrochem. Soc.*, Vol. 120, p. 1529 (1973).
33. Matsuzawa, A., Itoh, T., Ishikawa, Y. and Yanagida, H., *J. Vac. Sci. Technol.*, Vol. 17, p. 793 (1980).
34. Sugano, T., *Thin Solid Films*, Vol. 92, p. 19 (1982).
35. Ho, V.Q. and Sugano, T., *IEEE Trans. Electron. Dev.*, Vol. ED-27, p. 1436 (1980).
36. Ho, V.Q. and Sugano, T., ibid., Vol. ED-28, p. 1060 (1981).
37. Reisman, A. and Berkenblit, M., unpublished work.
38. Jennings, T.A. and McNeill, W., *Appl. Phys. Lett.*, Vol. 12, p. 25 (1968).
39. Locker, L.D. and Skolnick, L.P., ibid., Vol. 12, p. 396 (1968).
40. O'Hanlon, J.F., ibid., Vol. 14, p. 127 (1969).
41. O'Hanlon, J.F. and Pennebaker, W.B., *Appl. Phys. Lett.*, Vol. 18, p. 554 (1971).
42. Laska, L. and Kodymova, J., *Czech. J. Phys.*, Vol. B26, p. 157 (1976).
43. Labunov, V., Parkhutik, V. and Tkharev, E., *J. Crystal Growth*, Vol. 45, p. 399 (1978).
44. Musil, J., Zacek, F., Bardos, L., Loncar, G. and Dragila, R., *J. Phys. D: Appl. Phys.*, Vol. 12, p. L61 (1979).
45. Fromhold, A.T., Jr. and Baker, M., *J. Appl. Phys.*, Vol. 51, p. 6377 (1980).
46. Moruzzi, J.L., Kiermasz, A. and Eccleston, W., *Plasma Physics.*, Vol. 24, p. 605 (1982).
47. Fromhold, A.T., Jr., *Thin Solid Films*, Vol. 95, p. 297 (1982).
48. Kiermasz, A., Eccleston, W. and Moruzzi, J.L., *Solid State Electronics*, Vol. 26, p. 1167 (1983).
49. Ng, K.K. and Ligenza, J.R., *J. Electrochem. Soc.*, Vol. 131, p. 1968 (1984).
50. Perriere, J., Siejka, J. and Chang, R.P.H., *J. Appl. Phys.*, Vol. 56, p. 2716 (1984).
51. Fu, C.Y., Mikkelsen, J.C., Jr., Schmitt, J., Abelson, J., Knights, J.C., Johnson, N., Barker, A. and Thompson, M.J., *J. Electronic Mats.*, Vol. 14, p. 685 (1985).
52. Ligenza, J.R. and Spitzer, W.G., *J. Phys. Chem. Solids*, Vol. 14, p. 131 (1960).

53. Spitzer, W.G. and Ligenza, J.R., ibid, Vol. 17, p. 196 (1961).
54. Ligenza, J.R., *J. Phys. Chem.*, Vol. 65, p. 2011 (1961).
55. Ligenza, J.R., *J. Electrochem. Soc.*, Vol. 109, p. 73 (1962).
56. Powell, R.J., Ligenza, J.R. and Schneider, M.S., *IEEE Trans. Electron. Dev.*, Vol. ED-21, p. 636 (1974).
57. Reisman, A., Berkenblit, M., Chan, S.A., Kaufman, F.B. and Green, D.C., *J. Electrochem. Soc.*, Vol. 126, p. 1406 (1979).
58. Panousis, P.T. and Schneider, M., 143rd Spring Meeting, Electrochem. Soc., *Extended Abs.*, Vol. 73-1, Chicago, ILL. (May 13-18, 1973).
59. Tsubouchi, N., Miyoshi, H., Nishimoto, A. and Abe, H., *Jpn. J. Appl. Phys.*, Vol. 16, p. 855 (1977).
60. Tsubouchi, N., Miyoshi, H., Nishimoto, A., Abe, H. and Satoh, R., ibid., Vol. 16, p. 1055 (1977).
61. Miyoshi, N., Tsubouchi, N. and Nishimota, A., *J. Electrochem. Soc.*, Vol. 125, p. 1824 (1978).
62. Champagne, R. and Toole, M., *Solid State Technology*, p. 61 (Dec. 1977).
63. Private conversations with M. Toole, Gasonics Corp., 250 Caribbean Drive, Sunnyvale, CA 94089.
64. Razouk, R., Lie, L.N. and Deal, B.E., *J. Electrochem. Soc.*, Vol. 128, p. 2214 (1981).
65. Katz, L.E., Howells, B.F. and Adda, L.P., Thompson, T. and Carlson, D., *Solid State Technology*, p. 87 (Dec. 1981).
66. Katz, L.E. and Kimmerling, L.C., *J. Electrochem. Soc.*, Vol. 125, p. 1680 (1978).
67. Tsubouchi, N., Miyoshi, H., Abe, H. and Enomoto, T., *IEEE Trans. Electron. Dev.*, Vol. ED-26, p. 618 (1979).
68. Ho, C.P., Plummer, J.D. and Meindl, J.D., *J. Electrochem. Soc.*, Vol. 125, p. 665 (1978).
69. Irene, E.A. and Dong, D.W., ibid, Vol. 125, p. 1146 (1978).
70. Fuoss, D. and Topich, J.A., *Appl. Phys. Lett.*, Vol. 36, p. 275 (1980).
71. Hirayama, M., Miyoshi, H., Tsubouchi, N. and Abe, H., *J. Electronic Mat.*, Vol. 11, p. 919 (1982).
72. Hirayama, M., Miyoshi, H., Tsubouchi, N. and Abe, H., *IEEE Trans. Electron. Dev.*, Vol. ED-29, p. 503 (1982).
73. Wu, T.C., Stacy, W.T. and Ritz, K.N., *J. Electrochem. Soc.*, Vol. 130, p. 1563 (1982).
74. Deroux-Douphin, P. and Gonchond, J.P., ibid., Vol. 131, p. 1418 (1984).
75. Lie, L.N., Razouk, R. and Deal, B.E., ibid., Vol. 129, p. 2828 (1982).
76. Deal, B.E. and Grove, A.S., *J. Appl. Phys.*, Vol. 36, p. 3770 (1965).

Index

Acetylene - 275, 280
Adatom migration - 345
Adatoms - 351
Alkyls - 237
AlN - 323
Al_2O_3 - 361
Alumina - 323
Aluminum - 277, 323, 325
Aluminum and aluminum silicon alloys - 108
Aluminum oxide - 163, 277
Aluminum oxynitride - 277
Angular distribution - 375
Anodization - 7
Antistatic coatings - 15
Arsine - 279
Atmospheric-pressure CVD (APCVD) - 9
Auger electron - 204
Auger electron spectrometer - 172
Auger electron spectroscopy (AES) - 357
Autodoping - 66

Backside transfer - 55
Basic MBE process - 170
Beam assisted CVD - 319
Beam induced thermal processes - 333

Beam neutralization - 382
Bell jar reactors - 85
Bias sputtering - 6, 303
Bismuth - 278
Boron - 278
Boron nitride - 108
Buried layer pattern transfer - 62

Cadmium - 278
Catalytic poisoning - 275
Chase - 307
Chemical beam epitaxy - 212
Chemical reduction plating - 12
Chemical vapor deposition (CVD) - 2, 8
Chromium - 278, 325
Cleanliness - 252
Cluster beam deposition - 7
Compositional control - 201
Conformal coverage - 299
Continuous sources - 190
Control of variables - 43
Copper - 278
Cracking cells - 188
Cylinder reactor - 34

DC magnetrons - 299
DC sputtering - 297
Deep level transient spectroscopy - 207

Defects in epitaxy layers - 56
Deposition characteristics - 154
Deposition equipment - 306
Deposition mechanism - 120
Deposition technologies - 1
Diacetoxyditertiarybutoxy-
    silane (DADBS) - 106
Diamond - 280
Dichlorosilane - 274
Dielectrics - 276
Diethyltelluride - 279
Difluorosilane - 274
Dimethylselenide - 279
Dimethylsilane - 275
Dimethylzinc - 279
Diode sputtering - 5
Directional etching - 311
Disilane - 274
Dopant flow control - 47
Dopant transitions - 69
Doped LPLTO - 107
Doping control - 200

ECR plasma CVD system - 166
ECR plasma deposition - 148
ECR plasma generation - 147
Edge crown - 57
Electroless plating - 12
Electrolytic anodization - 12
Electron beam alloying - 336
Electron beam annealing - 334
Electron beam assisted processing - 318
Electron-beam deposited dielectric films - 327
Electron beam heated sources - 191
Electronic displays - 14
Electron impact dissociation - 301
Electrophoretic deposition - 12
Electroplating - 12
Elevated pressure oxidation - 402
Epitaxy - 345
Epitaxy process chemistry - 31
Epitaxy specifications - 28
Equipment for plasma deposition - 172
ESCA (Electron Spectroscopy) - 172
Etch pits - 58

Excimer lasers - 271, 282, 286

Faceted growth - 57
Facetting - 313
Film nucleation - 370
Film properties - 299
Film purity - 299
Fomblin - 285

GaAs on silicon - 210
Gallium arsenide - 279, 360
Gas control system - 44, 84
Gas flow control - 46
Gas manifold design - 254
Gas phase chemical processes - 8
Gas phase chemical reactions - 251
Gas source MBE - 212
Gas Sources - 191
Germanium - 278
Glow discharge etching - 312
Glow-discharge technologies - 5
Growth conditions - 244
Growth mechanisms - 249
Growth procedure - 199

Hall effect - 206
Hard surface coatings - 15
Haze - 57
Heating power supplies - 51
Hetero-epitaxy - 26
Hexachlorodisilane - 274
Hillocks - 270
Hollow-cathode - 366
Homoepitaxy - 2, 26
Horizontal reactor - 34, 83
Hydride MBE - 212
Hydride sources - 243
Hydrides - 237
Hydrogen incorporation - 323

Immersion plating - 13
Implantation sources - 191
Indium oxide - 279
Indium phosphide - 279
In-line system - 308
In situ analysis - 204
In situ metallization - 203
In situ processing - 216

Insulators - 276, 300
Integrated circuits - 181
Interrupted growth - 203
Intrinsic resistivity - 56
Ion assisted deposition (IAD) - 367, 373, 380
Ion beam deposition - 364
Ion beam etching - 312
Ion beam gun - 295
Ion beam probing - 367
Ion beam sputter deposition (IBS) - 373
Ion beam sputtering - 6, 295
Ion flux - 298, 369
Ionic mechanism - 123
Ion implantation - 11, 298
Ion incidence effects - 159
Ionized cluster beam deposition - 344
Ionized cluster beam epitaxy - 217
Ionizing sources - 190
Ion plating - 317
Ion sources - 378
Irradiation assisted MBE - 218

Kaufman ion sources - 364
Knudsen cells - 187

Laser and electron beam recrystallization - 338
Laser assisted CVD - 320
Laser assisted processing - 318
Laser CVD system - 282
Laser-deposited metallic films - 325
Laser-induced chemical vapor deposition (LCVD) - 10
Leak integrity - 252
Leak testing - 45
Lewis acid reactions - 251
Lift-off - 314
Line source recrystallization - 339
Liquid-phase epitaxy (LPE) - 13, 173
Load-lock - 171
Load-lock deposition systems - 135
Load-locked sources - 190
Low pressure CVD (LPCVD) - 10, 80

Low pressure MO-CVD - 260
Low temperature epitaxy - 75
Low-temperature oxide (LTO) - 100
LPCVD equipment - 81
LPCVD Processing - 88

Magnetic films - 14
Magnetron discharges - 299
Magnetron sputtering - 6
Material characteristics - 15
Materials purity - 245
MBE deposition equipment - 183
MBE devices - 176
MBE sources - 186
Mechanical methods - 13
Mercury discharge lamps - 271
Mercury-telluride - 279
Metal-organic chemical vapor deposition (MO-CVD) - 174, 234
Metal-organic-molecular beam epitaxy (MO-MBE) - 212, 237
Methylsilane - 275
Microwave electron cyclotron resonance deposition (ECR) - 7
Microwave electron cyclotron resonance plasma CVD - 147
MO-CVD sources - 237
Molecular beam epitaxy (MBE) - 5, 170
Molybdenum - 130, 278, 325
Monomethylsilane - 275
Multi-chamber systems - 184

Nucleation density - 351

Operating parameters - 138
Optical coatings - 14
Optical data storage devices - 15
Optoelectronic devices - 180
Orange peel - 57
Organometallic compounds - 238
Organometallic source packaging - 241
Oval defect reduction - 210
Oxygen gettering techniques - 253
Oxynitride - 328

Patterning - 311, 314
Penning reactions - 119
Phantom p-type layer - 56
Phosphorus-doped silicon - 106
Phosphorus nitride - 277
Photo-activated CVD - 271
Photochemical vapor deposition - 270
Photo-CVD equipment - 280
Photodissociation - 273, 325
Photo-enhanced chemical vapor deposition (PHCVD) - 10
Photo-enhanced MO-CVD - 237, 264
Photoluminescence spectroscopy - 207
Photolysis - 286
Photox - 276
Pits - 57
Planarizing - 295
Planarizing process - 304
Plasma - 297, 393
Plasma-assisted chemical vapor deposition - 112
Plasma-assisted oxidation - 396
Plasma-enhanced chemical vapor deposition (PECVD) - 2
Plasma-enhanced MO-CVD - 237, 264
Plasma etching - 313
Plasma extraction - 150
PMMA - 330
Polysilicon - 90
Polyvinylidene fluoride - 284
Pressure oxidation - 393
Principles of low pressure CVD - 87
Process repeatability - 299
Process technology - 17
Production equipment - 214
Pulsed electron beams - 330
Pumping considerations - 186
Pyrolysis - 8

Radiation damage - 319
Radiation defects - 395
Radio frequency discharges - 367
Reaction chamber - 256

Reaction kinetics in plasma - 117
Reactive evaporation - 317
Reactive ion etching - 313
Reactive sputter deposition - 302
Reactive sputter etching - 313
Reactive sputtering - 6
Reactor geometries - 33
Reactors - 9
Recondensation - 285
Reflection high energy electron diffraction (RHEED) - 172, 204, 357
Refractory metals - 278
Residual gas analysis - 205
RF magnetrons - 301
RHEED oscillation control - 209
Rotation control - 193
Rutherford backscattering spectroscopy (RBS) - 357

Safety - 59, 208
Sample mounting - 192
Sample transfer techniques - 186
Sapphire substrates - 278
Secondary ion mass spectrometry (SIMS) - 172, 204
Selenium - 279
Semi-insulating polysilicon (SIPOS) - 108
Sensitizer - 271
Shallow pits - 58
Silane - 274
Silicides - 278, 336
Silicon - 278
Silicon dioxide - 159, 361
Silicon epitaxy - 26
Silicon films - 130
Silicon nitride - 95, 124, 155, 277, 328
Silicon-on-insulator (SOI) - 338
Silicon oxide - 128
Silicon oxynitride - 126
Silicon tetrachloride - 274
Single wafer reactors - 86
Slips - 58
Source shutters - 192
Spikes and hillocks - 58
Spin-on deposition - 2

Sputter etching - 311
Sputter yield - 373
Sputtering - 5, 163, 291
Sputtering deposition - 2
Sputtering yield - 295, 298
Stacking faults - 58
Step coverage - 299
Sticking coefficient - 351
Stoichiometry - 303
Strained-layer superlattices - 213
Substrate cleaning - 198, 199, 370
Substrate preparation - 198
Superlattice buffer layers - 214
Superlattice device structures - 214
Superlattice structures - 213
Surface mobility - 351
System automation - 194
System design and construction - 252

Tantalum - 163
Tantalum silicide - 130
Target integrity - 299
Temperature control - 192
Temperature measurement - 53
Tetraethylorthosilicate (TEOS) - 106
Thermal evaporation - 3
Thermal evaporation sources - 187
Thermal oxidation - 11
Thermal transient - 200
Thermionic emission - 366
Thin-film manufacturing equipment - 19
Tin - 278
TiO_2 - 361

$TiSi_2$ - 337
Titanium - 278
Titanium chloride - 278
Titanium silicide - 130
Toxic gases - 218
Trenching - 313
Triethylgallium (TEG) - 279
Triisobutylaluminum (TIBAL) - 277
Trimethylgallium (TMG) - 279
Trisilane - 275
Tungsten - 107, 130, 278
Tungsten silicide - 108, 130

Uniformity - 299

Vacuum chemical epitaxy - 217
Vacuum evaporation - 2
Vacuum MO-CVD - 237
Vacuum system construction - 183
Vacuum systems - 84
Vapor-phase epitaxy (VPE) - 10, 174
Vertical reactor - 35, 85

WSi_2 - 337

X-Ray lithography - 330
X-Ray photoelectron spectroscopy (XPS) - 172, 204

Zinc - 278, 279
Zinc oxide - 277
Zinc selenide - 279